Land Snails
in Archaeology

Studies in Archaeological Science

Consulting editor G. W. DIMBLEBY

Other titles in the Series

The Study of Animal Bones from Archaeological Sites
R. E. CHAPLIN

Ancient Skins, Parchments and Leathers
R. REED

Methods of Physical Examination in Archaeology
M. S. TITE

Frontispiece. Alfred Santer Kennard, A.L.S., F.G.S. "With a strong leaning towards the antique—old books, ancient buildings, obsolete customs, primeval men—it was no wonder that he was attracted to the study of those long departed lives so well represented around London by fossil Mollusca" (Wrigley, 1948).

Land Snails in Archaeology

With special reference to the British Isles

JOHN G. EVANS
*Department of Archaeology,
University College, Cardiff, Wales*

1972

SEMINAR PRESS · London and New York

SEMINAR PRESS INC. (LONDON) LTD
24–28 Oval Road,
London NW1

United States Edition published by
SEMINAR PRESS INC.
111 Fifth Avenue
New York, New York 10003

Copyright © 1972 by
SEMINAR PRESS INC. (LONDON) LTD.

All Rights Reserved
No part of this book may be reproduced in any form by photostat, microfilm,
or any other means, without written permission from the publishers

Library of Congress Catalog Card Number: 72-9081
ISBN: 0 12 829550 3

PRINTED IN GREAT BRITAIN BY
WILLIAM CLOWES & SONS, LIMITED, LONDON, BECCLES AND COLCHESTER

Preface and Acknowledgements

This book is the result of five years' research, done largely between 1964 and 1969, on the relevance of land snails to archaeology with particular reference to their use as indicators of the environment of ancient man; some of the material has been reported in the author's Ph.D. thesis (Evans, 1967). Certain broad conclusions have emerged about the impact of prehistoric farming on the landscape, particularly in Britain, and in Chapter 11 I have briefly summarized our knowledge of habitat development in the calcareous regions of this country over a somewhat wider time range, including the Late Weichselian and Post-glacial periods as a whole.

I am greatly indebted to many people who have given advice and material help during the course of this work. In particular I would like to thank Dr M. P. Kerney who initially provided a reference collection of shells and spent many hours explaining the techniques of analysis and identification. His help and advice throughout have been invaluable. I am grateful to Dr I. W. Cornwall, who originally suggested the project, and to Professor G. W. Dimbleby, who read the manuscript and made a number of valuable suggestions; both have been a constant source of encouragement. On the archaeological side, I would like to thank Dr Isobel Smith, who has taken considerable interest in this work from the start, and made possible the collection of material for analysis from many of the sites in the Avebury area.

To the following archaeologists, grateful acknowledgement is made for their co-operation in giving me access to their excavations or in making available samples for analysis, and in giving of their time in explaining the context of various features and horizons: Mr L. Alcock, Dr J. Alexander, Professor R. J. C. Atkinson, Mr D. Benson, Mr P. J.

Fowler, Dr M. G. Jarrett, Mr T. Manby, Professor S. Piggott, Mrs Edwina Proudfoot, Mr D. D. A. Simpson, Mr W. G. Simpson, Dr Isobel Smith, Mrs M. E. Robertson-Mackay, Mrs Faith de M. Vatcher, and Dr G. J. Wainwright.

I am grateful to the staff of the British Museum Research Laboratory, and in particular to Mr R. Burleigh, for doing the radiocarbon determinations from South Street, Ascott-under-Wychwood, Cherhill and Northton.

I would like to acknowledge the help given by the technical staff in the Department of Archaeology, University College, Cardiff, and in particular Miss Gloria Stephenson who drew a number of the diagrams and undertook the tedious work of lettering the histograms. Mrs Gaye Booth kindly printed all the photographs (with the exception of Fig. 145).

Latterly, Mr S. P. Dance of the National Museum of Wales, has assisted in loaning a number of specimens for drawing, and kindly lent the photograph of A. S. Kennard used as the Frontispiece.

The following gave their permission for various illustrations to be reproduced, and I am most grateful for their co-operation: the Hutchinson Publishing Group Ltd., Fig. 51, from N. W. Runham and P. J. Hunter "Terrestrial Slugs". The Conchological Society of Great Britain and Ireland, Figs 17, 40, 42 and 45. Blackwell Scientific Publications, Figs 25, 41, 46, 47, 50 and 52 from the *Journal of Animal Ecology* and the *Proceedings of the Malacological Society of London*. The Leicester University Press, Figs 87, 88 and 90. Mr R. Baker, Figs 41, 47, 50 and 52. Dr R. A. D. Cameron, Fig. 46. Mrs Mary Farnell, Fig. 145. Dr M. P. Kerney, Figs 43 and 45. Mr B. W. Sparks, Figs 2 and 25. Mr C. Saunders, Fig. 117. Mr J. W. Stephenson, Figs 40 and 42. Dr B. Verdcourt, Fig. 17.

I am also much indebted to the many people whose work I have used in compiling this book.

The research work was done in the Department of Human Environment, Institute of Archaeology, London University and financed by two grants from the Science Research Council—a Research Studentship (1964–1966) and a Research Assistantship (1967–1969), the latter under the direction of Professor G. W. Dimbleby.

Note on the radiocarbon dates. It has recently been shown through the C-14 assay of wood dated dendrochronologically that there is a discrepancy between conventional radiocarbon dates and calendar dates which in the Neolithic and Bronze Age periods amounts to between

300 and 800 years (Renfrew, 1970). In this book conventional radiocarbon dates have been used, but these can be converted to dates in calendar years by reference to Fig. 1 (p. 6), kindly constructed by Mr P. J. Ashmore.

September, 1972 J.G.E.

Contents

Preface and Acknowledgements   v

PART I: The History, Principles and Methods of Snail Analysis

1
History   3

2
Principles   17
Preservation 22
Deposits containing shells 24
Other environmental evidence 35
Dating 39

3
Methods   40
Field recording of soils and sediments 40
Sampling 41
Extraction 44
Identification 45
Graphical presentation of results 79

PART II : The Snails

4
Factors Controlling the Distribution and Abundance of Snails 87

Shelter and humidity 92
Temperature 99
Food 103
Predators and parasites 105
Competition 108
The interaction of environmental factors 109
Lateral variation in snail faunas 111
Changes in snail populations with time 118

5
Distribution, Habitats and History 133

6
Ecological groups 194

PART III : The Stratification of Shells in Calcareous Soils and Sediments and their Relevance to Archaeology

7
Soils 207

The incorporation of shells into calcareous soils . . . 207
Variations on the basic rendsina profile. 214
Fossil rendsina soils 223
Time scale 228

8
Representative Soil Profiles   232

Modern soils 232
   PROFILES MS I–MS V 233–240
   COOMBE HOLE 240
Neolithic soils 242
   WINDMILL HILL 242
   BECKHAMPTON ROAD 248
   ASCOTT-UNDER-WYCHWOOD 251
   SOUTH STREET 257
   HORSLIP 261
   WEST KENNET 263
   WAYLAND'S SMITHY II 265
   SILBURY HILL 265
   AVEBURY 268
   MARDEN 274
   KILHAM 277

9
Colluvial Deposits and Colluvial Soils   280

Ploughwash 282
Scree and limestone rubble 287
Solifluxion débris 289
Loess 290
Blown sand 291
Tufa and travertine 297
Freshwater deposits 305

10
Caves, Dry Valleys, Lynchets and Ditches   308

Caves 308
Dry valleys 311
   PINK HILL 312
Lynchets 316
   FYFIELD DOWN I 317
   OVERTON DOWN XI/B 320

Ditches 321
   SOUTH STREET 328
   HEMP KNOLL 332
   ROUGHRIDGE HILL 335
   BADBURY EARTHWORK 337
   ASCOTT-UNDER-WYCHWOOD 343
Ditches on river gravel 344
   MAXEY 346
   ARBURY ROAD 349

11
Habitat change in the calcareous regions of Britain   351

Appendix: Tables   369

Glossary   393

References   401

Index   418

Part I

The History, Principles and Methods of Snail Analysis

1
History

Students of early man have long been interested in his environment, for there is no doubt that factors such as climate, vegetation and soil type have been fundamental in controlling his economy, his use of the land and his place of settlement. Moreover, it is now well known that these factors, far from being static, are ever changing, and that what we see today in our surroundings is generally quite different from the conditions experienced in the past. There are two reasons for this. In the first place, since the end of the Ice Age, various natural changes have taken place—fluctuations in temperature and rainfall, the isolation of Britain from the Continent brought about by the Post-glacial rise of sea level, the spread of forests and the development of nutrient-rich soils. Secondly, there is the effect of man himself on the environment, and ever since the recognition by Iversen and Godwin in the early 1940s of Neolithic forest clearance phenomena this has held a fascination for prehistorians and biologists alike. As a hunter and food-gatherer, man made small attempt to control his environment, though his influence may not have been altogether negligible (Simmons, 1969; Smith, A. G., 1970). But it is as a farmer that man's impact on the landscape has been most strongly exerted, and one of the prime concerns of environmental archaeology is the way in which agricultural communities have destroyed the ancient forests of these islands and brought about their replacement, through successive stages of land use, by the predominantly open landscape of today. Thus any investigation of the environment of farming communities is closely tied to a study of their economy and land use as well.

The method of approach to past environments depends on several factors, but most of all it depends on the available techniques. Prior to the invention of pollen analysis, we were dependent on macroscopic plant fossils preserved in peat bogs, and, largely through the work of

J. Geikie, F. J. Lewis and G. Samuelsson in the later nineteenth and early twentieth centuries, a sequence of Post-glacial environmental change was put forward for several places in Scotland based on this kind of evidence. This, in fact, was an extension of the more fundamental scheme established by Blytt and Sernander in the nineteenth century for Scandinavia in which the terms Pre-boreal, Boreal, Atlantic, Sub-boreal and Sub-atlantic were first used to describe the climatic phases of the Post-glacial (Godwin, 1956).

The discovery that the microscopic spores and pollen grains of plants were preserved in peat and lake sediments, and that they could be extracted and identified, sometimes to the species level, revolutionized the study of Post-glacial environmental history. In the first place, since only a small quantity is needed for the extraction of pollen, samples can be taken at close intervals and much more detail about vegetational change obtained than was possible using macroscopic material alone. Secondly, it was quickly realized that not only were the vegetational zones established by pollen analysis of regional significance, corresponding over wide areas of countryside, but that the periods formerly established by the use of macroscopic remains were of only local application. Pollen analysis of layers of tree stumps in peat bogs failed time and again to demonstrate the peaks of abundance which would have been expected on the macroscopic evidence alone. The application of radiocarbon assay, also, has shown that sequences such as those described by Lewis are of essentially local significance (Godwin, 1966).

There is thus a basic distinction between the two kinds of evidence. Pollen can tell us about the general environment over wide areas of countryside but gives us less indication of the local structure of the vegetation—the arrangement of different tree species in a forest, the size of clearings, etc.—though this deficiency is now being remedied by detailed work in restricted areas, the construction of three-dimensional pollen diagrams (Turner, 1970) and the study of pollen from buried soils. Macroscopic plant remains on the other hand are most suitable for the detailed reconstruction of local habitats, and apart from giving us a general idea of the character of the climate of a locality —e.g. whether it be temperate or sub-arctic—are of little value in its overall characterization, owing to the powerful influence of local environmental factors in controlling the immediate structure of the vegetation. Where macroscopic plant remains *can* be used as indicators of climatic change is in the mapping of former distributions. For example, the northward extension of the hazel in Scandinavia during the late Boreal and Atlantic periods beyond its present-day range indicates the prevalence of a temperature during those periods higher than that of today (Godwin, 1956: Fig. 13). To be accurate, such evidence must be based on macroscopic remains and not pollen which is susceptible to the caprice of long-distance wind transport.

Similar criteria apply to a greater or lesser degree to the use of animal fossils—vertebrates, insects and molluscs. They are best used as indicators of local environment, but, like plant macro-fossils, can be applied to the reconstruction of former climates if the influence of local environmental control on the structure of their communities is realized. And here again, as with plant macro-fossils, changes of distribution are best brought out by mapping, and as such may be important indicators of climatic change as has been shown in the case of the land snail *Pomatias elegans* (Fig. 43) (Kerney, 1968a).

The distinction between the use of macro-fossils in the reconstruction of local environments, and the use of micro-fossils for climate and other widespread environmental parameters is not exclusive, but must be appreciated as a general rule, particularly in dealing with single localities.

A fundamental defect of pollen analysis is that it is not applicable to large tracts of chalk or limestone countryside in southern, central and eastern England due to the absence of peat deposits from these areas. The East Anglian fens and the Somerset Levels are important exceptions to this, and sites close to the Chalk have been investigated in Kent (Godwin, 1962), Lincolnshire (Smith, A. G., 1958b) and Sussex (Thorley, 1971). But on the whole these are marginal to the main masses of chalk and limestone downland. Pollen analysis has been able to tell us virtually nothing about the environmental and ecological history of the chalklands, and this is particularly unfortunate in view of their importance as areas of settlement and agriculture in the prehistoric and Roman periods. Clearly an alternative technique must be applied and the shells of land snails which occur in such abundance in calcareous soils and sediments would seem to be an obvious choice. They are small, they are plentiful and they are easily extracted and identified. There are a manageable number of species and they cover a wide range of habitats. Some prefer open ground, others shade and moisture; and they are particularly sensitive to changes of land use, some groups living in arable land, some in tall grass or scrub, and others in closely cropped downland pasture.

Among the first to notice and comment upon the presence of snail shells in an archaeological context was General Pitt-Rivers (Lane-Fox, 1869, 1876) when in 1869 he remarked on the abundance of *Cyclostoma* (*Pomatias*) *elegans* and other species in the ditch of a hill-fort at

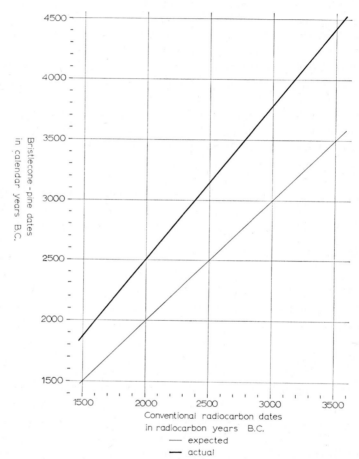

Figure 1. Regression line of radiocarbon years against calendar years. Calculated by P. J. Ashmore from data in Renfrew (1970: Fig. 1).

Cissbury in Sussex, suggesting their use as indicators of the past environment there.

But in Britain, most of the early work on land and freshwater Mollusca from archaeological sites was done by A. S. Kennard (Frontispiece). Alfred Santer Kennard began working on fossil molluscs at the end of the nineteenth century, and early on published a number of regional studies, some in conjunction with B. B. Woodward, and many of which were of a purely geological nature (Kennard, 1897, 1923, 1924; Kennard and Woodward, 1901, 1917, 1922). These two workers were also particularly interested in archaeological sites of the later prehistoric periods in the south and east of England where the calcareous subsoils of the Chalk and younger limestones cause conditions very favourable to the preservation of shells. During the period between the two World Wars a large number of investigations was made, and these were generally published as appendices to excavation reports. One of the earliest concerned the Neolithic flint mines at Grimes Graves in Norfolk where snails were used not only as indicators of the past environment, but as evidence for the Post-glacial origin of the mines whose age was then in dispute (Kennard and Woodward *in* Clarke, 1915: 220; *in* Peake, 1919: 91; Kennard *in* Armstrong, 1934: 393; Clarke, 1917; Armstrong, 1927). In Wessex, Kennard is best known for his work with Mrs M. E. Cunnington—for example at Woodhenge, The Sanctuary and Yarnbury Castle (Kennard and Woodward *in* Cunnington, 1929: 70; 1931; Kennard *in* Cunnington, 1933a: 207)—and with J. F. S. Stone, especially on the series of excavations at Easton Down (Kennard *in* Stone, 1933: 235; *in* Stone and Hill, 1938, 1940). Mrs Cunnington (1933b) and her nephew R. H. Cunnington (1935) used the evidence of climate, derived from land-snail faunas, to argue a Bronze Age date for Stonehenge.

This early work generally comprised the extraction of assemblages of subfossil shells (faunules as Kennard preferred to call them) from spot samples of soil usually taken from visibly shell-rich horizons, e.g. the *Cyclostoma elegans* zone referred to by Pitt-Rivers. The results of analysis were presented as lists with either the numbers of each species or a graded series of frequency estimates such as "abundant", "common", "frequent" and "rare". Environmental interpretation was based on comparison of the subfossil faunule with that of present-day populations.

In some respects however, the work of Kennard and Woodward is unsatisfactory (Evans, 1970). In the first place, the collection and extraction of shells was often on a partial basis, which, taken with the tendency to select shells by hand-picking during excavation, resulted in a bias towards the larger and otherwise readily recognizable species (Figs 2 and 105). This inevitably led to faulty interpretation of the assemblage as demonstrated by Sparks (1961: Fig. 1). Secondly, insufficient attention was paid to the stratigraphy of the deposits from which shell collections were made, interpretations being based largely on the results of analysis. Kennard seldom visited the archaeological sites whose land-snail faunas he studied, depending on the excavators to provide him with samples. The stratigraphy of soils and sediments is often extremely fine, pronounced faunal changes taking place within

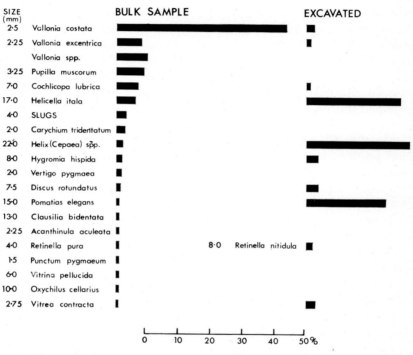

Figure 2. Comparison between faunas derived from the soil beneath a Bronze Age barrow at Arreton Down, Isle of Wight, by bulk sampling and visual selection. (After Sparks, 1961: Fig. 1).

a small thickness of deposit. If, therefore, mixed and ecologically meaningless assemblages are to be avoided, samples must be taken on an equally fine basis, and until recently this has not been done.

Thirdly, the interpretation of assemblages was often faulty, and not only due to the analytical deficiencies mentioned above. For instance, a subfossil assemblage of shells, even if extracted from a single homogeneous horizon, may not be an exact reflection of a former live population owing to *post mortem* changes such as the differential destruction of the shells of certain species or to the sorting action of earthworms (p. 212). Furthermore, interpretation was often biased by attempts both to date deposits and to make deductions about past climatic conditions. The basis of many of Kennard's interpretations was climatic, and generally in terms of the prevailing humidity as controlled by rainfall, less frequently in terms of temperature. The overriding influence of local habitat conditions in controlling the composition of land-snail populations, which we now know to be the case, was hardly recognized at all by Kennard. Looking back over his publications, which span half a century, they seem surprisingly monotonous, all the more remarkable when one recalls that during this time, pollen analysis was discovered and set on a firm quantitative basis.

The large size of snail shells in comparison with pollen grains, and the ease with which results could be obtained by simple techniques of extraction were no doubt factors in the uncritical and quasi-quantitative approach to snail analysis in the early days. Pollen analysis from its inception was the province of the professional researcher because of the more exacting methods of extraction and the need for a fully quantitative approach if results were to be meaningful at all. Moreover, the strong appeal of pollen analysis, in giving an apparently direct reconstruction of former vegetation types and its unquestioned value both as a chronological tool and as an indicator of former climates led to its rapid and successful establishment.

Kennard, working as an amateur in a large City warehouse, was without the benefit either of a university training or of a scientific environment. In addition he was essentially a geologist, imbued with the concept of using zone fossils in correlating geological formations over wide areas. The idea of palaeoecology, whereby ancient environments and faunas are studied for their own sake, is a relatively recent development (Ager, 1963). Indeed even the science of animal ecology itself, as the study of populations and communities of animals in relation to their environment, was then young. C. S. Elton's pioneer and classic work "Animal Ecology" was not published until 1927 by which time Kennard had thirty years of research behind him. It must be remembered too that in these early years the efforts of pollen analysts were directed largely towards the establishment of a chronology for Post-glacial time, the basis of which was climatic. Only since the last war, and particularly after the recognition in the pollen record of the influence of early man on vegetation, has there been a shift of emphasis towards the reconstruction of local environments. Nor could

Kennard benefit from the stimulus of other researchers on sub-fossil Mollusca for there were few, and none in the archaeological field.

It is against this background that Kennard's work must be assessed. The idea of using organisms as indicators of local environment was foreign to the research trends of the time, which was all the more unfortunate in the case of land snails as they are most suitable for this purpose—perhaps even more so than pollen.

However, even at the time of the Cunningtons' work on Stonehenge, other archaeologists were beginning to read alternative conclusions into Kennard's results. A footnote by J. G. D. Clark to the report on the snails from the Thickthorn Down Long Barrow (Kennard in Drew and Piggott, 1936: 94) called attention to the fact that differences between two faunas might be explained in terms of habitat differences rather than climate. Piggott (1954: 6), too, was clearly not satisfied with some of Kennard's conclusions regarding the climate of the Neolithic and Bronze Age periods. And Kennard himself in some of his later reports expressed unease in placing too much emphasis on climate as the main factor in controlling molluscan faunas; for example, "The whole series indicates a scrub growth and certainly damper conditions than now exist . . . Whether this is due to a difference in age or whether it arises from local conditions one cannot say" (in Stone, 1931: 363).

Kennard died in 1948 at the age of 78, and since then snail reports have been less prominent in the archaeological literature. Those which have been published, however, show a greater tendency towards interpreting results in terms of local habitat conditions than was previously the case (e.g. Sparks in Alexander et al., 1960: 299; Connah and McMillan, 1964; Evans in Smith, I. F., 1965a: 44; Evans, 1966a). Others dealing with shells hand-picked during excavation are of less value.

The earliest graphical record of subfossil assemblages of land snails in Britain was made by Burchell (Burchell and Piggott, 1939: Fig. 2) using a technique known as the pie-dish or sector diagram (Fig. 3). Four species, each selected as representing a particular facies of the environment, and as being characteristic of a specific period of the Post-glacial, were plotted as segments of a circle. The size of each segment (or sector) is a direct reflection of relative abundance, and the total area of the circle is 100%. This method is unsatisfactory in ecological studies due to the use of only a part of the fauna, with the result that the reconstruction of the ancient environment is incomplete. Its application should be restricted to faunal elements whose abundance can be safely assumed to be controlled by climate or other widespread environmental factors. With land snails, at any rate in the Post-glacial, one can rarely be certain that this is the case with any species.

In two later papers, Burchell (1957; 1961) attempted a synthesis

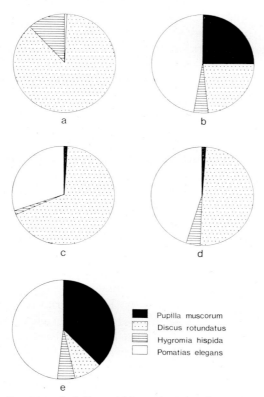

Figure 3. Sector diagrams of shell assemblages from Neolithic and Bronze Age sites. (a) Peaty alluvium, Ebbsfleet. (b) Sub-aerial loam, Bean Valley. (c) Bronze Age alluvium, Ebbsfleet. (d) Bronze Age peaty silt, Brook Vale. (e) Turf line beneath Julliberries Grave. (After Burchell and Piggott, 1939: Fig. 2).

of the Late Weichselian and Holocene land Mollusca of south-east England, and suggested that the successive stages of these periods could be characterized by the composition of their land-snail faunas, an attempt no doubt inspired by the success of palaeobotanists in their pollen analytical studies. It is true that there have been changes in the overall composition of the land-snail fauna in Britain during the Post-glacial which can be attributed to climate, the restriction in the distribution and abundance of *Pomatias elegans* since the Bronze Age being a case in point (Fig. 43) (Kerney, 1968a). But in terms of the individual habitat, with which the archaeologist is generally concerned, local conditions may easily override the influence of climate, and *P. elegans* can occur in extraordinary abundance on the very edge of its present-day range, and, if conditions are suitable, far beyond in isolated colonies. As Kerney (1963: 239) pointed out, "To correlate percentage frequencies from one deposit to another is misleading and is only to correlate possible similarities in environment, not in age."

In Britain, the analysis of land and freshwater molluscs was put on a firm quantitative basis by B. W. Sparks who applied the technique of serial sampling to shell-bearing deposits and presented the results of analysis in histogram form (Sparks, 1957a; Sparks and West, 1959). Individual species and groups of ecologically or climatically related species were plotted as histograms of relative abundance in a manner similar to that used in presenting pollen analytical data. Vertical changes in abundance through the deposits were interpreted in terms both of local environmental and climatic change. Since then, Sparks has made a number of detailed studies, mainly on freshwater deposits of Pleistocene age (Sparks, 1964a; Sparks and West, 1964, 1968, 1970).

Similar techniques have been applied to Late Weichselian and Post-glacial deposits in south-east England (Kerney, 1963; Kerney *et al.*, 1964; Evans, 1966b), investigations which are of direct relevance to archaeology in that they deal with terrestrial deposits; the latter two demonstrate the impact of prehistoric agricultural communities on the landscape of the Chalk. In the last five years a number of papers have been published which deal directly with the land-snail fauna from archaeological sites (Fowler and Evans, 1967; Evans, 1968a, 1969a, 1970, 1971b) and are concerned largely with landscape changes on the calcareous soils of Britain. Kerney (1966b) has discussed the various ways in which man has altered the land-snail fauna; and in the case of a few species, has shown that the Post-glacial thermal decline was responsible for changes in their distribution and abundance (Kerney, 1968a).

An interesting development in the past few years has been the recognition of changes in the frequency and distribution of various forms of the polymorphic snail *Cepaea nemoralis* during the Post-glacial (Cain and Currey, 1963a; Currey and Cain, 1968; Cain, 1971). This snail is the largest to occur on prehistoric sites, and is well-known to excavators on the Chalk. It occurs in a variety of forms, the main variations being the background colour of the shell and the number and disposition of darker bands of colouring which may vary from none to five. Cain and Currey, using subfossil *Cepaea nemoralis* shells from tufa sites and archaeological excavations, have demonstrated considerable alterations in southern England and Wales in the proportions of various morphs from the altithermal to the present day, changes which are consistent with climatic selection (pp. 43 and 174).

The technique of snail analysis has now been refined to the extent that we are able to recognize and differentiate between the effects of climatic, local environmental and anthropogenic influences on ancient snail populations. But we still need to know a good deal more before realistic reconstructions of ancient habitats can be made. In the first place, precise information on the modern geographical ranges of many species is not always available even within the British Isles, though this situation is being remedied by the Conchological Society's mapping scheme, and a similar scheme for Europe is being devised. Then there is the need for more palaeontological work, particularly from deposits in which pollen and snails are preserved together where we can get direct information about the contemporary plant communities with which ancient land-snail faunas were associated. But the most urgent requirement is for quantitative data on modern populations from habitats whose physical, chemical and biological attributes are fully described. Until we can obtain this information—and it is lacking for most species and habitat types—the interpretation of subfossil assemblages can be made only at the most general level. This is particularly critical for archaeological work where the interest is in local environment rather than climate, and I feel quite sure that were the data available the interpretation of faunas could be pushed beyond the simple level of "grassland" or "shaded ground" to the point where we could define what sort of grassland was present—uniform, tussocky, grazed by sheep, grazed by cattle or long reverting to scrub.

With the notable exception of Central Europe, and in particular Czechoslovakia (Lo^{Z}ek, 1964), little comparable work on land snail assemblages from archaeological and Pleistocene deposits has been done outside Britain. I have briefly discussed work which has been done in Central Europe, the Middle East, Africa and North America (Evans, 1969a), and the interested reader should consult this paper for further references.

In Central Europe and North America, molluscan analysis is of considerable relevance to Pleistocene studies, in some ways much more so than in Britain, owing to the widespread occurrence in these areas of deposits of calcareous loess. This material (p. 290) accumulated during periods of cold climate and is often rich in shells. Intervening warm periods are reflected by soil horizons which, when not decalcified, also contain Mollusca. These have been used successfully in Czechoslovakia to characterize the successive warm and cold periods of the Quaternary in the same way that pollen is used in this country to define the various interglacial periods. Not only can the distinction between interglacial, interstadial and glacial stages be recognized but in many cases the faunas are sufficiently diagnostic to characterize specific periods of time. This is in many ways due to the fact that Central Europe is situated between two mountain masses—Scandinavia and the Alps—on which ice-caps formed during the last three major glaciations. As these encroached from both north and south, many species of snail were exterminated; during the succeeding warm periods, the barrier of the Alps prevented northward movement from the Mediterranean area of species which might otherwise have been able to tolerate the climate of Central Europe. It is thus to some extent a matter of chance in the case of certain species as to whether they were present or not in any particular interglacial. In North America, on the other hand, where the main mountain ranges run north–south, movements of land snails are not thus restricted and there is far less chance of a species becoming extinct. Differences between the fauna of the various interglacials should therefore be less pronounced and depend more on climatic differences from one interglacial to the next than on chance. Owing however to the paucity of investigations in North America and the difficulty of isolating true interglacial horizons (apart from the Last, Sangamonian, Interglacial) this has not been established as a fact.

An important adjunct to the development of snail analysis as a branch of Quaternary science has been the increased awareness and investigation of a variety of calcareous deposits, for a knowledge of the way in which these have formed is essential if shell assemblages are to be interpreted correctly. For example, wind-lain sand, tufa and hillwash are deposits which are widely distributed in Britain (Figs 7 and 112) yet this is not generally appreciated, largely, one suspects, because they have been but rarely exploited commercially. In contrast to peat, sections through such deposits are comparatively uncommon, yet from what we know about the archaeological and environmental content of the few which have been investigated in detail, they can undoubtedly yield equally revealing results. Certainly in the south and east of England where great areas of countryside are totally devoid of peat or lake sediments, their study by archaeologists and Quaternary biologists alike has been delayed too long.

In addition to deposits of this sort which have formed by subaerial accumulation, there are those former land surfaces which have developed by *in situ* weathering of the subsoil, and these we call soil horizons. They are best preserved when buried by field monuments and are frequently encountered thus in the course of archaeological excavation. Less commonly they occur beneath colluvial deposits of natural or semi-natural origin of the type mentioned above such as loess or hillwash. And as with these, there is in Britain no great tradition of their study. This is in marked contrast to the situation in the Low Countries and Central Europe where thick and extensive deposits of wind-blown silt or sand have buried and preserved, often without any concomitant disturbance of the profile, soil horizons of many ages. The investigation of these and their molluscan faunas has greatly aided the establishment of a scheme of climatic phases for the Pleistocene, particularly during the periods of periglacial climate when loess formation was rife (Klima *et al.*, 1962; Ložek, 1965a, 1967).

In this country, the study of ancient soils is of recent origin. Cornwall (1953) and Dalrymple (1958) applied strictly pedological techniques in their investigations; Dimbleby (1955, 1961, 1962) applied pollen analysis to heathland soils; and I have made a study of buried soils on calcareous substrata from the point of view of their molluscan faunas (Evans, 1967). Archaeological investigations have been comparatively rare (O'Kelly, 1951; Atkinson, 1957; O'Kelly, 1969).

At the 1968 British Association meeting in Dundee, a study group including archaeologists, biologists, geologists, soil scientists and a radiocarbon expert, met to discuss buried soils—their potential value in various aspects of research, and the methods of investigation that could be applied to them (Dimbleby and Speight, 1969). The results of this meeting made it eminently clear that the study of buried soils in Britain can provide important information about former environments, phases of land use and habitat change.

This book is essentially a study of the terrestrial ecosystem of early man in the calcareous territories of Britain, in which the main evidence is derived from snail shells and buried soils. The technique of snail analysis has enabled environmental archaeology to be applied to the Chalk and other lime-rich areas in Britain (Fig. 7)—vast tracts of land which were vitally important in the settlement and livelihood of early farming communities from the earliest Neolithic to the end of the Roman period, and whose history was formerly obscure. The founder of this study was A. S. Kennard whose results have been left to us in over 200 papers—largely published as appendices to archaeological reports—and in his note-books, now in the British Museum. These have formed the basis of our present-day research. Today, some of Kennard's techniques appear a little unrefined and his conclusions are often at variance with current ideas; but his energy and enthusiasm for the subject maintained over a period of 50 years, coupled with his achievements, are testimony to a remarkable man. As students of Britain's ancient snails, our debt to A. S. Kennard is very great indeed.

2
Principles

The basic principles of snail analysis are these: snails are small invertebrate animals whose soft parts are enclosed within a hard exoskeleton, the shell, which is composed largely of calcium carbonate. As such it is readily preserved in lime-rich soils and sediments. Shells can be extracted from ancient deposits, identified, generally down to species level, and an estimate made of their former abundance. By comparing these ancient faunas with their nearest present day analogue, whose composition in terms of species is governed to a large extent by climatic and local habitat factors, we can get some idea of the past environment in which they lived. The usefulness of the method in the study of ancient human environments depends on our ability to date the subfossil fauna, on the degree to which we can assess its true composition in life, and on the accuracy with which the former environment can be reconstructed from it.

Snail faunas from archaeological sites can be used to interpret past climates and local environments. In this book the emphasis is on the latter, for the majority of sites discussed fall in the second half of the Post-glacial when climatic changes were of minor intensity and the all-powerful influence of man generally overriding. Moreover it is the detection of man's activities in altering the habitat and landscape which is of most interest to archaeologists and ecologists; climate change, while very relevant to the study of man's past, is more difficult to detect in the fossil record and is an aspect about which land-snail faunas have yielded little information for the past 6000 years, i.e. from Neolithic times onwards.

Difficulties fall into two main categories: stratigraphical and ecological.

The analysis of a soil sample for land snails yields what is known as a subfossil assemblage (Kennard's faunule)—a collection of shells of various species and of varying antiquity and states of preservation—which differs from the live population (or populations) from which it is derived in several respects. Thus within a community of living snails at any one time, there will be found variation in the age of the individuals (some being adults, others half grown, and others newly hatched), variation in their horizontal distribution over the ground, and variation in their vertical distribution. These aspects are not preserved in their entirety in the fossil record. For example, let us consider vertical distribution. In life land snails occupy a range of habitats above and within the soil. Some species live close to the soil surface in leaf litter or at the base of grasses; others are more frequent some distance above the soil on the stalks of herbaceous plants or beneath the bark of fallen branches. Yet others may be found well above the ground on trees or stone walls. There are also some which spend all or part of their life within the soil itself at various depths, some immediately beneath the surface, others at a depth of 2 m (Fig. 55). Each species has an optimum habitat on this vertical scale but many range widely above or below it. Thus many species which normally live on vegetation may aestivate or hibernate within the soil where, if weather conditions are severe, they may die and become entombed. *Limax maximus*, the Great Slug, copulates suspended from

Figure 4. Copulation in *Limax maximus*. (After Adam, 1960: Fig. 6).

a branch of a tree on a thread of slime (Fig. 4) but spends most of its life on the ground.

At death, all snails on or above the ground are accumulated together at its surface by the force of gravity, while those within the soil remain *in situ*. There is thus an immediate alteration in the structure of a snail population with the implication that shells can never be fossilized in their position of life. What was once a three-dimensional distribution in life becomes a two-dimensional one at death and such fascinating details of molluscan behaviour as are depicted in Figs 4 and 5 are not preserved in the fossil record. This collection of shells is known as a *death assemblage*.

The discrepancy between the structure of a death assemblage and the living community is greatest in a forest environment where the

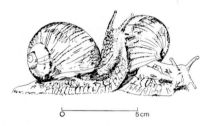

Figure 5. Courtship in *Helix pomatia*.

vertical range of habitats is large, and least in an environment such as short-turfed grassland where it is minimal.

Thus even if no further changes occur in the composition of the death assemblage, the structure of the living population from which it derives can only be surmised by reference to the habits and behaviour patterns of modern populations. But further changes generally do occur.

The periostracum of a shell is destroyed within a year of death, leaving the calcium carbonate exposed to the agencies of physical weathering and solution. Unless quickly buried, a shell is soon destroyed, and many are probably lost in this way. During its incorporation into a soil or sediment, a death assemblage of shells is subjected to a variety of processes, physical, chemical and biological, which further distort its composition, and which it is necessary to understand fully in order to reconstitute the original structure of the living population. These will be considered in Part III. They concern the way in which shells are incorporated into a soil, the differential destruction of shells of varying degrees of resistance to weathering, and the segregation of shells of different sizes (and species) by worm action. These processes change the composition of the death assemblage so that what we extract from the soil is one stage further removed from the living population. This we call a *subfossil assemblage*. The relationship between the soil sample, the subfossil assemblage, the death assemblage, the living community and the environment are shown in Fig. 6.

I shall now turn to ecological problems. The abundance of each species in a community is controlled partly by hereditary factors congenital to the species, and partly by the environment (Fig. 6).

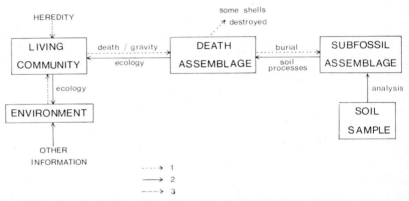

Figure 6. 1 = stages in the formation of a subfossil assemblage. 2 = stages in analysis and interpretation. 3 = factors controlling the composition of snail communities.

Some species, such as *Vertigo pygmaea*, seldom occur in abundance however suitable the habitat; others, such as *Carychium tridentatum* and *Discus rotundatus*, are usually prolific. Thus the ratio of different species in a community is to some extent a function of their powers of reproduction and survival. Otherwise, the composition of a population is controlled by the environment. Some factors determine the distribution and abundance of certain species on a regional basis, such as climate and the lime content of the soil. *Pomatias elegans* is instructive in this respect. This species is an obligatory calcicole, and the way in which its distribution follows the chalk downland and other limestone outcrops of southern Britain is strikingly apparent (Fig. 43). The few occurrences in the West Country on localized outcrops of Devonian Limestone and wind-blown shell sand demonstrate in a convincing manner this particular environmental control (cf. Fig. 7). Other factors, such as slope, exposure and shade operate on a more local scale and in general it is these which are of relevance to archaeological problems.

One of the difficulties of using modern snail populations as a standard of comparison for the past is that the necessary quantitative data is generally lacking. Observations have been made on the life histories and seasonal changes in abundance of about half-a-dozen species (Morton, 1954; Baker, 1965, 1968, 1970; Chatfield, 1968), and we are beginning to find out, through the Conchological Society's mapping scheme, quite a lot about the distribution of land snails in Britain (Figs 43 and 45) (Kerney, 1967). The work, largely of Cameron, on *Cepaea* and *Arianta* in which ecological observations have been tested against laboratory experiments on their behaviour under various controlled environments has given us an impressive body of information about these species (Cameron, 1969a and b, 1970a and b; Cain *et al.*, 1969; Cameron and Palles-Clark, 1971). Almost all other ecological records have been of a qualitative kind, and there has been little synecological work on the molluscan population of an individual habitat in Britain. Notable exceptions are the work of Mason (1970) and Chappell *et al.* (1971).

Another problem is that certain species appear to have changed their habitat requirements with time, so that direct comparison with present-day faunas, even if we do have adequate data about their ecology, is not always valid. As I shall discuss in Part II (p. 130), this is not such a serious problem as it appears to be, and in a number of cases I suspect that it is not the snail species which have changed their habitat preferences but the habitat itself which has changed. For example, *Vallonia costata* and *Helicella itala* were prolific in prehistoric arable habitats (e.g. Wayland's Smithy, p. 265, South Street, p. 259 and Badbury Earthwork, p. 341), in which the soil was intensively disturbed by cultivation; today they are absent from such places. On the face of it these two species have undergone a change of habitat preference. But can we equate the modern arable environment with that of the past? For a variety of reasons I think not. Even supposing that the immediate appearance and properties of cultivated land today and in the prehistoric period were similar (which they probably were not, owing to different methods of tillage), there have been changes of climate, faunal changes due to the introduction by man of species of snail which have competed with the indigenous population, and changes of soil type due to prolonged cultivation. Ancient landscapes and habitats are not only fossilized; in certain cases they are extinct as well.

In the absence of an exact present-day parallel to a subfossil fauna, its structure as a whole may yield some useful results about both the contemporary climate and the local environment. On the climatic side, broad changes in the distribution of certain species, if related to a particular isotherm or isohyet and meaningful in terms of their physiological requirements, may be indicative of climatic change (Fig. 43). On a more local scale, the structure of a fauna, not in terms of species so much as the distribution of numbers among the various species present, may be a useful guide to the favourability of a habitat for molluscan life (Figs 35 and 132); e.g. the degree of disturbance, the water-retaining capacity of the soil and the range of temperature fluctuation. For example, a fauna in which 90% of the snails are of one species with a few others in low abundance reflects a severe habitat such as a salt-marsh, ploughed field or sand-dune. At the other extreme, the fauna of an undisturbed, well-established woodland may contain upwards of fifty species with none amounting to more than 15% of the total fauna.

The particular value of the histogram as a method of presenting the results of snail analysis is that it shows the pattern of change. Often it is not possible to reconstruct the exact character of an environment at a given point in time, but one can deduce from it what factors are changing and the direction of change; i.e. we can talk in terms of a decreasing soil-moisture content or an increase in shade, but without knowing the values for these factors in any one instant.

In view of these various difficulties of interpretation, it is vitally important that as much additional information as possible about the habitat of ancient snail populations be obtained from other sources (p. 35).

Preservation

In life, the shell is covered with a thin proteinaceous coat called the periostracum, which is usually destroyed within a year of death, unless the shells are rapidly preserved in anaerobic conditions. Such conditions obtain in peat and in the waterlogged deposits of ditches (Arbury Road, p. 349 and Maxey, p. 346). The turf-stack of Silbury Hill (p. 266) was also anaerobic and there it was possible to separate out, by virtue of their preserved periostraca, those shells of animals which were alive or had only recently died when the mound was built, from those of more ancient origin (Fig. 93). In general however, the deposits in which shells occur are aerobic, and their preservation depends on the maintenance of a calcareous environment.

The crystalline form of calcium carbonate in the shells of most species is pure aragonite. The internal shells of the *Arion* and limacid slugs are exceptions in being composed of the more stable form, calcite. In both groups, the shells are subject to little change either in their crystalline or chemical composition, and for this reason are described as subfossil rather than fossil.

The degree of preservation varies considerably but in most terrestrial deposits which are highly calcareous, such as buried soils and hillwashes, it is good, the finest details of shell sculpturing being preserved. This applies to deposits derived from the Chalk and softer Jurassic limestones. In the case of the Magnesian and Carboniferous Limestones which are harder and weather more slowly, the derived soils may be barely calcareous and the shells in a poor state of preservation or absent altogether. This is by no means always the case, there being considerable variation from site to site depending on local conditions.

Shells from ditch deposits on river gravel (Maxey, p. 346 and Arbury Road, p. 349) are generally less well-preserved, being fragile and eroded. Sparks (1969) notes that in well-aerated, oxidized deposits, shells may be affected by leaching, and "... even though the shell itself may not be destroyed, the fine detail of ornamentation, often a diagnostic feature, may suffer severely."

On neutral and acid soils shells are never preserved. Indeed on extremely acid soils where microbial activity is slight, I have seen instances in which the calcium carbonate fraction of the shell has been destroyed and the periostracum preserved.

Different species of snail vary in the resistance of their shells to physical and chemical destruction, thin-shelled species such as *Oxychilus* being more readily destroyed than the more robust shells of *Cepaea*. Some shell apices, notably those of *Pomatias elegans* and *Clausilia*, become enlarged by the accretion of calcium carbonate, and appear to remain in the soil indefinitely. And the undersides of weathered apices of *Acanthinula aculeata* (Fig. 23) and, occasionally, *Cochlicopa*, bear calcareous granules, not present in living or well-preserved shells. This problem of the differential preservation of shells is discussed in Part III (p. 212). It is a phenomenon which can be easily recognized and dealt with when it arises, there being no question of its seriously invalidating the results of analysis. There are perhaps only two species, *Hygromia subrufescens* and *Monacha granulata*, which, because of the extremely fragile nature of their shells, rarely occur subfossil. In comparison with the gross discrepancies which prevail in the preservation of pollen grains, where whole groups are never represented, the problem is a minor one indeed.

Deposits Containing Shells

Soils and sediments containing the shells of land molluscs are abundant and widespread in the British Isles, and occur in association with a variety of geological solids (Fig. 7). Sites on the Chalk are generally rich in shells. Areas covered with sandy drift or Clay-with-flints are exceptions; for example much of the Chiltern Hills (particularly the dip slope), the North Downs and the East Anglian Breckland. The chalk of the Yorkshire Wolds is extremely hard and brittle in places, often giving rise to non-calcareous soils (e.g. Kilham, p. 277); and in the south of England decalcified "chalk heath" soils (p. 216) may develop if their content of siliceous minerals is high, even though the chalk parent material be no more than 30 cm below the surface (Perrin, 1956). Even so some of the activities of prehistoric man, such as the construction of barrows and the mining of flint, led him to dig through the drift skin and bring up to the surface the underlying chalk which was spread around thus creating a calcareous environment more suitable for molluscan life and the preservation of their shells. This is well exemplified at Grimes Graves in Norfolk (Fig. 8). Here, while the ancient soil is a non-calcareous brown earth on sand, the modern soil and fill of the mine shafts are highly calcareous being derived from the dumps of chalk waste brought up in the process of mining flint.

In general, soils derived from Jurassic rocks are also rich in shells, though as with the Chalk, conditions vary from site to site. Rocks of Portland Limestone, Corallian, Cornbrash, Greater and Inferior Oolite and parts of the Lias all provide parent materials suitable for the preservation of shell (Fig. 7).

On the Carboniferous and Magnesian Limestones, one is less certain of obtaining shells, but there is no doubt that they can occur in profusion given the right conditions. Stone cairns, ditches and caves are generally satisfactory due to the high concentration and constant

2. PRINCIPLES

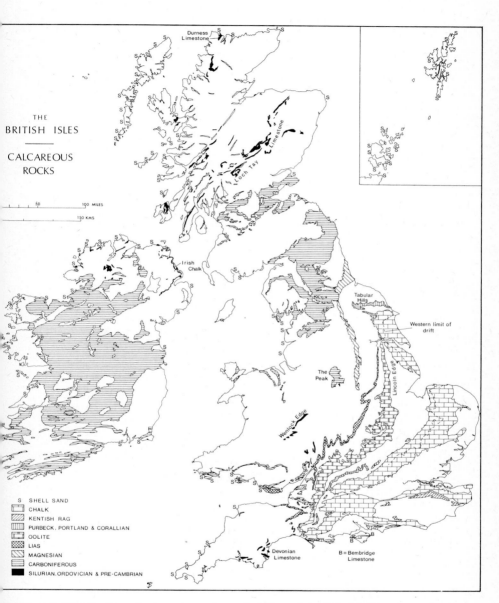

Figure 7. Distribution of calcareous rocks and shell sand in Britain. (Based on the Ordnance Survey 10 miles : 1 inch map, with the sanction of the Director General: Crown Copyright Reserved.)

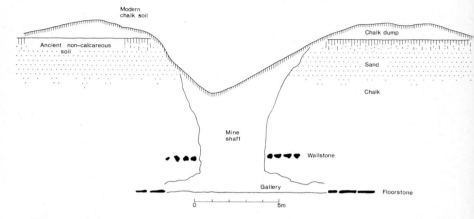

Figure 8. Grimes Graves, Norfolk. Section through flint-mine shaft and surrounding dumps.

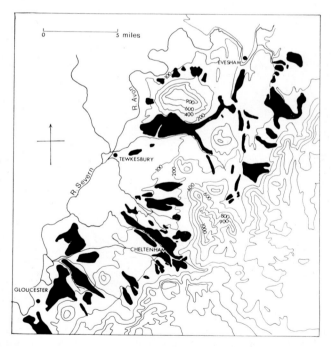

Figure 9. Distribution of fan gravels (black) at the foot of the Cotswold escarpment. (After Arkell, 1947: Fig. 43.)

replenishment of calcium carbonate in the environment. *In situ* soils on these parent materials are less suitable for they are often decalcified due to the slow rate of weathering of the parent rock (Fig. 64); indeed it is a general rule that on limestones other than the Chalk, colluvial deposits are more calcareous and thus richer in shells than soils formed *in situ*. Much of the Carboniferous Limestone, particularly in the north of England and in Ireland, is overlain by blanket peat and boulder clay.

There are too a number of older limestone formations such as the Loch Tay Limestone in the Central Highlands of Scotland, the Durness Limestone in the north west, the localized outcrops of Devonian Limestone in south-west England and the Silurian of Wenlock Edge whose soils are calcareous and suitable for molluscan analysis.

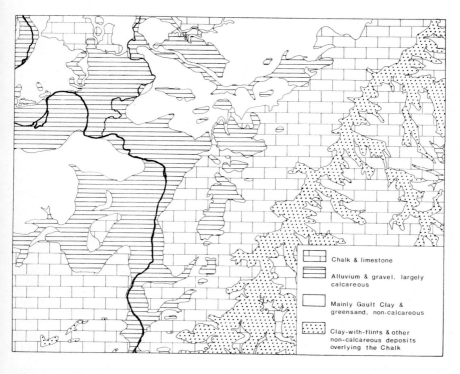

Figure 10. Distribution of drift deposits associated with part of the Chiltern escarpment, Oxfordshire. (Based on the Ordnance Survey 1 mile:1 inch geological drift map, sheet 254, with the sanction of the Director General: Crown Copyright Reserved.)

Derived from these geological solids is a variety of lime-rich drift types which give rise to calcareous soils capable of supporting a rich molluscan fauna and of preserving their shells when dead. River alluvium and gravel, and slopewash deposits of solifluial origin are the most widespread of these, and in the present context their chief importance is that they extend the range of the use of snail analysis away from the immediate vicinity of calcareous rocks into regions which would otherwise be non-calcareous. The fans of solifluxion gravel beyond the foot of the Chiltern and Cotswold escarpments are

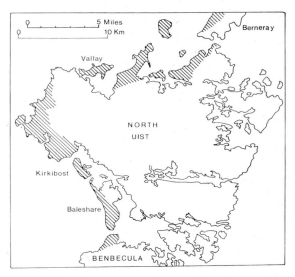

Figure 11. Distribution of machair (shaded) in North Uist, Outer Hebrides. (After Sissons, 1967: Fig. 40.)

good examples of this phenomenon (Figs 9 and 10). Even more striking are the thick deposits of calcareous wind-blown sand which overlie igneous rocks and boulder clay in many places along the western and northern shores of Britain (Figs 7 and 11). "In many of the western islands blown sand is very widespread and often consists almost entirely of comminuted shell fragments." In Scotland "it gives rise to the sandy plains, or machair, that lie between the coastal dune belt and the hills and bogs of the interior," and is derived from "wide sandy beaches exposed at low tide . . ." which ". . . supply abundant material for the winds to carry landwards" (Sissons, 1967: 229). The soils of the machair, or fixed-dune pasture as it is known in other parts of Britain, are often highly calcareous, supporting an abundant land-snail fauna. "Blown sea sand ... often enriches an otherwise inhospitable shore and is ... one of the reasons why the immediate neighbourhood of the sea is apt to be much more prolific than the ground half a mile inland" (Boycott, 1934: 11). It is also to be pointed out that these deposits of lime-rich drift are important from the point of view of human settlement, in providing tracts of fertile, well-drained soil suitable for agriculture and grazing stock (Fig. 12) in

Figure 12. Benbecula, Outer Hebrides. The machair.

areas which would otherwise be barren and hostile wastes (Fig. 13). That they are also rich in the shells of land snails gives them an added significance in the study of ancient human environments.

The construction of buildings and walls of limestone also constitutes an extension of the calcareous habitat beyond its natural limits. Today, hundreds of miles of limestone field walls in western Britain provide very good living conditions for a number of species of land snail which would otherwise find existence precarious in the predominantly agricultural and often acid landscape (Boycott, 1929a, 1934; Kerney, 1966b). The Roman town of Caerwent in South Wales is sited on non-calcareous sandstone and gravel; but the walls and bastions are built largely of Liassic Limestone. Tumbled rubble at the base is rich in the shells of land snails, and indeed the founders of

Figure 13. Taransay, Outer Hebrides. Hillsides which once supported birch forest now laid bare.

many of these snail colonies may have been brought to the site in the first place on the stone from which the walls were built (cf. Fig. 111).

Clearly, in assessing the likelihood of shells being present on an archaeological site, neither the solid nor drift geological maps, nor indeed the soil map will necessarily give us the required information. As Boycott (1934: 10) pointed out, there are unexpected areas of calcareous ground such as the many patches of cornstone in the Old Red Sandstone of Herefordshire and the patches of limestone among the schists of Inverness; and man, by his building, quarrying and agricultural operations, has upgraded the lime content of many places. On the whole of course, the south and east of Britain are more suitable for the application of molluscan analysis than the north and west. But this is only a general rule. Each site must be examined individually, for shells may be found in abundance on the northern shores of Scotland and yet be absent from our southern Weald.

Land snails occur in a variety of archaeological deposits, not all of which, however, are of value in yielding information about former environments. It is important that the kind of data to be expected and required from a deposit be intelligently assessed *prior* to sampling. The richer horizons are not necessarily the most useful from the environmental point of view, the "*Cyclostoma elegans* zone" of the early excavators being a case in point (p. 6). On terrestrial sites, shell-bearing deposits fall into four categories:

1. Soil horizons formed *in situ*
2. Deposits in hollows such as ditches, pits and wells
3. Ploughwash and other colluvial sediments
4. Occupation horizons and building débris.

Ancient soil horizons are preserved in a number of ways, the most usual being beneath a man-made structure such as an earthwork, stone wall, building or road. They are also frequent within the fill of ditches, when they are generally sealed by ploughwash or collapsed building débris, e.g. from the rampart of a hillfort. A further type of situation is when a soil is buried beneath a natural or semi-natural deposit such as tufa, hillwash or blown sand; examples are discussed in Part III. But unless in a direct archaeological context this latter type of situation is less satisfactory due to the difficulties of dating. Archaeologically, a scheme of environmental change, however elaborate, is of little value unless it can be tied in to a cultural horizon, and this is generally possible only when there is a direct link between the environmental and archaeological evidence.

In considering soil horizons, one of the main problems is in estimating the area of environment reflected by a shell assemblage from a soil sample. This we shall discuss in Part II. Here, all that need be said is that by taking a *series* of samples from a soil horizon rather than a spot sample we can get an idea of the environmental changes which are taking place, without necessarily being able to assess their exact value; it is reasonable to assume that a similar *sequence* of events would be recorded from other parts of the same soil horizon, even though the general facies of the fauna throughout the sequence be somewhat different. In theory, the only sure method of checking the significance of a fauna is to analyse several samples, laterally separated, from a single stratigraphical horizon. In practice however, experience has shown that the fauna from a buried soil reflects the environment of an area, the size of which can be considered in terms of human land use. Certainly the environment of a soil is uninfluenced by the extreme micro-habitat conditions which obtain in a pit or on the bottom of a ditch where the fauna is a reflection of those conditions alone, and not of the surrounding landscape.

Because it is protected from such weathering processes as chemical solution and ploughing, the buried soil beneath an earthwork is often preserved at a higher level than that of the adjacent, modern soil—a phenomenon known as differential weathering (Fig. 14). Discrepancies

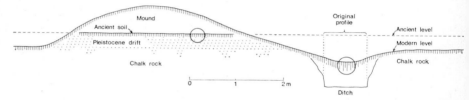

Figure 14. Transverse section through an earthwork of bank and ditch type. Critical horizons, likely to yield the most valuable environmental information, are ringed.

of about 30 cm have been recorded between Bronze Age and modern levels (Atkinson, 1957). Because of this, features are preserved in and beneath a buried soil which from the modern, active profile, have been long since destroyed. In many cases superficial deposits of Pleistocene origin are present only beneath an ancient earthwork, elsewhere, beyond its protective bounds, having been eroded away (Fig. 14) (Evans, 1968b). Land snails too, once sealed in a buried soil may remain preserved indefinitely, while in an unprotected and actively weathering soil they are gradually destroyed (p. 229). In a modern soil one does not find an unbroken sequence of snail faunas representing its entire history from its origin at the beginning of the Post-glacial to the present day; indications are that in chalk soils, only the last 200 or 300 years are represented in the active zone, though in exceptional circumstances features of greater antiquity may be preserved (e.g. Coombe Hole, p. 241). Hence the importance of a buried soil is the information it can yield about the environment at a specific time in the past: once sealed by an ancient monument, a soil, its associated drift deposits, shell assemblages and cultural paraphernalia remain preserved.

From a buried soil in the kind of situation just described, information can be obtained about phases of environment and land use before and right up to the construction of an earthwork. The main evidence is derived from morphological features of the profile, from molluscan analysis, from radiocarbon assay of organic inclusions, and from cultural remains (e.g. Evans, 1971b). Similar studies using pollen and soil analysis have been made on acid soils (Cornwall, 1953; Dimbleby, 1955, 1962); the basic idea, however, is the same.

Deposits occurring in concave situations such as ditches, pits and wells, features which were originally excavated by man, are another important source of environmental information and are often rich in shells. The fill of these features accumulates in a variety of ways and may yield different types of information according to its origin (Part III: Ditches). Buried soil horizons within a ditch fill represent stand-still phases in its accumulation—periods of surface stability when an unbroken vegetation cover obtained. Often they contain cultural material, and, like their analogues beneath earthworks, repay careful study. Of this group of features, ditches are the most useful from the point of view of obtaining information about the environment of a site as a whole. Where a site comprises a bank and ditch, the sequence in the ditch fill may continue—with only a short break while the earthwork was being constructed—the story already obtained from the buried soil beneath the bank, thus enabling the environmental history of the site to be worked out over a considerable length of time. Today, when so many of the sites being excavated have been ploughed flat, ditches constitute the only source of environmental data relating to their history, and it is thus vitally important that the origin of their soils and sediments be completely understood (p. 321).

Post-holes, small pits and graves have not been investigated in detail by snail analysis, but it is felt that generally they would be unsatisfactory due to the heterogeneous nature of their fill and uncertainties about the way in which it accumulates. In the case of a post-hole, for example, the post-hole pit is first occupied by the post, some packing stones and part of the excavated material, the latter containing shells from the contemporary soil. When the post decays, the cavity is filled up with soil from round about, also containing shells but of a later vintage than those in the original packing material and possibly, therefore, from a different environment. If the post-hole is a large one, a hollow remains after the natural filling processes have ceased, which may then be colonised by denser vegetation than that in the surrounding area, and consequently by a different snail fauna. Thus shells from three different sources, each reflecting a different environment, may be found in close proximity in the fill of a post-hole. The same complexities apply to the infilling of small pits which may have one or more of several functions, none of which may be at all clear, and to graves.

Nevertheless some interesting results may be obtained. For example, shells of the carnivorous snail *Oxychilus cellarius*, were found in the skulls and amongst the bones of Neolithic burials in the mortuary house of the Waylands Smithy I Long Barrow (Kerney, personal communication), the inference being that the bones had been buried with flesh adhering, and the mortuary house left open long enough for the snails to crawl in. The occurrence of two species of marsh snail in post-holes of the Neolithic timber monument known as the Sanctuary (Kennard *in* Cunnington, 1931; Wainwright and Longworth, 1971: 371) in an area which is excessively well-drained, suggests the use of rushes on the site for flooring or for thatch.

Deposits of ploughwash are widespread and common in areas of lime-rich soil and are a frequent agency in the burial (and preservation) of archaeological sites (e.g. Evans, 1966c). A layer of ploughwash is a usual feature in the upper fill of ditches and also comprises a substantial component of cultivation terraces or lynchets. Its snail fauna is generally of open-country type and differs from that of *in situ* soil horizons in that the shells are culled from a wider area, possibly reflecting a number of different minor habitats, and thus being a truer representation of the environment as a whole. Ploughwash and other colluvial deposits are discussed in detail in Part III.

Occupation horizons and building débris, though by no means uncommon on prehistoric sites, are more usual on those of Roman and medieval age. Very little is known about their land-snail faunas but their study might well be worthwhile in connection with the origin of the snail species which today live in close association with man—

Preserved grain and artefacts associated with cereal cultivation may indicate the extent to which a community was dependent on crops, and an estimate of the amount of arable land needed to support the community might be made. The presence of fragments of pottery in a soil, if not directly associated with occupation, may be an indicator of manuring, and that the soil had once been cultivated; such fragments are frequently found in ploughwash deposits. The vast expanse of "Celtic field" systems and medieval ridge and furrow are perhaps the most spectacular and direct fossil evidence of formerly arable land.

Dating

For a knowledge of former environments and land-use phases to be of value in the study of early human communities, they must be linked to them, and this can be done in several ways. Most satisfactory is the direct association of cultural (archaeological) remains such as pottery and other artefacts with environmental data—for instance, occupation material on or in a calcareous soil containing snail shells—a situation which applies to the majority of sites discussed in Part III of this book. Radiocarbon assay of organic material within these soils can be of value in providing a rough time-scale for a site, and in correlating cultural horizons from one site to another (Evans, 1971b; Evans *in* Wainwright and Longworth, 1971: 329), (Part III, p. 228).

3
Methods

There are several papers in which the methodology of snail analysis is discussed. The most useful are by Sparks (1961; 1964a), Kerney (1963) and Ložek (1965b). Here I want to discuss particularly the analysis of terrestrial soils and sediments, as being of most relevance to archaeological problems.

Field Recording of Soils and Sediments

Before samples for molluscan analysis are taken, a detailed record of the deposits to be analysed must be made. The field recording of soils and sediments is to be discussed by Miss Limbrey in her forthcoming book in this series. Excavators should be familiar with the soil maps and Soil Survey Memoirs of their area if available (e.g. Roberts, 1958; Avery, 1964).

First of all the position of a deposit in relation to archaeological structures or local geomorphology must be assessed. To do this it is often necessary to expose and record a considerable length of section, and in this way too, the extent of a layer—whether it be local or widespread—can be judged. This can be an important criterion in the recognition of an *in situ* fossil soil horizon, which can be expected to maintain a relatively uniform morphology, in contrast to the patchy distribution and variable thickness of slumped or solifluxed humic matter such as turf.

In recording, scale drawings are made of the section, showing in detail the stratigraphy of the point to be sampled, and in more general terms the relationship of the deposits to their surroundings (e.g. Figs 87 and 89). A scale of 1:10 or 1:5 is necessary for the detailed recording of soil profiles. A photographic record and written description of the profile succession should also be made. In this book I have followed the terminology of Avery (1964: 41, 205).

Sampling

Sampling is done at a point where the stratigraphy is most complete and most representative of the deposits as a whole, remembering at the same time that the exact location will be reflected in the composition of the snail fauna. For example, in the investigation of a dry-valley fill, the most complete picture of its environmental history will be obtained from deposits in the valley axis where they are thickest. Towards the valley sides where the deposits become thinner, the pattern of environmental change will be compressed. But at the same time one would expect the fauna from the valley axis to contain an element more hygrophilous and suggesting greater surface stability than that from the valley sides. This principle applies particularly to soil horizons and less so to colluvial deposits which generally contain shells taken from a variety of micro-habitats. Sparks (Sparks and Lambert, 1961; Sparks, 1964a) has shown that the same principle applies to freshwater deposits as well.

The best method of sampling is to cut samples from the face of a vertical section. The section is thoroughly cleaned back to remove loose débris, and samples are then cut out with a pointing trowel, starting at the base of the section and working upwards (Fig. 15). A useful sample size is 1·0 kg. Each sample is put in a stout polythene bag (30 × 20 cm) and labelled, the depth below the modern soil surface being a convenient datum for this purpose. The interval at which samples are taken depends on a variety of factors, but 10 cm is a useful standard. The main factor which governs the sample interval is the rate of build up of a deposit; the faster the rate, the less likelihood there is of critical faunal changes taking place within a small thickness of deposit, and therefore a large sample interval (10–15 cm) may be used. This applies to deposits such as flood loam, wind-lain sand and hillwash. Where sedimentation is slow, or where soils have formed during standstill phases, a smaller sample interval of 3–5 cm is preferable. Another factor governing the sample interval is the texture of the deposit: a fine-grained, compact sediment can be sampled at closer intervals than a coarse-grained material.

Samples must not be taken across a pronounced stratigraphical boundary such as the surface of a buried soil. Such a boundary

Figure 15. Arbury Road, Cambridge. Section through a Roman pit to show the columns from which samples have been taken for insects and seeds (left-hand column), pollen (narrow right-hand column) and molluscs (broad right-hand column).

generally constitutes a marked environmental break and as such will be reflected by the molluscan fauna, the faunas above and below the boundary reflecting different environments and comprising different snail populations.

As an alternative to cutting out individual samples a monolith sampler can be used. This is a three-sided metal container, length 60 cm, two sides 10 cm and back 12 cm, with or without closed ends, which can either be hammered forcibly into the section, or slotted on to a column of sediment produced by cutting away material from either side. (Air holes are necessary in the back if the ends are closed.) The column of sediment is then removed by slicing down behind the sampler with a spade. The advantage of this method is that an entire sample of the section *in situ* can be removed and taken back to the laboratory for careful study.

An auger or borer can be used for obtaining samples when a section is not readily available, but for terrestrial deposits this is not to be recommended as it does not allow a critical investigation of the deposits being sampled, nor does it enable the recovery of archaeological material, important for dating. It is generally a simple matter to dig a soil pit two to three metres deep, either by hand or with a mechanical excavator, and although still far from satisfactory this does enable a better section of the profile to be examined. The use of a borer may be justified where the deposits are exceptionally deep, or where they are waterlogged and unlikely to remain intact as a section. The use of a percussion corer for deep Pleistocene deposits is recommended by Sparks (1961) who finds that serious contamination occurs very infrequently.

To obtain a general idea of the fauna of a soil horizon or sediment, or to obtain a quantity of shells of a particular species, a *spot sample* can be taken. It should be emphasized, however, that the amount of ecological information which can be got from the analysis of such a sample is slight.

Fully-grown shells of the larger species such as *Cepaea* and *Succinea* are rarely present in quantity in soil samples and should therefore be collected by hand-picking during the course of excavation. There are two reasons for doing this. First, identification of some of the larger species is not easy from juvenile or fragmentary material. And second, large collections of adults of the polymorphic snails, *Cepaea* spp., *Helicella itala* and *Cochlicella acuta* are of interest to population geneticists who are studying the variations which have taken place during the Post-glacial in the colour patterns of these shells. Professor A. J. Cain has asked me to put in a plea for large collections of *well-dated* subfossil *Cepaea* which should be sent to: P.O. Box 147, The Department of Zoology, University of Liverpool,

Liverpool L69 3BX. Some of the results of his work are discussed on p. 174.

It is worth pointing out that these two stages of analysis—the field recording of soils and sediments and the sampling procedure—are the only ones in which the subjective judgement of the investigator is involved. Subsequent stages are largely mechanical, and, while the interpretation of the results of analysis is a matter of individual taste, the results themselves are incontravertible. The moral is obvious. Not only must these two stages be carried out with care and precision; they must be preceded by an intelligent assessment of all the available evidence presented by a site, and a rigorous appraisal of the problems which snail analysis is likely to answer.

Extraction

A standard weight of air-dried material is used so that counts can be compared, if necessary, on an absolute frequency basis; between 0·5 kg and 2·0 kg is usually sufficient. The material is placed in a bowl of water and allowed to collapse; the process can be speeded up by gentle stirring. A large number of shells generally float to the surface and these are poured off into an eight-inch diameter sieve, mesh size 0·5 mm (30 holes per inch), and oven dried at about 90°C. Generally a series of graded sieves with mesh sizes of 2·0 mm, 1·0 mm and 0·5 mm is used for convenience of working. The sludge is then washed through a second series of sieves until all humic material has been removed. Resistant soil crumbs can be broken up with 100 volume hydrogen peroxide (General Purpose Reagent, not Analar). The washed débris is then oven dried and all shells extracted from it using a hand lens or low-power stereoscopic microscope; a small paint brush and forceps made from tinfoil (commercially known as "pelican forceps") are most suitable for manipulating the shells. Plastic petri dishes make admirable containers for the washed and dried débris.

Small-mammal bones, ostracod valves and wood-charcoal fragments can be collected at the same time, though the majority of ostracods are too small to be trapped by a 0·5-mm mesh. This method of extraction is unsuitable too for seeds and insects, not only because many are lost through a 0·5-mm mesh, but because oven drying cracks and shrivels insect cuticles and seed coats.

All the shells extracted from the sample are counted, but to avoid including the same specimen more than once only shell apices are

3. METHODS

used for this purpose. In cases where apertural fragments are more diagnostic than apices, these may be extracted and counted instead. This applies to the genera *Cochlicopa*, *Clausilia*, *Vertigo* and *Cepaea*. It is wise to be constantly on the look out for non-apical fragments of species otherwise not represented in the sample.

Identification

There is no standard work on British land and freshwater Mollusca that can be used for the identification of subfossil shells. All the available text books deal with the identification of adult individuals only and not with the juvenile and fragmentary shells so common in subfossil material. Moreover, the identification of living animals is frequently based on features which are lost in the subfossil state, such as the soft parts of the body—colour of the mantle and morphology of the genitalia—or the periostracum, its colour, and whether or not it bears hairs or spines. One snail, *Oxychilus alliarius*, gives off a strong smell of garlic when molested and this can be used to distinguish it from the similar *O. cellarius* when live.

Nevertheless, the identification of living Mollusca is very adequately covered by a number of easily available works of reference which should be consulted, for it is essential that the intending student of subfossil shells should also be acquainted with the living animals. The most useful are listed below:

Ellis, A. E. (1926, reprinted with notes, 1969). "British Snails". Clarendon, Oxford. The standard work on British snails.

Ellis, A. E. (1964). Key to land shells of Great Britain. *Conchological Society of Great Britain and Ireland: Papers for Students*, No. 3.

Adams, L. E. (1896). "The Collector's Manual of British Land and Freshwater Shells". (2nd edn) Taylor Brothers, Leeds.

Taylor, J. W. (1894–1921). "Monograph of the Land and Freshwater Mollusca of the British Isles." 3 vols. + 3 parts (unfinished). Taylor Brothers, Leeds.

Turton, W. (1857). "Manual of the Land and Freshwater Shells of the British Islands". (New edition, with additions by J. E. Gray.) Longman, London. Recommended for its true to life water-colour sketches of most snail species.

Quick, H. E. (1949). "Synopses of the British Fauna. No. 8. Slugs (Mollusca) (Testacellidae, Arionidae, Limacidae)". Linnean Society, London.

Quick, H. E. (1960). British Slugs (Pulmonata: Testacellidae, Arionidae, Limacidae). *Bull. Br. Mus. nat. Hist. Zool.* **6**, No. 3, 103–226.

The following two works deal with freshwater species only, and are listed for their illustrations of the slum species (p. 200).

Macan, T. T. (1969). Key to the British fresh- and brackish-water gastropods. *Freshwater Biological Association Scientific Publication*, No. 13, 3rd ed.

Ellis, A. E. (1962). "Synopses of the British Fauna, No. 13. A Synopsis of the Freshwater Bivalve Molluscs." Linnean Society, London.

Most of the Late Weichselian land snails of Britain are illustrated by photographs, including some of juveniles, in two papers by Kerney (1963: Kerney et al., 1964). Apart from these however, British subfossil shells are illustrated only occasionally: e.g. in Kerney (1959)— *Acicula* spp., *Azeca* spp. and some of the interglacial clausiliids now extinct in Britain; and in Sparks (1957a)—various interglacial species such as *Vallonia pulchella* var. *enniensis* and *Belgrandia marginata*, and slug plates.

Two continental works contain illustrations of a high standard, of modern and subfossil shells many of which occur in Britain. One of these by W. Adam (1960) covers the land and freshwater fauna of Belgium and is concerned largely with modern species. The other, by V. Ložek (1964), deals with the Quaternary Mollusca of Czechoslovakia. Both are well worth consulting.

In fact, identification is more tedious than difficult. The total number of land and freshwater mollusc species both living and extinct in Britain is not more than 200, and on terrestrial sites, fewer than fifty of these can generally be expected. With a little practice a good working knowledge of the fauna can be acquired, and identification can then become a matter of rapid routine. It is essential, nevertheless, to have a reference collection of named specimens of subfossil shells in various stages of growth and degrees of preservation for this is the only secure means of identifying unknown material. The following notes and drawings are not intended as a substitute for this but are designed simply to act as a guide to some of the key features used in identification. They cover the majority of British land snails likely to be met with on archaeological sites of Post-glacial age.

Identification is best done using a low-power stereoscopic microscope with interchangeable × 10 and × 25 lenses. It is advisable to use a model in which the drive-housing and optics of the microscope are fixed to a long-arm stand, and with no stage. The shells are contained in a shallow tray which is set directly on the laboratory bench. In this way one can obtain maximum manoeuvrability of the object so that a large number of shells can be rapidly compared.

Some of the more important descriptive terms are illustrated in Fig. 16 with reference to an adult shell of *Vertigo pygmaea*.

In the section which follows, I have adhered as closely as possible to the order of species given in Chapter 5, which is based on their taxonomic relationships. Although this has been compiled largely through a consideration of the anatomy of their soft parts (Watson, 1943), it does provide, with a few exceptions which will be mentioned as they arise, a realistic basis for dealing with the identification of their shells also.

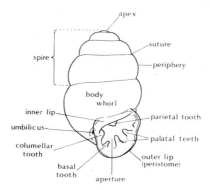

Figure 16. Shell of *Vertigo pygmaea* to illustrate some of the terms used in shell description.

POMATIIDAE

Pomatias elegans

Readily identified by its large size, conical form and well-rounded whorls (Fig. 19). Its apex is similar to that of *Ena montana* but is easily separated by its larger size (Fig. 20). Broken apices at the one-whorl stage are close to those of *Cepaea*, but smaller and more concave (Fig. 28); only with much battered and minute specimens is confusion at all likely. Shell fragments of *P. elegans* are characterized by strong spiral and transverse ribbing and by their ochreous inner surface.

P. elegans is one of the two British land operculates; the other is *Acicula fusca*. In *P. elegans*, the operculum is strong and calcareous, and frequently preserved, either whole or fragmentary (Fig. 19).

When recording *P. elegans* shells, it is important to note whether they are adult or relatively complete, or are battered and worn apices. The latter are resistant to destruction by comparison with those of most other species, and thus tend to accumulate in ploughwash and at the base of soils, leading to a false impression of abundance (p. 212).

ACMIDAE

Acicula fusca

The shell characteristics in the adult and juvenile stages are distinctive, and confusion with other species is unlikely. In size and form, though only as small juveniles, *Acicula* comes closest to *Carychium* (Fig. 21), from which it can be distinguished by the absence of columellar and parietal folds, the sharply incised vertical striae, the shallower suture and less tumid whorls; the apex is more obtuse. From *Cecilioides*, *Acicula* apices are distinguished by their more depressed whorls.

The Pleistocene species of *Acicula* are illustrated by Kerney (1959).

ELLOBIIDAE

Carychium

There are two British species of *Carychium*, *C. minimum* and *C. tridentatum* (Watson and Verdcourt, 1953). The only species with

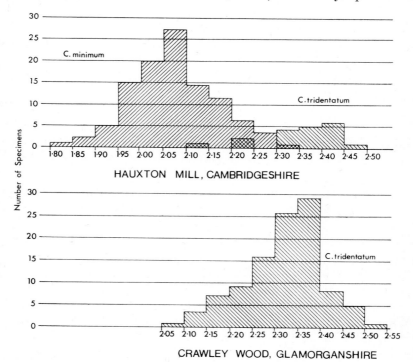

Figure 17. *Carychium*. Ratio of height to minor diameter. (After Watson and Verdcourt, 1953.)

which *Carychium* might be confused is *Acicula fusca*, and the separation of these two, as discussed under the latter, presents no problems. But the determination of the two species of *Carychium* is less easy, though it should be pointed out that the occurrence of *C. minimum* on terrestrial (non-marsh) archaeological sites is unlikely. As adults, there is a number of characteristics which may be used. *C. tridentatum* is taller and narrower, with a deeper suture and more shouldered whorls (Fig. 21). The ratio of height to the minor diameter of the last whorl is generally greater in *C. tridentatum*, although there is some overlap (Fig. 17). The penultimate whorl of *C. tridentatum* is generally broader than in *C. minimum*. In both species the whorls below the protoconch bear fine vertical ridges which in *C. tridentatum* are more strongly developed and closer together than in *C. minimum*. The aperture of *C. tridentatum* is smaller than that of *C. minimum*, and the lip is thicker and more strongly developed. The form of the parietal fold within the body whorl differs markedly in the two species, being more elaborate in *C. tridentatum*. This latter characteristic can sometimes be seen through the shell, but in doubtful specimens in which the shell is opaque, the outer wall can be broken open with a scalpel tip to expose the folds within. With juvenile shells, separation is more difficult, though with well-preserved material, characters such as the tumidity of the whorls, depth of suture and development of striae enable identification to be made with a certain degree of confidence.

LYMNAEIDAE

Lymnaea truncatula
In terrestrial contexts, the only confusion likely is with *Succinea*, from which it is, however, easily separated by its smaller apex and deeper suture (Fig. 18).

SUCCINEIDAE

The separation of the species of *Succinea* on their shells alone is not easy, even with adult specimens. With subfossil, fragmentary material (and *Succinea* shells are unusually fragile and susceptible to fragmentation) it is often impossible. Useful photographs are given in Stelfox (1911), Quick (1933), Sparks (1957b), Kerney et al., (1964) and Ellis (1969).

The key features in their identification are the depth of the suture, tumidity of the whorls, shape of the aperture, and heights, relative

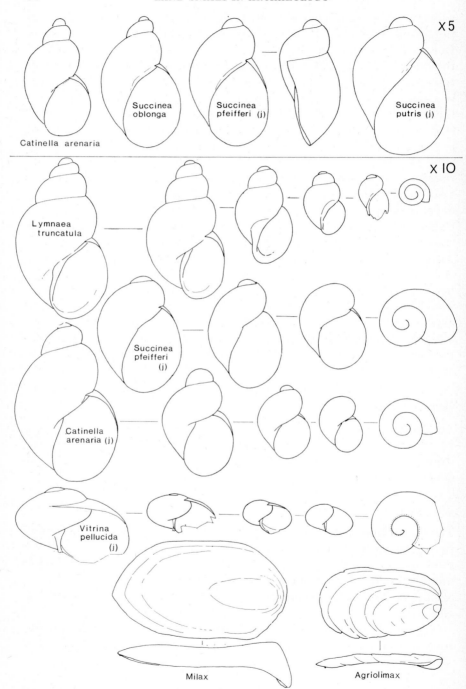

Figure 18. Succineidae, *Lymnaea truncatula*, *Vitrina pellucida*, Limacidae. (j) = juvenile.

to each other, of the spire, body whorl and aperture. The following notes are taken largely from Ellis (1969).

Catinella arenaria and *Succinea oblonga* are separated from the other three by the relatively taller spire and shorter body whorl. They resemble *Lymnaea truncatula* from which they can be readily separated, however, even as very small juveniles, by the larger and more swollen protoconch (Fig. 18). The differentiation of *C. arenaria* and *S. oblonga* is difficult, particularly as the former varies, specimens sometimes approaching *S. oblonga* extremely closely (Sparks, 1957b). *C. arenaria* is less elongated, has a deeper and less oblique suture, a blunter apex, a rounder (less oval) aperture which tends to have a squared-off (truncate) appearance at the base, and more inflated whorls even at the apex (Sparks, 1957b).

The other three can be arranged in a series:

S. sarsi
S. pfeifferi
S. putris

S. sarsi is the most elongate, with the highest spire, and deep regular striae. *S. pfeifferi* has whorls more inflated than *S. sarsi* but less so than *S. putris*; the height of the spire is intermediate between the two. In *S. putris* the spire is short, the suture less oblique, and the whorls relatively broader (Fig. 18).

COCHLICOPIDAE

Azeca goodalli

Similar in size and form to *Cochlicopa*, but distinguished in the adult stage by the presence of at least three apertural denticles (Fig. 19). As juveniles, *Azeca* is separated from *Cochlicopa* by the smaller apex, tighter coiling, almost non-existent suture and barely convex whorls (Fig. 20).

The diagnostic features of the Pleistocene species, *Azeca menkeana*, are illustrated by Kerney (1959).

Cochlicopa

There are two British species of *Cochlicopa*, *C. lubrica* and *C. lubricella* (Quick, 1954). As a genus, the only possible confusion is with *Azeca* (discussed above), and *Ena obscura* from which *Cochlicopa* is readily distinguished by its smaller and shiny apex (Fig. 20), that of *E.*

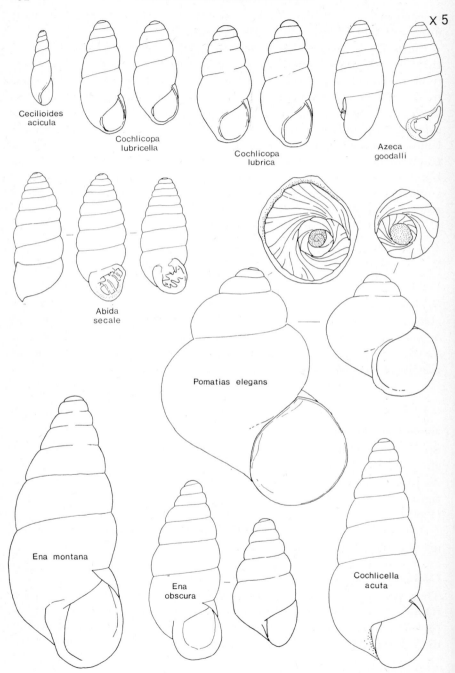

Figure 19. *Cecilioides acicula*, Cochlicopidae, *Abida secale*, *Pomatias elegans*, Enidae, *Cochlicella acuta*.

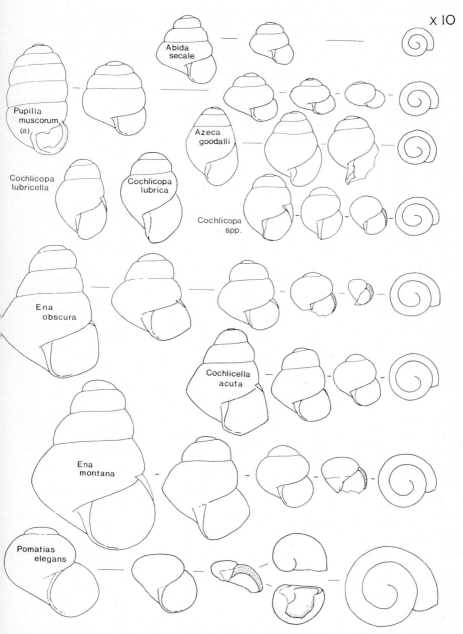

Figure 20. *Abida secale, Pupilla muscorum,* Cochlicopidae, Enidae, *Pomatias elegans.*

obscura being mat. With very small and battered specimens there is also the possibility of confusion with *Pupilla*, but in the latter the whorls are more compressed (Fig. 20) and the apex has a mat texture.

The differences between the two species of *Cochlicopa* are discussed by Quick (1954) and illustrated in Figs 19 and 20. As a rule, *C. lubrica* is the larger, with more rounded whorls and a relatively more acute apex; *C. lubricella* tends to be fusiform in shape with a relatively blunt apex. As juveniles, they are difficult, if not impossible to separate, though *C. lubricella* is generally the smaller (Fig. 20).

VERTIGINIDAE

Pyramidula rupestris

The apex of this species is smaller than that of *Discus rotundatus* and larger than that of *Punctum pygmaeum*. It is similar in size to that of *Vallonia*, but is distinguished by its more conical form and dark coloration (Figs 22 and 23).

The other members of the Vertiginidae differ from *Pyramidula* in being taller than broad, cylindrical, sub-cylindrical or fusiform. Four other species, *Abida secale, Acanthinula aculeata, Acanthinula lamellata* and *Euconulus fulvus* can usefully be considered here as their apices bear resemblances in size and form to those of certain Vertiginidae. Two groups can be recognized:

	LARGE	
	Diameter (mm)	
	1 whorl	1½ whorls
Euconulus fulvus	0·64	0·92
Lauria cylindracea	0·64	0·76
Pupilla muscorum	0·62	0·76
Lauria anglica	0·60	0·68
Acanthinula aculeata	0·52	0·68
Abida secale	0·52	0·60
	SMALL	
Vertigo pusilla	0·48	0·56 ⎫ Sinistral
V. angustior	0·44	0·52 ⎭
V. substriata	0·44	0·52
V. alpestris	0·44	0·52
V. pygmaea	0·44	0·52
Columella edentula	0·44	0·52
Truncatellina cylindrica	0·36	0·52

Acanthinula lamellata (1 whorl, 0·48, 1½ whorls, 0·6 mm) is transitional between the two groups.

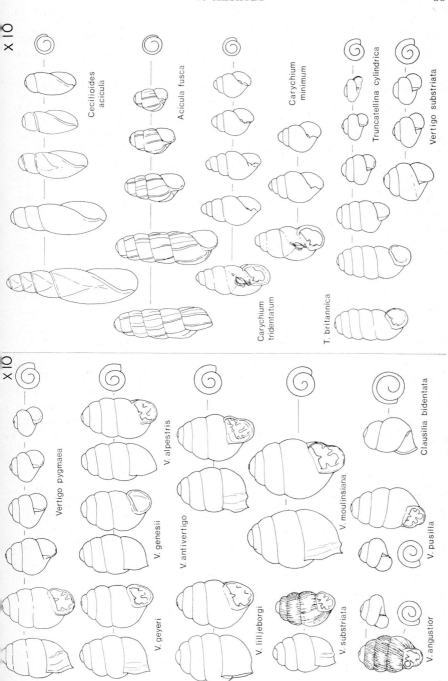

Figure 21. Vertiginidae, *Cecilioides acicula*, *Acicula fusca*, *Carychium*.

Columella

The shell of adult *Columella* differs from that of *Vertigo* in being taller and slightly broader, and can be readily distinguished by its more or less parallel-sided form and edentulate aperture (Fig. 22). From *Truncatellina*, which it resembles in form, it can be distinguished by its larger size and far weaker striations—virtually absent in *C. edentula*. The apex of *Columella* at the one- and $1\frac{1}{2}$-whorl stages is the same size as that of *Vertigo*, but it is less conical and the whorls perhaps slightly less shouldered. At the two-whorl stage and thereafter, the apex is broader and flatter. From *Pupilla*, *Columella* apices are readily distinguished by their smaller size (Fig. 23).

The two species of *Columella*, *C. edentula* and *C. aspera*, are readily separated as adults (Fig. 22). *C. edentula* is longer and narrower, both relatively and absolutely, and the last whorl is often slightly wider than the preceding whorl. The shell is ornamented with fine *irregular* striae. Length 2·2–2·8 mm, width 1·23–1·31 mm (6–7 whorls). *C. aspera* is shorter and fatter. There are 5–6 whorls which are usually more convex and broader, and with a deeper suture. The shell is ornamented with fine *regular* striae, which are more strongly developed than in *C. edentula*. Length 1·85–2·13 mm, width 1·26–1·36 mm (Paul, 1971).

The Pleistocene species, *C. columella* (Fig. 22), is distinguished from *C. edentula*, which it most closely resembles, by its larger size, more parallel-sided form, deeper suture and more tumid whorls. The diameter of the penultimate whorl is often less than that of both the antepenultimate and the body whorl, giving the shell a markedly waisted appearance.

Truncatellina

Truncatellina is distinguished from *Vertigo* by its narrower and more cylindrical form at all stages of growth (Fig. 21). The shell is closely rib-striate, but can be separated from *Vertigo substriata*, which is similarly marked, by its smaller size, more pronounced ribbing, and dull texture. (The two, however, are unlikely ecological associates.)

Truncatellina britannica was considered by Kennard and Woodward (1923) to be a species distinct from *Truncatellina cylindrica*, but is now thought to be at most a subspecies (Ellis, 1969). The main difference between the two is in the aperture, which in *T. britannica*

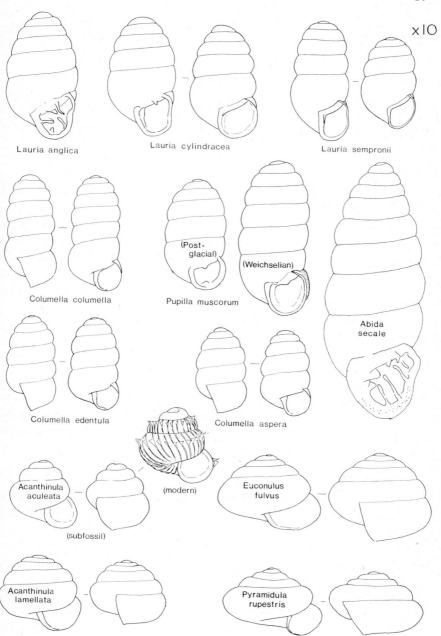

Figure 22. Vertiginidae, *Abida secale*, *Acanthinula*, *Euconulus fulvus*, *Pyramidula rupestris*.

bears two denticles while that of *T. cylindrica* is edentulate. Moreover, *T. britannica* is more cylindrical in form than *T. cylindrica* (Fig. 21).

Vertigo

Adult, the species of *Vertigo* are fairly distinct (Fig. 21). Juveniles and apical fragments are more difficult, and it is impossible to separate those of *Vertigo genesii, V. geyeri, V. pygmaea* and *V. antivertigo*. In instances where two or more of these species are present, apertures rather than apices must be counted.

Vertigo pusilla and V. angustior

These two species are the only two sinistral *Vertigo*. The apex of *V. angustior* differs from that of *V. pusilla* in that the whorls are narrower, as viewed from the side, and are coarsely striate whereas those of *V. pusilla* are smooth. From *Clausilia bidentata* and *Balea perversa*, the smallest of the Clausiliidae which are also sinistrally coiled, *V. pusilla* and *V. angustior* differ in their smaller size (Fig. 21).

Vertigo antivertigo

Characterized by its smooth shell and the presence, in the adult, of two parietal denticles (Fig. 21).

Vertigo substriata

Characterized by its coarsely striate shell, tumid whorls, and, in the adult, the presence of two parietal denticles (Fig. 21). The coarse striations serve to separate this species from all other *Vertigo*. From *Truncatellina*, which is similarly striate, the apex is distinguished by its larger size (Fig. 21).

Vertigo pygmaea

The adult is distinguished from *V. geyeri* and *V. alpestris* by the presence of five rather than four denticles (the additional one being a small basal fold between the columellar and palatal denticles) and by the presence of a strong callus or thickening behind the aperture where the palatal denticles are inserted (Fig. 21).

Vertigo geyeri

This species is distinguished from the extinct *V. genesii* by the presence of four apertural denticles, a less polished shell with irregular striations, a *slight* thickening behind the aperture, more tumid whorls and a deeper suture (Kerney et al., 1964). From *V. alpestris* it is distinguished by its more convex, barrel-shaped form, more tumid whorls and less pronounced striations—though in neither species are these as strong as in *V. substriata*; and from *V. pygmaea* by the weak rather than strong callus behind the aperture (Fig. 21).

Vertigo genesii

This species is narrower and more cylindrical than *V. geyeri*, and generally edentulate, though occasionally there is a small parietal tooth (Fig. 21) (Kerney, et al., 1964).

Vertigo moulinsiana

Distinguished from other species of *Vertigo* at all stages of growth by its larger size (Fig. 21).

Vertigo alpestris

The shell is faintly striate and cylindrical in form, features which serve to distinguish it from *V. pygmaea* and *V. geyeri* both in the adult and juvenile stages (Fig. 21). The adult is larger (2 mm high as against 1·5 mm) than *V. geyeri*, more cylindrical, and with less tumid whorls.

Vertigo lilljeborgi

This species resembles *V. moulinsiana* but is distinguished by its smaller size, blunter apex, and more tumid and glossy whorls. From *V. pygmaea* it is distinguished by its more swollen, glossy and striate whorls and deeper suture; and from *V. antivertigo* by its more swollen and striate whorls and deeper suture (Kevan and Waterston, 1933: 309; Dance, 1972) (Fig. 21).

Vertigo angustior

(See *V. pusilla*).

Pupilla muscorum

Distinguished on a size basis from *Vertigo, Truncatellina, Columella, Abida* and *Acanthinula* in being invariably larger at the single-whorl stage. It differs from *Lauria cylindracea* and *Euconulus* in its more rounded whorls, deeper suture and mat surface at the one-whorl stage, from *Lauria anglica* by its larger size, mat surface and absence of denticles, and from all three by its more open umbilicus (Figs 22 and 23). From *Cochlicopa* it differs in being squatter (less globose)

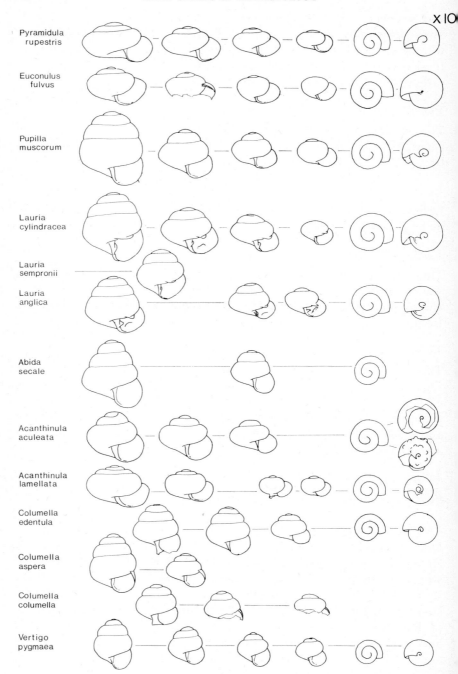

Figure 23. *Pyramidula rupestris, Euconulus fulvus,* Vertiginidae, *Abida secale, Acanthinula.*

(Fig. 20) and in having a mat texture. With practice, the smallest apical fragments of *Pupilla* can be identified.

Lauria cylindracea and *L. anglica*

L. cylindracea is readily separated from *L. anglica* at the one-whorl stage by its curiously flattened apex and lack of suture; the apex of *L. anglica* is slightly smaller at both the one- and $1\frac{1}{2}$-whorl stage (Fig. 23). The parietal denticle and columellar fold appear at an early stage in the development of these two species thus serving to distinguish them from other genera of the Vertiginidae and from *Euconulus* to which there is a superficial resemblance (Fig. 23). It should be noted that the small outer denticle which is invariably present in the young shell of *L. anglica* (Fig. 23) is sometimes present in that of *L. cylindracea*, although absent from the adult of the latter (Fig. 22) (Adam, 1960: Fig. 67), and is therefore no basis for the separation of the two species. In the juvenile stage, the umbilicus of *Lauria* is more open than that of *Euconulus* and less open than that of *Pupilla*, a useful diagnostic feature when the aperture is obscured by sediment. The adults of *L. cylindracea* and *L. anglica* are readily separated (Fig. 22).

Lauria sempronii

The differences between this rare species and *L. cylindracea* are discussed by Kerney (1957b). Essentially, *L. sempronii* is more nearly cylindrical and with more tumid whorls; in the adult stage it is edentulate (Fig. 22). The western form of *L. cylindracea*, var. *anconostoma*, is intermediate between the two, and sometimes edentulate (Fig. 22).

CHONDRINIDAE

Abida secale

The apex is distinguished from that of *Pupilla* by its smaller size, and from that of *Acanthinula* by its pyramidal form—a striking feature of *Abida* apices—and flatter whorls. The calcareous nodules which are present on the underside of weathered *A. aculeata* apices (Fig. 23) are absent from *Abida*. Apertural fragments are particularly resistant to destruction, and often occur in chalk soils when all other trace of the shell has been destroyed. Adult, *Abida* is readily recognized (Figs 19 and 22).

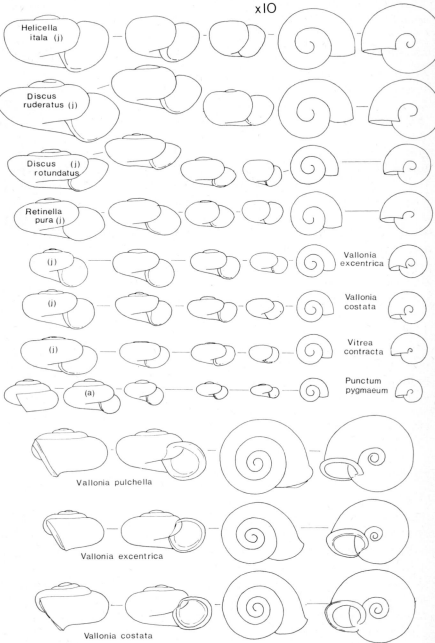

Figure 24. *Vallonia*, Endodontidae, *Vitrea contracta, Helicella itala.* (j)=juvenile; (a)=adult.

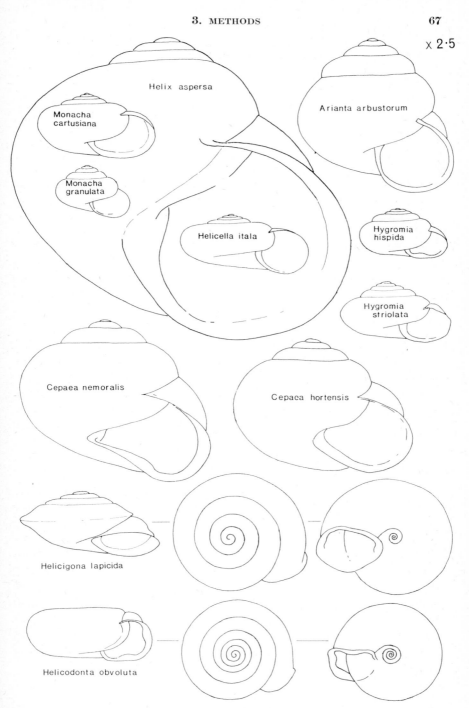

Figure 27. Helicidae, adults.

Hygromia hispida
H. striolata, Monacha cantiana, Helicella
Helicigona lapicida
Arianta arbustorum
Cepaea
Helix aspersa
H. pomatia

Fruticicola fruticum

The shell resembles that of *Monacha cantiana*, "but is more finely and regularly striated, and has also numerous delicate spiral lines, and the whorls more swollen" (Ellis, 1969: 193).

Helicodonta obvoluta

Young individuals can be recognized by their involute apex, rather like that of a planorbid (ram's-horn) snail, by their deep suture, and by the presence of hair pits; in size the apex is similar to that of *Hygromia*. The adult shell is unmistakable (Fig. 27).

Helicigona lapicida

The apex is intermediate in size between that of *Arianta* and *Helicella*, and is characterized by its flattened form, relatively shallow suture and mat surface. Adults and juveniles are unmistakable by reason of their strongly keeled shell (Figs 27 and 28) and reticulate sculpturing; the latter feature is diagnostic, thus enabling the smallest shell fragments to be recognized.

Arianta arbustorum

The apex is similar in size and form to that of *Cepaea*, but is slightly smaller, with more rounded whorls and more open umbilicus. These differences, however, are somewhat tenuous, and unless the shells are well preserved and in considerable abundance, it is generally not worthwhile attempting to separate them from *Cepaea*. Adults and shell fragments, on the other hand, are readily recognizable (Fig. 27); thus the lip is white and deflected at right angles to the body of the shell and there is a small umbilicus, absent in adult *Cepaea*; the shell is incised with fine, spirally orientated, wavy lines which appear at about the two whorl stage, and shell fragments fracture cleanly with many right-angle breaks—rather in the manner of a broken hen's egg.

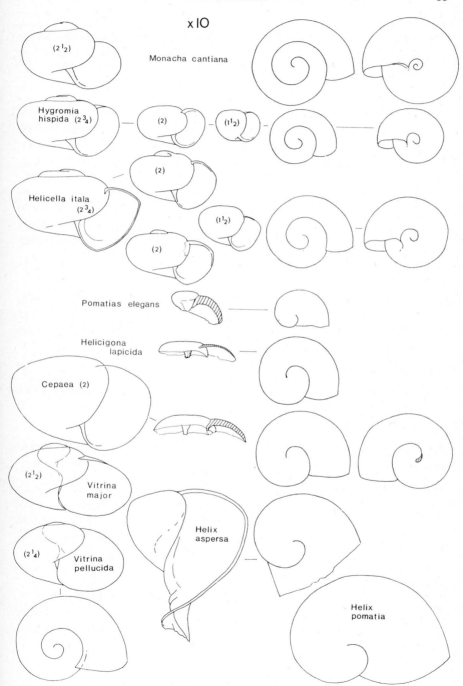

Figure 28. Helicidae, juveniles. *Vitrina*. Number of whorls in parentheses.

Theba pisana

"The shell is sculptured with transverse and spiral striae, the latter being very noticeable ..." and the whorls " ... are squarish or shouldered above in a characteristic manner" (Ellis, 1969: 222).

Helix (Cepaea) hortensis and H. (Cepaea) nemoralis

Apices of these two species are the largest to occur in prehistoric assemblages in Britain, being exceeded only by *Helix aspersa* and *H. pomatia* which are almost certainly Roman introductions. As juveniles, it is virtually impossible to separate the two species, but adults (Fig. 27) and lip fragments are readily identifiable. Thus the lip of *C. nemoralis* is generally pink, and this colour is retained in the sub-fossil state in well-preserved material, while that of *C. hortensis* is white. However, white-lipped individuals of *C. nemoralis* occasionally occur (p. 174), so that identifications of *C. hortensis* are unsatisfactory on lip fragments alone.

Helix aspersa

The apex is intermediate in size between *Cepaea* and *Helix pomatia* (Fig. 28). Adults (Fig. 27) and shell fragments are readily identified, the latter characterized by a coarsely wrinkled or puckered surface unlike that of any other species.

Helix pomatia

This species is the largest British land snail, and has the largest apex (Fig. 28). At the one-whorl stage, the shell can be covered by five or six adult *Carychium tridentatum*.

Hygromia limbata and *H. cinctella* are introductions of very recent origin, living in south Devon, and will not therefore be discussed.

Hygromia subrufescens

The shell of this species differs from that of other indigenous species of *Hygromia* in the absence of periostracal hairs and the minute umbilicus. The shell is extremely thin and feebly calcified, and is thus seldom preserved (p. 176).

Hygromia striolata

Juvenile shells of this species can be distinguished from those of *H. hispida* by their more regular transverse striations, slightly larger apex and fewer hair pits. The adult shell is non-hispid, and has a keeled periphery (Fig. 27).

Hygromia hispida

This is the commonest species of *Hygromia* to occur in archaeological deposits (Fig. 27). The shell is characterized by its open umbilicus, serving to distinguish it from *H. subrufescens*, *H. liberta*, *H. subvirescens* and species of *Monacha*, by its irregular rather than striate sculpturing and by the presence of numerous hair pits. Juvenile shells are readily separated from those of *Helicella itala*, a commonly associated species of similar size, by the presence of hair pits, the slightly smaller apex and the more rounded profile of the underside of the whorls—i.e. around the umbilicus (Fig. 28).

Hygromia liberta

Distinguished from *H. hispida*, which it closely resembles, by its smaller umbilicus.

Hygromia subvirescens

An introduction, though of somewhat earlier date than *H. limbata* and *H. cinctella*, living in the extreme south-west of England and Wales as a coastal species (p. 178); it is most unlikely to occur in prehistoric deposits, and certainly not inland. I have not examined apices, but according to Ellis (1969: 212), "the shell is somewhat globose, very thin, of a green hue, transversely striate and clothed with periostracal hairs. There are 4 to $4\frac{1}{2}$ convex whorls with a deep suture, the spire is slightly raised and the peristome a trifle thickened and partly reflected over the very narrow umbilicus".

Monacha

Juvenile shells of *Monacha* are characterized by their bulbous apex, disproportionately large in relation to the size of the succeeding whorls, and by their small umbilicus, features which serve to distinguish them from *Hygromia* (Fig. 28).

Monacha granulata

Seldom preserved subfossil due to its feebly calcified shell. The apex is extremely swollen and covered with hair pits. The suture is shallow at the one-whorl stage, a feature which serves to distinguish apices from those of *M. cantiana*, and the umbilicus minute. The adult shell (Fig. 27) is readily distinguished from the other two species of *Monacha* by its smaller size.

Monacha cartusiana and *M. cantiana*

"Apical fragments of *M. cantiana* may be distinguished from those of associated *M. cartusiana* by the presence of hair pits, and by the relatively larger size of the nepionic whorls." (Kerney *et al.*, 1964: 172) (Figs 27 and 28).

Helicella

Shells of *Helicella* are characterized by the absence of hair pits and the presence of fine spiral striae. The separation of apices within the genus is not easy, but fortunately only *H. itala* is at all likely to occur in prehistoric deposits of Post-glacial age. The shell features of the introduced species are described by Ellis (1969) and will not be repeated here.

Helicella itala

Young shells have a slightly more open umbilicus than *Helicella caperata*, *H. gigaxi* and *H. virgata*. The differences betweeen young *H. itala* and *Hygromia hispida* are illustrated in Fig. 28, the main feature being the more angular profile of the periphery and underside of the whorls around the umbilicus in *Helicella itala*. In addition, the absence of hair pits and the regular striations in *H. itala* prevent any real confusion. In well-preserved shells, spiral bands of pigment may be seen.

Helicella striata and *H. geyeri*

The diagnostic features of these two extinct species are described and illustrated by Sparks (1953b).

Cochlicella acuta

This species (Figs 19 and 20) can only be confused with *Ena obscura* as already discussed (p. 64).

ENDODONTIDAE

Punctum pygmaeum

The apex is smaller than that of any other species of similar form, i.e. *Vallonia* and *Pyramidula* (Fig. 24).

Discus ruderatus

Shells can be distinguished from *D. rotundatus* by their more widely spaced and stronger ribs, less keeled periphery and more concave form (Fig. 24). The mottled pattern characteristic of *D. rotundatus* is absent, and the apex is larger, being similar in size to that of *Helicella itala*. The latter, however, can be separated by its deeper whorls.

Discus rotundatus

The shell apex is larger than that of *Vallonia costata* and smaller than that of *Helicella* and *Hygromia* (p. 66). It is closest in size and form to *Retinella pura*, but can be distinguished from this species by its slightly deeper suture, more rounded form as viewed from above, and more shouldered whorls (Fig. 24). By the $1\frac{1}{2}$- to 2-whorl stage, the characteristic striations become apparent, serving to distinguish it from all other species.

ARIONIDAE

Geomalacus maculosus

Unlike the other Arionidae, the internal shell of *Geomalacus* is a flat plate, 3×5 mm.

Arion

The internal granules of the Arion slugs are " . . . small ovoidal or subspherical calcareous granules, mostly about 0·5 to 1·5 mm in maximum diameter. They have a radial crystalline structure, and sometimes show crystal facets on their outer surfaces" (Kerney, 1971a; 7). They

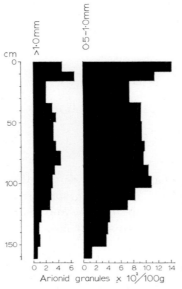

Figure 29. Arionid granules from a thirteenth century ditch fill at Middleton Stoney, Oxon.

are not specifically determinable, but granules greater than 1·0 mm are more likely to belong to the *Arion ater* aggregate (p. 186) than to smaller species. In Fig. 29 changes in the frequency of two size grades of granules are shown from the fill of a medieval ditch at Middleton Stoney, Oxon. The differences in the two graphs probably reflect changes in the species composition of the slug fauna.

ARIOPHANTIDAE

Euconulus fulvus

The shell is characterized by a shallow suture and minute umbilicus, the latter serving to distinguish it from *Lauria cylindracea* (Fig. 23). The apex is intermediate in size between *Vitrea*, which is smaller, and *Retinella pura*, which is larger. In the absence of the umbilicus, small apical fragments are separated from *L. cylindracea* by their flatter form, and slightly smaller size. As adults, they are readily identified (Fig. 22).

ZONITIDAE

The Zonitidae are characterized by thin-walled shells with a generally shiny surface and shallow suture, features which serve to distinguish them from the Helicidae, Endodontidae and Valloniidae. On a size basis, their apices can be arranged in the following series, beginning with the smallest (Fig. 31):

> *Vitrea diaphana*
> *V. contracta*
> *V. crystallina*
> *Retinella radiatula*
> *R. pura, Zonitoides excavatus, Z. nitidus*
> *Oxychilus alliarius*
> *Retinella nitidula*
> *Oxychilus cellarius*

The difference between any adjacent pair is often slight and sometimes non-existent, but of the commoner species, the apex of *Vitrea* is consistently smaller than the others, and that of *Retinella radiatula* and *R. pura* consistently smaller than *R. nitidula* and *Oxychilus cellarius*.

Two useful characters for separating *Oxychilus* from *Retinella*, and particularly *O. cellarius* from *R. nitidula* which are similar in size and commonly occur together, are the form of the umbilicus and texturing

3. METHODS

These are oval or subrectangular and there is considerable variation within each species, making certain identification virtually impossible. The shells are illustrated by Quick (1949; 1960) and tentative Pleistocene determinations by Sparks (1957a) and Hayward (1954).

Milax shells can be separated from *Limax* and *Agriolimax* by virtue of their bilateral symmetry (Fig. 18). Shells larger than about 7 mm probably belong to one of the larger species of *Limax* such as *L. maximus*, *L. flavus* or *L. cinereoniger*.

Graphical Presentation of Results

The purpose of graphical presentation is to give a clear and comprehensive picture of what are considered to be the more important aspects of the results. No final statement of the facts is intended, and it is unlikely that any two workers would produce identical graphs from the same set of figures. For this reason, it is essential in a definitive research paper that the results of analysis are also recorded in tabular form.

Broadly, the results can be presented in terms of either relative or absolute abundance. Kerney (1963: 231) in considering the Late Weichselian fauna of south-east England, puts forward arguments in favour of using absolute abundance as a basis for the histograms, which may be summarized thus: (a) The total number of shells is often inadequate for statistically valid percentages to be calculated; (b) The species do not fall into definite ecological groups; (c) An increase in one species may cause an apparent decrease of another when plotted on a relative basis, where in fact no decrease occurs; (d) The absolute abundance of Mollusca is often itself a reflection of climatic change.

When a greater number of species is involved and when these fall into clearly defined ecological groups, plots of relative abundance are more suitable.

The molluscan diagram is constructed in a manner similar to the pollen diagram (Fig. 116). The vertical axis is the depth below the modern soil surface or other convenient datum; the horizontal axis is the relative or absolute abundance of the species plotted. Alongside the vertical axis, a schematized stratigraphical column of the soils and sediments analysed is shown, and to the right of this, if plots of relative abundance are being used, the total number of shells in each sample on which the percentages are based. On the far right of each diagram is placed a brief summary of the general environment at each horizon.

The diagrammatic representation of soils and sediments in the stratigraphical columns needs little explanation. In general, vertical lines represent humic material, and the closer these are the more stable and long-standing is the soil which they represent. Chalk/limestone lumps and flints have been represented by open and blacked shapes respectively. Variations from this scheme are indicated in the individual diagrams.

The arrangement of species or groups of species in the diagram depends on the aspects of the pattern of change which it is intended to emphasize. Most of the faunas discussed in this book are terrestrial, and can be subdivided into three broad ecological groups: woodland or shade-loving species, intermediate or catholic species, and open-country species. In the histograms these are placed from left to right respectively. In Kerney's treatment of the Late Weichselian fauna of south-east England (1963), the arrangement is on a climatic basis, with climatically tolerant species on the left and relatively thermophilous species on the right irrespective of their habitat preferences. Freshwater species may be similarly classified either on an ecological or climatic basis as has been discussed by Sparks (1961; 1964a).

Percentages are generally based on the total number of shells in each sample though there are exceptions to this. Thus the terrestrial species, *Cecilioides acicula*, burrows to depths of up to 2·0 m, and the examples of this species are either represented as a percentage over and above the rest and plotted as an open graph, or excluded altogether. Percentages are corrected to the nearest whole number. Values less than 1·5 are plotted as a plus sign in the histogram. A plus sign in the tables indicates a non-apical fragment not included in the percentage calculations.

A second method of presenting the results of analysis, also making use of the histogram technique, has been found useful in comparing individual faunas from different sites. Percentages are plotted along the vertical axis and the species or groups of species along the horizontal axis (Fig. 105).

A third method makes use of the sector diagram in which groups of species are arranged radially in a circle, the size of each sector representing the relative abundance of each group (Fig. 3). The use of this method has already been discussed (p. 10).

Differences in faunal composition either between different levels on a site, or between different sites, can be checked statistically, but this is often not necessary for a variety of reasons. For example, if two shell assemblages are shown to be significantly different by means of a statistical test, while admittedly providing a basis for argument or further thought, it may still be uncertain whether this is of *ecological* significance or not. Thus in a histogram showing the faunal succession through a deposit, a change in the relative abundance of one or more species is usually due to one of three factors. A change in the environment, for example, generally brings about a change in the composition of the snail fauna which is then reflected in the fossil record. It is this kind of change in which we are primarily interested. But changes may also be due to stratigraphical and pedological processes; these are discussed in Part III and are usually easy to detect. Thirdly, changes may be of a statistical kind, due to the fact that percentage values are based on low total numbers; i.e. they are statistically insignificant. Statistical significance can be checked in various ways. Most conclusively, perhaps, a mathematical test can be applied, but to do this for each species at each level where a change occurs would be laborious. It would be useful if one could construct a histogram by means of a computer which at the same time automatically deleted all those changes which were not statistically significant. But the snail fauna cannot be considered in isolation. Other factors such as the stratigraphical and archaeological record must be taken into account, and as these are generally of a qualitative kind they cannot easily be fed into a computer or introduced into a mathematical test. A faunal change which is ecologically in line with a stratigraphical change has an added significance. As already pointed out (p. 22), it is important to consider other environmental data when assessing the significance of a subfossil fauna.

It is also desirable that marked changes of abundance be maintained over two or more samples if they are to be considered as of environmental origin. Alternatively the total numbers of shells on which percentage values are based should be high. The minimum number of shells in a sample needed to establish statistically significant values of abundance for each species depends too on the number of species and their relative abundance in the fauna, that is, on faunal structure. For example, in a fauna composed of 90% species a, 5% species b, 4·55% species c, and 0·05% species d, a count of one hundred shells will adequately reveal the broad structure of the fauna but may not reveal species b and c, and is unlikely to reveal species d. If the fauna is being studied to establish the local environment of a site, as is usually the case in archaeological work, then a count of 100 will be sufficient. If on the other hand, species d is an important climatic or zonal indicator, a larger count is necessary.

An experiment to determine the minimum number of shells needed

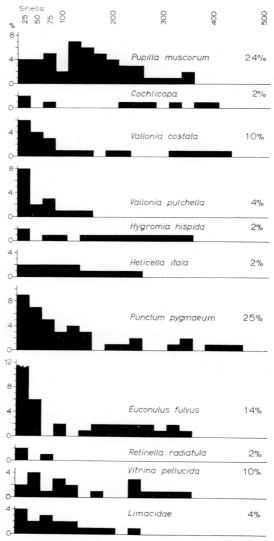

Figure 32. Percentage variation in a sample of shells at various values of abundance.

to obtain the true values of abundance for each species or group in a fauna of comparable composition to those described in this book was done as follows. The bottom of a wooden tray, 8·5 × 16·5 cm, was inscribed with 561 0·5-cm squares numbered randomly from 1 to 561. A sample of 500 shells comprising eleven species in known but varying proportions (Fig. 32) was put into the tray, spreading the shells as evenly as possible. Then shells were extracted 25 at a time, starting at square 1 and progressing to square 561, by which stage all the shells had been removed. After each group had been extracted, the counts for the various species were added to those of the previous counts and percentage values calculated. The variation from the true percentage was then plotted for each species (Fig. 32).

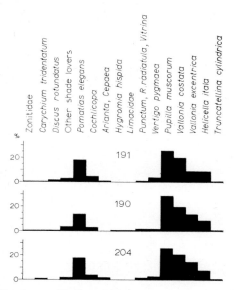

Figure 33. Durrington Walls. 0·5-kg samples from the buried turf line.

With 175 shells, variation from the true percentage had fallen to zero in five species, 1% in four species, 2% in one species and 5% in one species. It is thus felt that between 150 and 200 shells is a sufficient number to establish the broad composition of a fauna.

This is further substantiated by the results of three analyses of a soil sample from Durrington Walls in Wiltshire (Wainwright and Longworth, 1971: 329). The sample, which came from the turf horizon of the buried soil beneath the henge bank at Durrington

Walls was divided into three lots each weighing 0·5 kg, and each was analysed separately. The results (Fig. 33), both in total numbers (about 200) and in the composition of the fauna, are closely comparable.

We can now summarize the points to be noted in assessing the significance of an assemblage as follows:

> The total number of shells on which percentage values are based
> The number of species or groups plotted
> The maintenance of constant values over two or more samples
> Other environmental data

Perhaps the most relevant use of statistics is in the comparison of assemblages from different sites, particularly when the composition of the assemblages is felt to be a close reflection of formerly live populations. The faunas from the various Neolithic turf lines considered in Part III (Fig. 144) would be suitable material, but unfortunately the number of sites is extremely small. One would prefer to be dealing with hundreds rather than tens, and until more data can be obtained I feel that statistics will be of little help in assessing the results of snail analysis.

Part II

The Snails

4
Factors Controlling the Distribution and Abundance of Snails

The basis for the interpretation of subfossil faunas in terms of ancient environments is the ecology of their present-day counterparts. As pointed out in Chapter 2, there are various difficulties, and these fall into two categories—stratigraphical and ecological. The former concerns *post mortem* changes brought about in the course of soil formation and sediment deposition as discussed at length in Part III. Here we are concerned with problems of an ecological kind, involving a consideration not only of factors controlling the composition of snail faunas today, but of the way in which the influence of these factors has changed with time.

Sparks (1969) has pointed out that " ... human beings usually take a wider view of environment than snails, so that conclusions drawn from the latter are valid only for small sections of the human environment." The ecology of a snail species may enable it to occur in a number of, what to man's eye are different, environments, but which to the snail are identical. For example, *Carychium tridentatum*, *Punctum pygmaeum* and certain zonitids, such as *Retinella pura* and *R. nitidula*, can live equally well amongst leaf litter in densely shaded woodland as at the base of tall grasses in grassland of *Zerna/Arrhenatherum* type. The important factors are a modicum of lime, a lack of disturbance and a soil whose moisture-retaining capacity prevents undue desiccation in the summer months. The gross structure of the environment—shaded woodland in one case, grassland in the other—is of no account. Likewise, the fauna of light woodland on a steep escarpment where there is a paucity of leaf litter and no herb-layer vegetation may closely resemble that of short-turfed grassland, species such as *Abida secale, Vallonia costata* and *Helicella itala* being common to both; here the critical factors are light, and a shallow soil with a poor water-retaining capacity.

In the Chiltern Hills, two contemporary vegetation seres have been recognized (Watt, 1934) each having its characteristic grassland, scrub, and woodland stages. The juniper sere is characterized by a short-turfed grassland stage with *Festuca rubra* as one of the important species, by juniper scrub and sanicle beechwood (Figs 75 and 79). The hawthorn sere is characterized by a grassland stage dominated by tall grasses such as *Arrhenatherum elatius*, by hawthorn scrub and mercury beechwood (Fig. 78).

Some of the properties of the two seres are listed below (From Watt, 1934).

HABITAT FACTOR	JUNIPER SERE	HAWTHORN SERE
Exposure to westerlies	More	Less
Slope	Steep—up to 34° 50′	Gentle—up to 29°
Chalk horizon	Middle and Upper—typical on Upper hard strata	Middle—typical on soft strata
Mean soil depth (cm)	31	51
% $CaCO_3$ in top 15 cm in mature beechwood	59·55	33·32
% loss on ignition in upper 15 cm of soil in mature beechwood	12·94	23·33
Leaf litter in mature beechwood	Sparse	2·5–5 cm thick
pH (upper 15 cm)	7·9	7·6
Soil water relations	Smaller water-holding capacity	Greater water-holding capacity
Nitrification	Less active	Active

As far as it affects the snail fauna, the occurrence of certain species appears to be governed not so much by the presence of grassland or woodland but by the factors which characterize the seres as a whole—the physical and chemical properties of the soil and ground surface, and, as is discussed later in connection with the ecology of the two species of *Vallonia*, the rate of nitrogen metabolism in the environment. From a rather cursory examination of habitats belonging to these two seres (MS I to MS V, p. 232) I gain the impression that species such as *Retinella nitidula*, *Carychium tridentatum* and *Vallonia excentrica* are associated with habitats of the hawthorn sere and others such as *Abida secale*, *Vallonia costata* and *Retinella radiatula* with habitats of the juniper sere.

Thus similar habitats, e.g. "grassland", may be occupied by different snail faunas, while dissimilar habitats, e.g. grassland and woodland of the same sere, may be occupied by identical faunas. This state of affairs presents problems only so long as the interpretation of a fauna is made in terms solely of vegetation. Once it is realized that other factors can be of overriding importance, then a more realistic assessment of the habitat can be made, and it is to be pointed out that some of these factors, such as the moisture-retaining capacity of the soil and the rate of nitrogen turnover, are often of greater relevance to the environment of early man than is the gross structure of the vegetation—at least as far as the potential of a habitat to support stock or yield crops is concerned.

A more serious difficulty of interpretation is that certain species can occupy different niches in the environment. This is not the same thing that we have just been discussing, in which a species may live in different habitats but occupy the same niche. Thus Boycott (1921d) pointed out that both phenomena were characteristic of certain species of snail: " . . . in looking through the literature I am impressed by the fact . . . that a species' habitat and habits vary in different parts of the British Isles." *Vallonia costata*, for example, can live in woodland in small numbers, thrives in abundance on stone walls—in both cases as a rupestral species, probably browsing on unicellular algae—lives in short-turfed grassland, again frequently in vast numbers, and is characteristic of the fauna of rubbish dumps, in association with synanthropic species such as *Oxychilus draparnaldi* and *Hygromia striolata*, scavenging on semi-decomposed organic matter. *Pupilla muscorum* is another species which can occur in a grassland or rupestral capacity, and again probably obtaining its food from a different source in each case.

The absolute abundance of shells in a soil or sediment is controlled by three factors: processes which destroy the shells; the rate of accumulation of the deposits; and the reproductive rate of the snails as controlled by the environment and their own hereditary mechanism. Destructive processes will be considered in Part III, and may generally be thought of as acting on shells after death. In soils, biological and chemical solution are the main processes, while in slopewash deposits physical smashing of shells is the main destructive agent. The faster the rate of sedimentation, the fewer the number of shells incorporated due to dilution of the sediment by inorganic matter—other factors such as reproductive rate being constant. In addition, rapid sedimentation generally implies an unstable surface, and this is deleterious to most (though not all) species of land snail. Thus the sedimentation rate controls the abundance of shells in a deposit in two ways, stratigraphically and environmentally, a rapid rate almost always resulting in low abundance, and vice versa.

There is generally, too, a direct relationship between abundance and

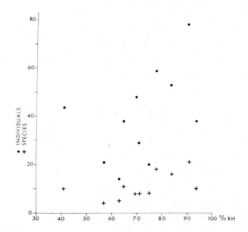

Figure 34. Numbers of individuals and species at various relative humidities, from sites in north-west Germany. The site at 41% is semi-dry calcareous grassland. (Data in Ant, 1963: Table 1.)

the suitability of the environment for molluscan life, so that the abundance of shells in a deposit can give us some idea of the contemporary environment, whether this be suitable or unsuitable for the various species of snail in the fauna and, when considered in conjunction with the sediment type, the rate at which deposition is taking place. In most cases, "suitability" implies high lime content and relative humidity, shelter and lack of disturbance, as Fig. 34 shows. But in calcareous grassland, other factors are clearly involved.

More information, particularly about the status of the ecosystem as a whole, can be obtained by examining the structure and diversity of the fauna. Two main types of ecosystem can be distinguished—generalized and specialized. Generalized ecosystems are characterized by a wide variety of plant and animal species, each of which (or the majority of which) is represented by a relatively small number of individuals; i.e. there is a fairly uniform distribution of numbers of individuals among the various species present. Specialized ecosystems, on the other hand, usually have a low diversity and are characterized by a small number of species, some of which are represented by a large number of individuals, as sometimes occurs, for example, in chalk grassland or arable habitats. This principle can be shown graphically as

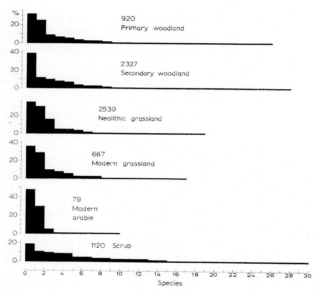

Figure 35. Diversity of faunas from different habitats. Subfossil except where stated.

in Fig. 35, where faunas from different types of ecosystem are compared (see also p. 335).

The ecological niche of most species of land snail is that of small herbivore, scavenger and carrion feeder, living close to the surface of the ground and preyed on, largely one suspects as a subsidiary food source, by birds, hedgehogs, shrews and other small beasts. Most favour the moister and less disturbed habitats of the ecosystem. None appear to be restricted to living on or in association with particular plant species, though the structure of the vegetation is an important factor in their livelihood. Most species have annual life cycles, though a few of the larger helicids may live for two or three years.

Boycott (1929b; 1934), who has given us much of our information about the ecology of our land-snail fauna, considered shelter and lime to be the two most important environmental factors controlling the distribution and abundance of snails. In the context of subfossil faunas however, where the preservation of shells is almost always a function of a lime-rich environment, this latter factor can be taken for granted.

Shelter and Humidity

With a few important exceptions, land snails have not evolved structural or physiological means of withstanding water loss from their bodies whilst active, and as a result, their habitats and behaviour patterns are controlled largely by the humidity of the environment. The majority of species feed and move around at night when the relative humidity of the air is high; during the day they confine themselves to moist and sheltered places. Under dry conditions, the shell can be a useful protection against water loss, being nearly impermeable to water in some species (Warburg, 1965; Machin, 1967). In some, a horny epiphragm is laid down across the mouth of the shell after it has withdrawn into it. *Pomatias elegans* and *Acicula fusca* have a calcareous operculum attached to the dorsal part of the foot, which seals the shell when the animal is inactive—in essence a permanent epiphragm (Fig. 19).

The humidity of the environment is controlled by the precipitation/evaporation ratio at the ground surface, primarily a function of climate (temperature and rainfall) secondarily of the structure of the vegetation cover, the water-holding capacity of the soil and the degree of disturbance of the habitat. Climatic variation across the British Isles is responsible for the restricted distribution of certain species, but local environmental factors are frequently of overriding influence.

There is, for example, a number of species whose main area of distribution is the north and west of Britain, which is possibly a function of the higher humidity in those parts as controlled directly by precipitation (Boycott, 1921b: 163) (Fig. 36). *Pyramidula rupestris, Vertigo substriata, V. alpestris, Lauria cylindracea, L. anglica, Acanthinula lamellata* and *Hygromia subrufescens* display a northern and western distribution to the most marked degree (Fig. 37). For some of these, winter temperature may be the factor controlling their eastern limit as has been suggested by Kerney in the case of *Lauria*

4. DISTRIBUTION AND ABUNDANCE OF SNAILS 93

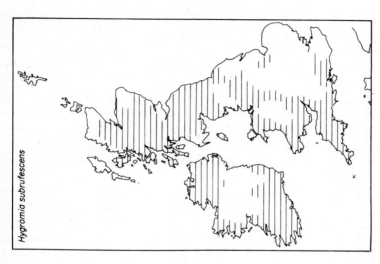

Figure 37. *Hygromia subrufescens* in Britain, a species whose distribution is probably controlled largely by rainfall. (After Ellis, 1951.)

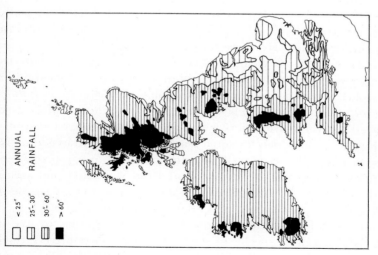

Figure 36. British Isles. Rainfall map.

cylindracea (1968a). But for others, humidity is probably the critical factor.

Sparks (1964a: 92) has suggested that the north-western distribution of *Acanthinula lamellata* and *Lauria anglica* may be "regarded as complementary to the distribution of the most intensively farmed lands of Britain". Both species were widespread in south-eastern England during the Atlantic period, and their present-day more restricted distribution probably had its inception in the Neolithic period when vast tracts of chalkland were cleared of forest and many suitable damp and shaded micro-habitats destroyed. The same processes, of course, took place in the highland zone of Britain, though perhaps to a lesser extent; here, however, the higher rainfall has enabled these species to survive (Boycott, 1921b: 165).

Most of the species which are fastidious with regard to their moisture requirements, however, have managed to remain common in the drier parts of Britain by confining themselves to the more sheltered, and thus moister, terrestrial habitats—hedgebanks, ditches, woodland and marsh. Nevertheless, in parts of East Anglia where the annual rainfall is less than 25 in., and where evaporation often exceeds rainfall, even some of the common species, like *Discus rotundatus*, are rare. Thus for a large number of snail species in Britain, and particularly in the south and east, shelter is an important factor in maintaining their abundance (Boycott, 1934).

For some species, a dry environment (as is provided by stone walls, chalk downland and sand dunes) is necessary and such species fall into two groups. First there are those for which such habitats are only one of a number in which they can live; i.e. they are not restricted to dry places but possess the necessary behavioural and physiological adaptations which allow them to do so. Examples are *Retinella radiatula*, *Euconulus fulvus*, *Punctum pygmaeum*, *Vitrina pellucida* and *Helix nemoralis*. Then there are others which are virtually restricted to dry places (Fig. 38), and which for some reason are unable to live in moist habitats. Some of these, as well as being confined to dry places, are inhibited by shade, and are thus absent from woodland and scrub. The most frequently occurring are *Pupilla muscorum*, *Vertigo pygmaea*, *Vallonia costata*, *V. excentrica* and *Helicella* species (Fig. 39).

It is not certain what adaptations enable these species to live in habitats which to most species of snail are hostile. A relatively thick shell (compared for example with that of the Zonitidae) which may help to impede water loss, is common to all. The shell lip of *Vallonia*

4. DISTRIBUTION AND ABUNDANCE OF SNAILS

costata is acutely reflected and flat in one plane so that when appressed to the substratum the interior of the shell is completely sealed (cf. Machin, 1967). The shell of this species is traversed by fine vertical ribs, one function of which may be to slow down the circulation of air

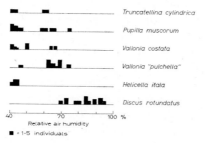

Figure 38. Dependence of certain land snails on relative air humidity, from sites in north-west Germany. (After Ant, 1963: Table 2.)

in contact with the shell and thus reduce the rate of water loss from it. The hairs on *Hygromia hispida* and *Monacha cantiana* may serve a similar purpose, particularly in the young, as suggested by Chatfield (1968: 244). *Pupilla* and *Vertigo pygmaea* have dark reddish-brown coloured shells, perhaps serving to absorb harmful ultraviolet light. *Vallonia* on the other hand has a white shell, which could be said to reflect heat, keeping the animal cool and thus preventing water loss.

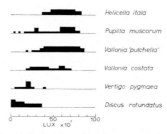

Figure 39. Relationship of certain land snails to amount of available light, from sites in north-west Germany. The smallest unit corresponds to 1–10 individuals. (After Ant, 1963: Table 3.)

Cain (Cain *et al.*, 1969) has suggested that in parts of northern Scotland there is an inverse relationship between the extent and intensity of banding on the upper surface of the shell of *Helicella itala* and the degree of insolation.

Light and low humidity are factors necessary for the occurrence of these species: they are inhibited by shade and moisture. There are several possible reasons for this. One is that their physiology is incompatible with shade and moisture; how these factors operate is not clear, and any hypothesis would have to explain how it is that the snails are directly inhibited. For example, Chatfield (1968: 234) noticed that juvenile mortality of young *Monacha cantiana* was especially heavy under conditions of "excessive moisture". Carrick (1942) states that waterlogged soil (above 80% saturated) drowns the developing embryos of *Agriolimax reticulatus*. Another possibility is that these species, by virtue of their being able to occur in open and dry places, are excluded from shaded and moist ones by other species (interspecific competition) as suggested by Boycott (1934).

Other species which occur in dry habitats are not restricted to open places, and their presence is generally controlled by the availability of a firm substratum such as is provided by stone walls, rocks, tree trunks and fallen branches. Included here are all the Clausiliidae, *Acanthinula aculeata*, *Vertigo pusilla*, *V. alpestris*, *Ena* spp. and *Helicigona lapicida*. Some of these have similar structural adaptations as are possessed by *Vallonia costata*—costae or spines (Fig. 22) and the shell aperture flat in one plane. Essentially their requirements are for humidity as provided by woodland or stone-wall habitats, coupled with dryness of the immediate substratum.

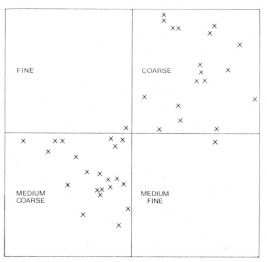

Figure 40. Distribution of *Milax budapestensis* in relation to soil texture. (After Stephenson, 1966: Fig. 1.)

Humidity is often controlled by soil texture, an important consideration in the case of burrowing species. A number of slug species, for example, bury themselves a short distance into the soil during the day or in periods of excessive dryness, and it has been shown by Stephenson (1966) that *Milax budapestensis* prefers coarser to finer soils in which to burrow. Fig. 40 shows the distribution after seven days of forty slugs offered a choice of four soil textures. The soils were arranged in boxes 1 ft square by 6 in. deep. It was noted that the finer soils tended to dry and form an unbroken surface, while the coarser soils, although dried near the surface, remained moist lower down

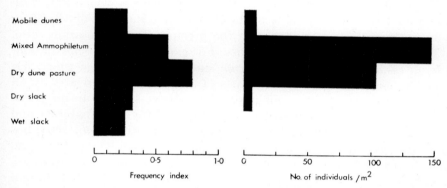

Figure 41. Abundance of *Helicella caperata* in different habitats, showing its restriction by increasing humidity in one direction and decreasing surface stability in another. (After Baker, 1968: Fig. 1.)

where the spaces between the moist particles provided an acceptable slug habitat.

Humidity is thus a powerful factor in controlling the distribution and abundance of Mollusca in a habitat, and it operates in various ways depending on the physiological requirements of the various species. It is because of this that our main ecological classification of non-marine molluscs into freshwater, marsh, woodland and xerophile species is based on the moisture régime of the environment. We may not always be able to explain in physiological terms what determines the ecological range of a particular species, but from the point of view of the palaeoecologist, the fact that certain species have upper and lower limits to their ecological requirements is invaluable (e.g. Fig. 41). Moreover, the variety of ways in which humidity is controlled—by climate (temperature and rainfall), vegetation structure, soil type and the degree of disturbance of the habitat (e.g. Fig. 41)—enables us to obtain other details of past environments in addition to their humidity *per se*.

As I have stressed throughout, humidity operates largely through the medium of the local environment. Rarely can we invoke rainfall as a determining factor in the distribution of land snails in the British Isles today, and the same would appear to be true of their later Postglacial history. The marked decline since the Atlantic period in the south and east of Britain of many species is due largely to anthropogenic processes such as forest clearance, agriculture and the draining of marshes (Boycott, 1934; Kerney, 1966b), and to a fall of temperature (p. 100). Nor are there any changes which can with certainty be linked causally to the climatic deterioration of 550 BC or any other subsequent change in the precipitation/evaporation ratio. This is not to deny their existence. For example, the decline of *Vallonia costata* and *Pupilla muscorum* since the Bronze Age which has taken place locally in many parts of Britain may be related to an increase in precipitation making for a moister substratum in grassland habitats, for both species tend towards the drier habitats and often occur in a rupestral capacity. Kerney (Kerney *et al.*, 1964) has tentatively suggested that certain changes in a hillwash fauna reflecting a shift from dryness to wetness can be equated with a recurrence surface recognized in a number of peat bogs as falling around AD 1. On the whole however we must be extremely careful, particularly in archaeological work where strong local humidity gradients are induced by man-made structures, before invoking rainfall as a causative agent in faunal change.

Earlier on, however, in the Late Weichselian and early Postglacial, changes in the land-snail fauna are more certainly associated with changes of precipitation. The abundance of *Hygromia hispida* and *Retinella radiatula* in zone III of the Late Weichselian probably reflects a more humid climate than obtained in zones I and II "either because of higher actual precipitation, or because a cover of dense cloud restricted evaporation" (Kerney, 1963: 244). The extinction of *Discus ruderatus* at the beginning of the Atlantic period may be linked to an increase of precipitation at this time (p. 184). It is probable too that the main episode of tufa formation in lowlying areas of Britain during the Atlantic period was responsible for an increase in the distribution and abundance of many hygrophile species, and its later cessation for their subsequent decline. This, however, is less certainly linked directly to precipitation than to the more congenial conditions extended their ranges far beyond their present eastern limits (Taylor, 1965: 606).

Food

Most species of land snail feed on dead and decaying plant and animal matter—i.e. they are scavengers and carrion feeders—and occupy the same trophic level in the ecosystem. There are, of course, exceptions, some snails and most slugs being herbivores, commonly eating live plants, generally seedlings; though leaves, roots and stems of adult plants are taken from time to time by one or other species. Many snails browse on the film of lichens and unicellular algae coating tree trunks, fallen branches and stones. Fungi are a favourite food with woodland species.

But no snail species is dependent on a particular species of plant, and this applies to the provision not only of food, but of shelter and substratum as well. Boycott (1934) was very insistent on this point and no evidence has been brought forward since his work to contradict this conclusion (e.g. Duval, 1971: 249). From the palaeoecological point of view this is disappointing for the composition of the vegetation in terms of species is an important aspect of the environment. The association today, for example, of *Helicodonta obvoluta* with beech woodland is due entirely to the fact that the only suitable habitats available to this species within its area of distribution happen to be beech woods (Cameron, 1972). There is no obligatory association of *H. obvoluta* with beech, and the occurrence of this snail in ancient faunas—as in the Neolithic flint mines of Sussex—is no indication of the former presence of beechwood there.

Snails of the family Zonitidae, and *Vitrina pellucida* are facultative carnivores. *Oxychilus cellarius* has occasionally been recorded from within the crania and amongst the bones of human skeletons buried in Neolithic long barrows, as for example at Wayland's Smithy I in Berkshire (M. P. Kerney, personal communication) where it was attracted by the rotting flesh of bodies. *Testacella* (not recorded fossil) eats earthworms, often chasing them some way along their burrows before catching and devouring them.

Boycott also insisted that food is not, or is only rarely, a limiting factor for land snails in controlling their distribution and abundance, and Runham and Hunter (1970: 135) have indicated that the same applies to slugs. Baker (1970) maintained that it was unlikely that seasonal changes in food supply could affect the mortality rate of *Cochlicopa lubrica* in a grass sward habitat since there was a plentiful supply of litter and vegetation throughout the year. Indeed this is held to be the case with herbivorous and scavenging species of animal as a whole (Andrewartha, 1961).

However, it is arguable whether this is universally applicable to snail populations, particularly in some of the very specialized ecosystems such as chalk downland, sand dunes or arable land with which prehistorians are particularly concerned. Thus the amino-acids necessary for protein synthesis must be acquired from vegetable (or animal) matter. Snails, like other animals, cannot utilize inorganic nitrogen compounds for this purpose, and are therefore to some extent dependent on the composition of the plant tissues which they consume. In a generalized ecosystem such as a woodland or rich grassland habitat where the number of plant species is large, the number and variety of amino-acids and organic nitrogen compounds is correspondingly great. But where the number of plant species is small, as is the case in the habitats mentioned above, is it not possible that the type and variety of organic nitrogen compounds available to Mollusca is reduced to the extent of becoming a limiting factor? I suggest this as a distinct possibility in controlling the composition of some of our grassland faunas.

Furthermore, might nitrogen turnover as a whole play some part? For example, it is well known that although sheep and cattle are not restricted to particular plant species for their food, sheep can be reared and *fattened* on grassland of a type (e.g. *Festuca/Agrostis*) on which cattle can only be *maintained,* and that this is related to the percentage of nitrogen in the leaves. The following table shows the

	NITROGEN	FIBRE
Alluvial pasture	3·0	21·5
Festuca–Agrostis grassland	2·1	24·6
Nardus grassland	1·6	31·2

percentage of nitrogen and fibre in dry matter from leaves of different types of grassland (Pearsall, 1968: 151). Cattle can make growth only on alluvial pasture.

Might not the same principle apply to closely related species of snail—for example, *Vallonia costata* and *V. excentrica*? I have already suggested that *V. costata* is characteristic of juniper sere and *V.*

excentrica of hawthorn sere habitats and with reference to Watts' figures (p. 88) it can be seen that in the latter, nitrification is more active and the organic content as a whole greater. On the available data we can take this line of enquiry no further, but there is a strong possibility that the abundance of the two species of *Vallonia* is in part related to the nutrient level and rate of nitrogen metabolism in the ecosystem, *V. costata* being able to survive in poorer, or more oligotrophic, environments than *V. excentrica*.

It must not be forgotten, however, that in certain circumstances, other factors may be of overriding influence—local temperature differences, human disturbance, particularly agriculture, and the texture of the substratum. Thus under conditions of optimum temperature, it is possible that both species of *Vallonia* might coexist, as they did in prehistoric times, in areas where today only one or the other is found (cf. p. 158). The interesting observations of Johnson and Lowy (1948) on the distribution of *Helicella caperata* and *Hygromia striolata* as controlled by leaf width might well apply to other species. They showed that *H. caperata* occurred mainly on narrow-leaved plants and *H. striolata* on broad-leaved ones, and that this association was conditioned not by humidity but by the width of the leaf itself.

Predators and Parasites

Boycott (1934) considered that neither predators nor parasites were effective in localizing the distribution of any species of land snail with the possible exception of *Balea perversa*. This is a geophobic species, generally living in habitats away from the ground such as trees and walls, and thought to do so as a means of escape from some (unidentified) destructive agency. There is however plenty of evidence—much amassed since 1934—that snails and slugs are preyed upon and parasitized by a wide variety of vertebrate and invertebrate forms. Wild and Lawson (1937) summarized the known enemies of land and freshwater Mollusca which included mammals such as the hedgehog, mole, voles and shrews, a long list of birds, two reptiles—the slow-worm and the viper—other land molluscs and a variety of insects.

As far as predators are concerned, Boycott felt that none is dependent on land molluscs, and though eating them "pretty freely" only do so as an item of mixed diet. Cameron (1969b) (Fig. 46) showed that thrushes turn to snails for food chiefly in January and February, and June and July, being a reserve food taken in quantity only when more palatable foods are scarce. The main species eaten are *Cepaea hortensis* and *Arianta arbustorum*. Cranbrook (1970) showed heavy predation by starlings, mainly of *Hygromia hispida* and *Cochlicopa lubrica*; and Chatfield (1968) noted the predation of larger specimens of *Monacha cantiana* by the glow-worm larva, *Lampyris noctiluca* (L.), birds and small mammals. But does Boycott's contention that these agencies are of little importance in the destruction of half- to full-grown snails still hold? Two lines of enquiry suggest not.

Since 1950 there has been a number of papers on the polymorphic snail, *Helix* (*Cepaea*) *nemoralis*, which have demonstrated conclusively that in many parts of Britain, frequencies of the different

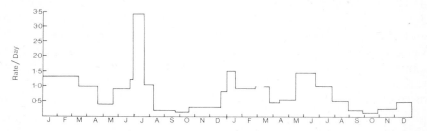

Figure 46. Variation in the rate of accumulation of shells of *Cepaea hortensis* on thrushes anvils over two years. (After Cameron, 1969: Fig. 1.)

morphs (or varieties) vary from place to place, mainly because predation by thrushes is visually selective, favouring the more cryptic varieties on any one background (Cain and Sheppard, 1950; Sheppard, 1951). For instance, in woodland where the carpet of leaf litter provides a uniformly brown background, the predominant morph is brown and unbanded; in grassland where the background is paler and more heterogeneous, the favoured morphs are yellow with black bands—a similar type of camouflage to that shown by the zebra and tiger. Predation is clearly a limiting factor controlling the distribution and abundance of the various morphs. As far as I know it has not been shown that predation is responsible for the restriction of any snail *species* to a particular habitat, but in the light of the *Cepaea* data it would seem to be a much stronger possibility than Boycott originally maintained. Taken with the evidence for seasonal peaks of predation (Cameron, 1969b) (Fig. 46), predators can at any rate be thought of as causing fluctuations in numbers if not irregularities in distribution.

The other major line of enquiry which perhaps necessitates a reconsideration of Boycott's conclusions is the predation and parasitism (in many cases the two have not been satisfactorily separated) of land molluscs by insects and other invertebrates—notably Protozoa, flatworms and nematodes. Most of the work on these associations has been done on slugs (Stephenson and Knutson, 1966; Stephenson, 1968) because of their economic importance, but there is no reason to suppose that snails should not be similarly affected, and indeed in some cases are.

Boycott appreciated that infestation of molluscs by other invertebates took place and cites a number of examples—*Helicella virgata* eaten by the glow-worm larva; *H. itala* as the intermediate host of the sheep fluke, *Dicrocoelium dendriticum*; and *Agriolimax reticulatus* as the intermediate host of the fowl tapeworm, *Davainea proglottina*. Ellis (1969: 195) considers *Helicella itala* "scarce on downs grazed by sheep" which may be linked to its parasitism by the sheep fluke, though no regulation has been demonstrated. Foster (1958a; 1958b) showed that infestations of *Milax sowerbyi* and *Agriolimax reticulatus* by trematodes led to severe necrosis of the "kidney"; infestation of *Agriolimax* occurred in March and April, reached 90 to 100% by June to July and then dropped sharply with a high death rate of parasitized slugs. The seasonal level of parasitism followed a regular pattern over five years and then lessened. Williams (1942) found that 20% of the slugs and snails from the Cardiff and Rhondda Valley areas of South Wales were infested with larvae of the nematode *Muellerius capillaris*. Knutson (Knutson et al., 1965) cites *Tetanocera elata* and three other sciomyzid flies as the only insects known to feed obligatorily on slugs, which are killed; *Agriolimax reticulatus* and *A. laevis* are the main victims. Stephenson and Knutson (1966) list 46 species of invertebrates (in North America and Britain) associated with 25 species of slugs, and 10 species of invertebrates known to kill 14 species of slugs—protozoans, flatworms, lungworms, lampyrid beetles and sciomyzid fly larvae being the more important enemies. Runham and Hunter (1970) consider the action of predators—parasites and diseases of the egg state, and occasional visits from flocks of birds—to be key factors in controlling the abundance of slugs.

Stephenson (1968) gives several estimates of the size of slug populations; these range from **13 000** through **50 000/70 000** to **275 000** per acre. Notwithstanding these vast numbers, he claims that the frequency with which infested and damaged slugs are found in nature suggests that biological control may be possible. Although these estimates are admittedly subjective, one is left with the impression that parasites and predators are of some importance in controlling the distribution and abundance of land molluscs, at any rate much more than Boycott allowed.

Competition

Competition between different snail species for food and living space—interspecific competition—is not considered to be a powerful influence in determining their distribution and abundance in a habitat (Boycott, 1934). "Land Mollusca cannot normally be thought of as 'filling' an environment in the sense of a plant community with its members mutually limited by competition, and in competition with other communities; the individual species may live in a variety of microhabitats overlapping within the same area, and many increase or decrease in numbers independently" (Kerney, 1963: 232). Indeed, the concept of competition among animals as a whole is a somewhat nebulous one, and a number of ecologists, notably Andrewartha, put forward a strong case for its non-existence. "In nature relatively few species ever become crowded relative to their 'living space' or food" (Andrewartha, 1961: 106).

However, in specialized ecosystems such as sand hills, arable land or heavily grazed chalk downland, where plant diversity is low and shelter sparse, interspecific competition for these assets may come into play. The number of species in such habitats is generally low, but as already pointed out (p. 90) those which thrive may do so to an extraordinary degree (Fig. 35). This property of xerophile faunas is further brought out in Fig. 34 where the numbers of individuals in habitats of varying humidity are compared. Under such circumstances where resources are limited but the number of individuals in a population high, it is likely that competition between different species will occur, and this has indeed been suggested on more than one occasion as, for example, in populations of *Cepaea hortensis* and *C. nemoralis* in sand dune and downland habitats (Oldham, 1929; Cain and Currey, 1963a) (cf. pp. 171 and 174).

I would suggest too that the data presented in this book with regard to the distribution and abundance of *Vallonia costata* and *V. excentrica* in both past and present populations indicates that the same might apply to these species (p. 158). It is difficult to understand the absence of *V. costata* from certain downland habitats at the present day (e.g. Overton Down and Windmill Hill in north Wiltshire) as being determined solely by environmental factors other than the restricting influence of other species. Competition is almost certainly not the whole answer, but until we know more about the physiological and nutrient requirements of these grassland species it is difficult to be more precise.

The Interaction of Environmental Factors

Boycott (1934) maintained that lime is one of the main factors controlling the distribution and abundance of the majority of land-snail species in Britain. In the context of subfossil faunas, where preservation is almost always due to high calcium carbonate content of the environment, we need hardly consider its relevance. Nevertheless, it is worth pointing out that lime is of supreme importance to Mollusca, and not only for shell building but also for certain physiological processes, which are dependent on calcium ions, so that when present in the environment, lime favours most species of land snail whether these be obligatory calcicoles or not. Thus a calcareous habitat is richer than one which is non-calcareous not only in species but also in numbers of individuals. In other words, lime acts as a compensatory factor, enabling a better and more complete use to be made of the environment. It is well to appreciate this point particularly when comparing subfossil faunas with modern faunas from non-calcareous habitats.

In the same way, other environmental factors may interact. All snails need food and most need shelter and generally they are not too fussy about how these are acquired. The example of *Cepaea hortensis* in north Wiltshire is instructive. Today, this snail occurs generally in valleys or lowlying areas around the Marlborough Downs, from which *C. nemoralis* is excluded by its intolerance of frost hollows (Fig. 57). In upland areas where conditions are more suitable for *C. nemoralis*, this species successfully competes with *C. hortensis* for food and shelter, excluding it from all but wooded and shrubby areas which are thus acting in a compensatory manner (Cain and Currey, 1963a). But in the Neolithic and Bronze Age periods both species were able to co-exist in open downland habitats on the uplands due probably to the compensatory action of the higher temperature at that time.

Cameron (1969a: 107) has shown that in grassland today, only one of the two species, *C. hortensis* and *C. nemoralis*, is generally present, as determined by height and locality. "Thicker and more heterogeneous habitats, especially woods, more often support mixed populations of the two species . . . Thus in any area the minority species will appear to be predominantly a woodland snail". This is an extremely important point, for although no work of this kind has yet been done on smaller species of snail, the same principles may well apply. For smaller species, too, such as *Vallonia*, *Pupilla* and *Vertigo pygmaea*, the scale of habitat differences which are relevant to their distribution and abundance may be less, so that instead of differences being created by macro-habitats such as woodland and grassland as in the case of *Cepaea*, we may find that different types of grassland—grazed, tall or tussocky—or different leaf widths (p. 105) create similar complementary distribution patterns among these species. This kind of concept would go some way to explaining the rather bewildering results from modern and ancient faunas relating to the habitats of *Vallonia costata* and *V. excentrica*.

In other words, the absence or low level of suitability of a particular factor may be compensated for by other factors if the value of these is optimum or above with respect to the species concerned. At the same time competition which might otherwise exist under more severe conditions, is, in very favourable habitats, relaxed, allowing several species which may be very close in their habitat requirements for food and shelter, to co-exist.

We are thus faced with a situation in which the immediate habitat of a species, e.g. woodland or grassland, depends upon the overall environment—climate, topography, etc. Clearly there are not such vast differences in all cases, particularly with the smaller species, but it is apparent that we cannot generalize from one particular set of circumstances. As Cain and Currey (1963a) pointed out, the behaviour and structure of populations of *Cepaea* on the Marlborough Downs differs from that in the area around Oxford; and similarly, the distribution of the two species of *Vallonia* in the Chilterns today cannot be used as a basis for their behaviour in the Chilterns in past times, or today on the Chalk as a whole.

Lateral Variation in Snail Faunas

The distribution of snail species in a habitat depends partly on the nature of the habitat and partly on the behaviour patterns of the snails. In a relatively homogeneous environment, such as a cultivated field or chalk downland, the distribution and abundance of a species in the area as a whole tends to be uniform. On a small scale, however, considering for example 2 m², snails occur more often in aggregations of a few individuals rather than singly (Mason, 1970). This is particularly the case in the breeding season as Baker (1968) has shown for

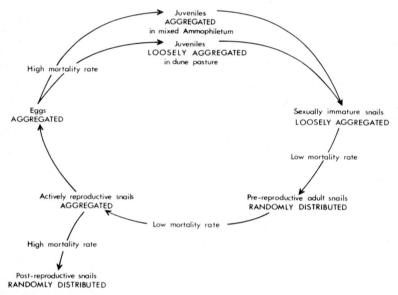

Figure 47. *Helicella caperata*. Distribution as determined by the stage in the breeding cycle. (After Baker, 1968: Fig. 5.)

Helicella caperata (Fig. 47) and Roscoe (1962) for *Cochlicopa lubrica*. In woodland, of many apparently suitable logs, rupestral species such as *Clausilia bidentata* may sometimes be found in abundance under only a proportion, and then in localized groups on one part of the log only. *Helix aspersa* hibernates in groups in places which are not apparently more suitable than closely adjacent loci, suggesting that it is not the habitat *per se*, but the gregarious habit of the snails which is responsible for their behaviour. Alkins (1928) found that colonies of *Clausilia cravenensis*, while uniform within themselves, showed considerable variation from one locus to another. There was no correlation between shell characters and the nature of the locus. The sharp boundaries between areas of constant morph frequency in *Cepaea nemoralis* found by Cain and Currey (1963a: Figs 5 and 6) in many places on the Chalk which appear not to have an environmental cause, provides additional evidence that discrete breeding populations can exist in close proximity, kept separate not by environmental factors, but by intrinsic properties of the populations themselves.

On the other hand, lateral variation is more often due to environmental heterogeneity, however subtle this might be. Boycott (1920), in studying the size distribution of specimens from colonies of *Clausilia bidentata* and *Ena obscura* in a beech wood, showed that from different areas in the wood, 50–300 yards apart, specimens could be distinguished from one another by differences in size. He also showed that the size variation in colonies of the two species ran parallel, and concluded therefore that it had an environmental rather than internal origin.

Woodland environments are more diverse than those of grassland and the number of ecological niches greater. Logs provide habitats for rupestral species such as *Clausilia*, *Acanthinula* and *Lauria*; piles of leaf litter on the woodland floor and banked up against the trunks of trees are favoured by moisture-loving species such as *Carychium*, *Retinella*, *Oxychilus* and *Vitrea*; while on the upright trunks may be found *Balea* and *Ena obscura*.

It is not known whether local variation on this scale is preserved in the fossil record. Sampling techniques probably do not enable the extraction of a single generation of snails from a sediment, and even if we could take very thin slices of soil (say *ca.* 2·0 mm thick, this being the average diameter of the largest number of shells generally found in an assemblage) it is most unlikely for stratigraphical reasons that these would derive from the population of a single generation, as discussed in Part III (cf. Sparks, 1964a: 89).

In addition to sampling difficulties, there is the fact that the heterogeneous distribution of snails in a habitat, whether this be environmentally or internally controlled, is constantly changing—extending at one point, contracting at another in amoeboid fashion. Snails are always moving from one micro-habitat to another, and although at any one time the faunal composition of a habitat may be different from that of an adjacent one, over a period of years the accumulative effect of diffusion from one to the other tends to blur such differences.

4. DISTRIBUTION AND ABUNDANCE OF SNAILS

And this is just the kind of picture we are getting from the fossil record—not simply a faunal reflection of the area within the measured limits of the soil sample, but of the ambient places as well. I have frequently been asked in connection with the analysis of faunas from buried soils, "How can you be sure that you are not taking samples from beneath a bush in an otherwise open environment, or from a small open space of a few square metres in a forest?" Apart from the slight chance of doing so, I feel that such a question is largely irrelevant owing both to the way in which snail populations wander, and to the blurred nature of the fossil record.

To take a specific example. In an area of open ground of about 100 m^2 surrounded by dense thorn scrub, in a Berkshire wood, I found several examples of *Acanthinula aculeata*. The soil surface was in places exposed and the vegetation of the clearing generally sparse—an unlikely habitat in fact for *Acanthinula*. But in the same woodland, in another piece of open ground, *Acanthinula* was totally absent despite prolonged search. This second site was similar to the first in all attributes except for area, which was 3000 m^2, thirty times that of the first. In other words the composition of the snail fauna, and the presence of *Acanthinula* was controlled not only by the character of the habitat, but also by its surrounds.

The question now arises as to how representative of the environment as a whole is the shell assemblage. What sort of distances do snails travel? Or, perhaps more significantly, what is the extent of the transition zone between two adjacent habitats (e.g. woodland and grassland) in which snails from each occur? Until the relevant quantitative work has been done on modern faunas, the answers to these questions will remain obscure. My own, qualitative, observations on faunas from chalk grassland habitats in various parts of southern England have led me strongly to believe that only when the habitat is very open and closely grazed, as on some of the barer downs, do faunas of the restricted open-country type, which are familiar from Neolithic and Bronze Age sites, occur. Once the grassland tends to become tussocky or unkempt and tall with species of *Zerna*, *Dactylis* and *Arrhenatherum* replacing the neater fescue sward, or once a few thorn bushes are introduced into the vegetation, then our mesophile species such as *Retinella nitidula*, *Discus rotundatus* and *Carychium* start to appear in the fauna. It is for reasons such as these that I believe the Neolithic and Bronze Age faunas discussed in this book, which are representative of dry grassland environments, reflect areas to be measured in hectares. In other words they are areas meaningful in terms of human land use.

Until more quantitative work has been done we cannot be more precise. Nor is it yet possible to devise a numerical standard such as the NAP/AP ratio used by palynologists to denote the degree of openness of an environment. This is not only due to our ignorance of the extent of the area of overlap between the faunas from two adjacent habitats, but also to the inherent difficulties of interpreting the environmental significance of woodland species any way, many of

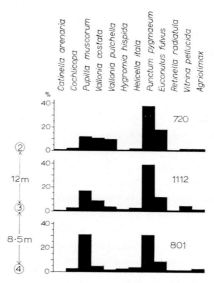

Figure 48. Pitstone, Bucks. Faunas from three points on a Late Weichselian land surface (samples 2·5 cm thick). (Data in Evans, 1966b.)

which, as we have already discussed, can occur in open habitats free from shade.

The analysis of samples taken at laterally spaced intervals from the same level along a soil horizon, and a comparison of their shell assemblages statistically, would help to clarify some of the problems of lateral variation, particularly if we could equate faunal changes with environmental changes along the profile as deduced by other means. For example, in Fig. 48 three assemblages from the surface of a buried soil of Allerød age at Pitstone, Bucks., are compared (data in Evans, 1966b: Figs 3a, 3b and 4). Although the general character of the fauna in all three is the same, differences are apparent,

SITE	DEPTH OF MARL (cm)	% PUPILLA	% EUCONULUS
2	60	12	18
3	47·5	17	12
4	25	31	9

particularly in the abundance of *Pupilla*. This can be related to the water-retaining capacity of the soil, in turn a function of the thickness of the underlying marl. *Pupilla*, at least in its more usual ecotype, is a xerophile, and in this instance clearly tends towards the drier habitats, being in greatest abundance where the underlying marl is least thick. *Euconulus fulvus* and *Vallonia pulchella*, species which favour more mesic habitats, show an opposite trend.

Kerney (1963: 241) pointed to differences in the snail fauna from the Allerød soil at two sites in the Medway Valley, and likewise suggested that these were due to varying local environmental conditions, such as the presence of bare chalk scree with scanty vegetation, or of more continuous plant cover, as deduced from features of the soil profile and local topography. But he goes on to say—and from the environmental point of view, this is the crux of the matter—"If we examine the vertical changes which take place within the soil in these two areas, we find that although the relative importance of species differs greatly, a strong parallelism exists . . ." Seldom can we be certain of the exact value of environmental parameters such as light intensity, relative humidity or the water-retaining capacity of the soil at any one time, but we can usually indicate with some confidence the general direction of habitat change.

As a topic of future research, the lateral variation of ancient land-snail faunas over the surface of buried soils is to be recommended, as a problem of both ecological and environmental interest. The Allerød soil at Pitstone would be ideal in that there is a long exposure, environmental changes are apparent along its length, and the stratigraphy of the site is remarkably clear (Fig. 71). The thin turf lines of Neolithic soils such as those at South Street and Durrington Walls would be equally suitable, and with care it is possible to take samples of 1 cm thickness, as was done at Avebury (Fig. 96), thus minimizing the time range involved.

We have already mentioned some of the micro-habitats—leaf litter, bare ground, tree trunks and fallen branches—which occur in woodland, and which impose their different properties on the molluscan fauna, creating a laterally heterogeneous distribution of species in the environment as a whole. On a larger scale, a fallen tree creates a clearing of perhaps 100 m² or more which might persist for a sufficient length of time to allow the development of its own characteristic fauna. Species such as *Oxychilus*, *Acicula fusca* and *Carychium tridentatum* having poor resistance to moisture loss would become less prominent, rupestrals more abundant, particularly as a fallen tree provides a wealth of suitable substratum close to the ground. *Vallonia costata* and *Abida secale*, heliophiles with woodland affinities, might also appear, and such changes could well leave their mark in the fossil record. Unfortunately we have little information about how such habitat variation in the woodland of the Atlantic period affected the snail fauna, due largely to the fact that the shell assemblages from such habitats are preserved in soil profiles, not in colluvial material. The creation of hillwash deposits, in which long records of change are preserved, is generally associated with forest clearance and cultivation, processes which necessarily destroy the forest fauna. Only in exceptional circumstances are woodland faunas preserved, and then usually in isolated pockets or subsoil hollows, seldom in soil profiles of sufficient length to enable us to examine the effect of lateral variation over the woodland floor. The problem is a relevant one in connection with the location of refugia, which must have existed in the Atlantic period, for the various open ground species which later became so common.

Topography is an important factor in creating differences in snail populations from place to place, as will be discussed in connection with the comparison of subfossil assemblages from ditches and their adjacent soil profiles (p. 325). It is clearly important that we should allow for the effect of local habitat differences when using ditch faunas to generalize about the environment of an area as a whole, and similar considerations apply to faunas from natural topographical features.

Thus at Brook in Kent, consideration of the results of snail analysis of Late Weichselian and Post-glacial deposits from six loci suggests some relationship between height above the level of the plain, or gradient, and the general composition of the fauna (Fig. 49). The highest position is the Rifle Butts section, at the head of a small scarp-face coombe; the lowest is Pit A on the Gault Clay plain, about 1·3 km from the scarp edge. The other four positions are at points intermediate between these two. The maximum recorded percentages at any one locus of various ecological groups of Mollusca are plotted in Fig. 49.

4. DISTRIBUTION AND ABUNDANCE OF SNAILS

This is a very crude assessment of the assemblages since the type of deposit and the environmental and climatic conditions under which they were laid down varies enormously. Nevertheless, the figures show a strong increase in freshwater and marsh species, a slight decrease in total land species and a strong decrease in xerophile species with decreasing height and gradient. These trends suggest, as is to be expected, that the general composition of the faunas is controlled by their position in relation to topographical features and not by differences of climate or land use through time. Thus changes in the water regime as controlled by precipitation can be assessed only when local environmental factors such as these are taken into account. For

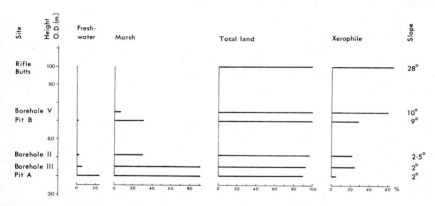

Figure 49. Brook, Kent. Variation in the maximum abundance of various ecological groups at five loci. (Data in Kerney et al., 1964.)

example, the abundance of marsh species at Pit B seems inconsistent with the general trend in this group shown by the other five sites, suggesting perhaps a climatic effect at this locus. And such is in fact the case, for the marsh fauna from Pit B comes from a calcareous tufa which formed during the Atlantic period when climatic conditions were wetter than previously or since.

Heterogeneity as imposed by climate—temperature and rainfall—has been discussed at the beginning of this chapter, and must be borne in mind when comparing the local environmental implications of faunas from different areas of the British Isles.

Lateral variation is engendered by faunal change, whatever its cause, and will persist until the full potential of the new conditions has been exploited. There is always a time lag—more so perhaps than in most creatures!—between an induced change in the fauna and its final manifestation. This, the time element, we shall now discuss.

Changes in Snail Populations with Time

Temporal changes in snail populations fall into a number of categories:

> Variations in abundance directly related to life cycles: variation in the relative abundance of juveniles and adults.
> Short-term fluctuations of a non-environmental origin.
> Response to environmental change:
>> Climatic.
>> Natural environmental.
>> Man-made, including the introduction of new species.
>
> Changes of ecological tolerance.
> Evolution.

Life cycles. Most snails and slugs are hermaphrodite, and some are protandrous—that is, they function as males when first sexually mature, and later as females. *Limax maximus* is said to be first female, then successively hermaphrodite, male and hermaphrodite, before finally returning to the female state (Ellis, 1969: 251). Populations of all species of *Vallonia* are made up largely of females (Watson, 1920:17) Mating may be preceded by courtship which in some species is elaborate and takes many hours. In some slugs, the partners circle for a time while touching each other with their tentacles or labial lobes; the Arionidae eat the mucus produced by the prospective partner, and some species copulate suspended by a thread of slime from the branch of a tree (Fig. 4). In some of the higher helicids, e.g. *Cepaea* and *Helix aspersa*, calcareous darts are fired into the flesh of the partner as a sexual stimulant. These, known as love darts, are occasionally found subfossil, as for instance by Woodward (1908: 84) in the sand deposits around Newquay, who called them *spicula amoris*.

Five species are viviparous; the rest lay eggs. The number of eggs laid varies from species to species, but is normally in the range of 50–300. They are deposited in loose damp soil, in shallow cracks, at the base of plants, under rotting vegetation and in other places having suitable moisture retaining powers. The eggs may be fairly large—in

Monacha cantiana, for example, being 1·5–2·2 mm diameter (Chatfield, 1968), in *Agriolimax reticulatus*, 3 × 2·5 mm (Runham and Hunter, 1970: 32)—and the shell is partly calcareous; because of this property, eggs are sometimes preserved subfossil particularly in anaerobic conditions. However, the calcium carbonate fraction is often confined to small granules which are embedded in a non-calcareous matrix (e.g. Chatfield, 1968). Thus when treated with dilute acid, subfossil eggs from a fourth century AD Roman well at Whitton, Glamorgan, effervesced for a short while and then ceased to do so, remaining intact in spite of the addition of more acid. When they were broken open, however, violent effervescence ensued due to the presence within the eggs of an embryonic shell, the protoconch.

In most species, the life cycle is completed within a year (9–15 months; Boycott, 1934: 4), but the larger helicids may live for several years. *Arion hortensis* takes nearly two years to mature when hatched late in the season, and *Milax budapestensis* usually has a biennial life cycle (Hunter, 1966; Stephenson, 1968: 171).

In many species, breeding takes place in spring. Xerophile species, however, frequently breed in the Autumn as an adaptation to life in a dry environment, the eggs and young thus avoiding the conditions of extreme dryness which prevail in the summer. Aggregations of adult snails may occur, and this allows as many members of the population as possible to mate (Fig. 47).

In the simplest type of life cycle, breeding takes place over a short period of time (weeks rather than months), there is a high mortality rate of post-reproductive snails immediately after breeding (Fig. 50), and at any one time there is only one generation of snails in the

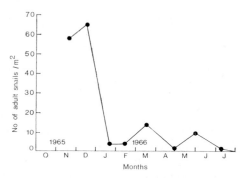

Figure 50. *Helicella caperata*, showing the sharp decline in adults following the November/December breeding season. (After Baker, 1968: Fig. 4.)

population which thus shows a unimodal distribution of size (i.e. age) classes. This was found to be the case in *Cochlicella acuta*, at the northern limits of its range, by De Leersnyder and Hoestlandt (1958). In the few other species which have been investigated, the situation is more complex. Thus in a population of *Cochlicopa lubrica* (Baker, 1970) at Braunton Barrows, Devon, young were present at all times, but were at their greatest numbers in May to August and November to January following the two main breeding periods, and in addition to which reproductive activity on a reduced scale continued all the time.

Monacha cantiana (Chatfield, 1968), *Carychium tridentatum* (Morton, 1954) and *Helicella caperata* (Baker, 1968) all showed a bimodal size distribution at some time in their life cycle over the period of

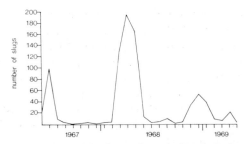

Figure 51. *Agriolimax reticulatus.* Numbers of newly-hatched slugs in field samples. (After Runham and Hunter, 1970: Fig. 18.)

study. In *Carychium*, this pattern persists throughout the year, though from July to November there was a tendency for younger individuals hatched in June to catch up with the main adult population, thus blurring the distinction between the two generations. In *Helicella caperata*, there were two breeding seasons, one in early summer and one in winter. After the summer breeding there was a high juvenile mortality due to the dry conditions of the habitat; but after the winter breeding, juvenile mortality was low. A similar pattern was observed in *Cochlicopa lubrica* (Baker, 1970) in a grass sward habitat, thus bearing out Boycott's contention (1934: 18) that autumn breeding is ecologically advantageous to snails living in dry places. In *Agriolimax reticulatus*, breeding takes place throughout the year but there are usually peaks of egg laying (Runham and Hunter, 1970: Fig. 18), as in *Cochlicopa lubrica* (Fig. 51); and as with *Monacha cantiana*, there may be several overlapping generations in a population at any one time, as these slugs may take up to a year to complete their life cycle.

Unless a subfossil assemblage is extremely rich in shells, it is generally impossible to obtain a count, which is statistically meaningful, for adult specimens alone; juveniles and apices of broken shells must be included. The age structure of a population is thus not recorded. This is probably of little consequence, for even if we were to distinguish between adults and juveniles it is unlikely that their proportions would reflect the age structure of a population at a point in time. In the first place, as discussed above, changes in abundance directly associated with the life cycle and breeding may be marked and rapid, and there is frequently more than one generation of individuals in a population at a given time. And secondly, the nature of the fossil record does not generally enable the preservation of such fine variation (p. 211). The closest we may hope to come to the ideal situation of extracting the shells of a living population is either in deposits which accumulate extremely rapidly, such as blown sand and loess, or where the burial of a soil surface is so sudden and massive as to create anaerobic conditions, thus preventing the further destruction of any shells, and preserving their periostraca, as took place at Silbury Hill. Under aerobic conditions, the periostracum is generally destroyed within a year of death, but at Silbury Hill (p. 265) where the soil had remained biologically inactive since burial, it was possible to separate out all the animals living within a year prior to the construction of the mound from those of greater age (Fig. 93).

Even when such factors as differential destruction and the confusion of juveniles with broken adults can be discounted, it is occasionally noticed that the proportion of adults to juveniles in a subfossil assemblage varies from species to species. For example, relatively large numbers of young *Oxychilus cellarius* and *Retinella nitidula* are sometimes recorded, indicating a high juvenile mortality in these species. Such a frequency, constituted largely of young individuals, may be a reflection of an environment unfavourable towards that species, but it is difficult to place much significance on this kind of variation until more is known about the different factors operating on the juvenile and adult stages of the life cycle. Boycott (1934) suggests that the main loss falls on the eggs and infant young since they are more susceptible to predation and moisture loss; " . . . I have watched *Hyalinia lucida* [*Oxychilus draparnaldi*] eating a hatching batch of *Helix aspersa* eggs with gusto" he remarks, hardly a situation which could arise with the adults. And this principle has been confirmed by Chatfield (1968: 242) for *Monacha cantiana*, in which the population density remained constant over a two-year period. "Although a high proportion of eggs may hatch, data from field and laboratory work confirm previous reports concerning land pulmonates of high mortality in the first month of life . . ." " . . . the increase in proportion of young snails is far below the level which one would expect should all the eighty to ninety eggs from each mature individual survive and grow." High *adult* mortality is often confined to immediately post-reproductive snails (Baker, 1968: 50; Chatfield, 1968: 243) (Fig. 50) and is presumably due to natural causes. Chatfield does not discount predation of *Monacha cantiana*—especially the larger specimens—by glow-worm larvae (*Lampyris noctiluca*), birds and small mammals. Runham and Hunter (1970) while allowing that parasites and diseases of the egg state may be of considerable importance in controlling slug populations, do not discount the possibility of control by occasional flocks of birds, particularly on surface-dwelling slugs.

Short-term fluctuations of a non-environmental origin. In a stable, unchanging environment (if such an ideal state exists in nature) which has a long-established snail fauna—for example climax wood-

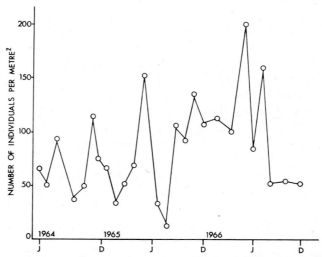

Figure 52. *Helicella caperata*. Variations in population density. (After Baker, 1968: Fig. 2.)

land—fluctuations in numbers of the various species of snail nevertheless probably occur. These may have no apparent or actual environmental basis but be due to chance—a characteristic, in fact, of most animal populations. Often the fluctuations are relatively short term (annual) in duration, and sometimes of considerable amplitude. They have been recorded for populations of *Catinella arenaria*, *Helicella caperata*, *Cochlicopa lubrica* (Baker, 1965; 1968; 1970) and *Monacha cantiana* (Chatfield, 1968) (Fig. 52). As with changes associated with life cycles, such short-term variations in abundance from year to year are unlikely to be preserved in the fossil record except in the same, unusual, circumstances. In general we are probably dealing with assemblages representing at one extreme, periods of five to ten years, and at the other, many centuries. In a buried soil, the assemblage of shells at the surface probably comes closest to reflecting a single population of snails; those at the base may be the remnants of ages, as was suggested in the case of the West Kennet Long Barrow (Fig. 91).

Response to environmental change. The initial effect on a snail population of an environmental change is in the abundance of certain species already present in the habitat. Some may increase, others may decrease and yet others may be eliminated altogether. On some, an environmental change may have no effect.

Then, species not present in the indigenous fauna may migrate into the area through being able to tolerate the new environment. The rate at which colonization takes place is a function of numerous factors such as the abundance of the colonizing species in the source habitat, the distance and type of land over which they have to travel and their rate of dispersal. Snails may not always occupy their full potential range, not because of the absence of suitable habitats, but because of their inability to reach them. As is suggested in discussing the fauna of ditches, particularly those of barrows isolated in arable land, this may be due to the presence of terrain hostile to snails—barriers to dispersal between refuges of life. The artificial spread of many species by man across ocean, mountain and climate barriers (for example of helicellids from the northern to southern hemispheres, shielded by him through the inhospitable tropics across which they could never pass by natural means (Quick, 1953)) is a further example of this phenomenon.

The introduction and spread of a species into a new habitat depend too on its *ability* to get a hold in and colonize the habitat, for this varies as Boycott (1934) pointed out: "Attempts at artificial colonization in apparently suitable places fail much more often than they succeed: transport is not necessarily followed by oekesis." The example of *Pomatias elegans* is a case in point. Boycott (1921d: 225) commented on the inability of this species to colonize old chalk pits, and both Bullen (1912) and Scharff (1907: 2) remarked on its absence from Ireland even though " ... dead shells have frequently been picked up on its shores."

The absence of a species is often due to the element of time alone, and, particularly when dealing with subfossil faunas, this is an important point to appreciate. As Boycott pointed out, "The age of a habitat may ... explain the absence of the appropriate Mollusca, and if a new habitat ... is quickly prolific, it may be assumed that the Mollusca found there were already present in small numbers in the neighbourhood." For example, *Pomatias elegans* had not spread as far as Ascott-under-Wychwood in the Oxfordshire Cotswolds by 2800 BC (p. 254) although common on the Chalk at this time not 20 miles away, and present at Ascott by the Roman period. *Helicella itala* does not appear in the Outer Hebrides until around the beginning of the Christian era (p. 295), though suitable habitats were available at least 3000 years previously; and *H. itala* was one of the first species to migrate into Britain at the end of the Last Glaciation.

Faunal changes in response to environmental change are thus a function of both the ability of the indigenous population to survive and adapt, and of the rate at which new species colonize, and their success in establishing themselves in new environments. We know very little about the dispersal of land snails, but the problems are discussed by Kew (1893), Scharff (1907) and Rees (1965).

The response of snail faunas to climatic fluctuations varies according to the nature and direction of change. A temperature increase is easier to detect than a temperature decrease. The climatic amelioration of the Allerød Interstadial, for example, is accurately reflected in much of south-east England by the introduction of the thermophile species *Helicella itala* and *Abida secale*. The onset of the Post-glacial warmth and attainment of the thermal maximum are reflected by the introduction and spread of thermophiles in at least two waves, first a group of species whose distribution in Europe is relatively northern, such as *Carychium tridentatum*, *Ena montana* and *Cepaea nemoralis*, which came in early on in the Post-glacial, and second, around

6000 BC, a group of species with a generally southern distribution, such as *Pomatias elegans*, *Lauria cylindracea* and *Azeca goodalli* (p. 361).

However, one cannot expect a pattern of change which is climatically controlled to apply rigidly over a wide area, on the one hand because of the slow rate of dispersal of snails, and on the other the ability of some to get a hold in favourable micro-habitats in an area prior to a general climatic amelioration. Thus *Helicella itala* and *Abida* were already present at Beachy Head on the south coast in zone I of the Late Weichselian, probably because of the proximity of the open sea and the induction of a climate milder than that of inland areas (Kerney, 1963). While at the other extreme, as already discussed, *H. itala* took over 8000 years to spread up the west coast to the Outer Isles.

A climatic deterioration is less clearly reflected. Thus Sparks (1964b) has shown that a number of thermophile species were able to survive into intemperate climatic conditions at the end of the Last Interglacial. The climatic worsening of zone III of the Late Weichselian resulted in no extinctions, and in calcareous sediments is detected more easily by lithological than by faunal changes. Recognition of the Post-glacial thermal decline is complicated by man-made changes, resulting in the creation of a host of habitats with their own microclimates, and making the assessment of faunal changes in climatic terms difficult. This problem can be got round by carefully selecting species which in Britain are on the edge of their range with regard to the factor being investigated, and whose distribution is controlled by known climatic factors, as was done for three species, *Pomatias elegans*, *Lauria cylindracea* and *Ena montana* (Kerney, 1968a). Detailed mapping of past and present distributions brought out changes which were compatible with a decline of winter and summer temperatures in southern and eastern England since Neolithic and Bronze Age times (Fig. 43).

Natural environmental changes other than climate include sea level, vegetational and soil changes. Britain was isolated from the Continent around 5000 BC, after which time further additions to our fauna by natural agencies were probably prevented by the absence of a land bridge. It is interesting, though possibly irrelevant, in this respect that a number of late arrivals into Britain, such as *Pomatias elegans*, *Truncatellina cylindrica* and *Azeca goodalli*, are absent from Ireland which was cut off from the rest of Britain at a much earlier stage.

The influence of vegetational changes, particularly the spread of forest in the Post-glacial, is discussed in Chapter 11 and the effect of vegetation as a whole on the distribution and abundance of land-snail populations earlier on in this chapter. The all-important factor is not species composition *qua* species, but structure, and this operates on a large scale through the major plant communities of woodland, grassland and arable, and on a small scale down to differences induced by variation in leaf width.

Pedological changes, prior to human interference, have not been recorded as controlling the composition of snail populations, but it is likely that where a brown earth developed from a rendsina there was a decline in the abundance of certain species and perhaps an extinction of obligatory calcicoles.

The various ways in which man has influenced the indigenous land-snail fauna of Britain have been discussed by several previous writers (Boycott, 1921b; 1934; Kerney, 1966b), and only a brief discussion of the main aspects is necessary here.

Man has brought about such drastic changes in the environment that few, if any, areas of the British Isles can today be thought of as entirely natural. The main changes have been an increased dryness and disturbance of the habitat, due largely to forest clearance, agriculture, grazing and settlement. The ground surface has been exposed to the full forces of sun, wind and rain; evaporation has been enhanced, and the water-retaining capacity of the soil and its nutrient status reduced. The sheer battering action of wind and rain is inimical to molluscan life. The effect of forest clearance on the molluscan fauna is well demonstrated at a number of sites (Pink Hill, Avebury and Ascott-under-Wychwood), and is one of the most striking features of later Post-glacial snail diagrams. Grassland and xerophile species have been favoured at the expense of woodland species, and the fauna of cleared areas has become depleted, as only a few species are able to tolerate the extreme conditions of an environment dominated by farming.

On many tracts of chalk downland there was once a cover of periglacial drift (p. 290) which may have brought about richer and more moisture-retentive soils. In places, this has now gone, destroyed by centuries of ploughing and erosion, leaving the soil poorer and more susceptible to drought. The development of *sols lessivés*, non-calcareous soils possibly created, at least in some instances, by forest clearance from soils which were once lime rich, undoubtedly led to the impoverishment of the fauna, if not in numbers of species, at least in individuals.

Another effect of forest clearance which has caused a change in the distribution pattern of certain snails is the frost hollow phenomenon. On downland where trees and hedgerows are sparse, cold air drains down the slopes from high land on still, clear nights, accumulating as pools and rivers in the valley bottoms. Cain and Currey (1963a) have suggested that this is the cause of the absence of the cold-susceptible *Cepaea nemoralis* from certain valley sites on the Chalk, where formerly it was common (Fig. 57). The frost hollow effect is unlikely to have been a serious factor in a forested environment owing to the blanketing effect of the tree canopy. It is possible that certain other species, such as *Vallonia excentrica*, are inhibited by frost hollows, and if this is the case, we might be able to use their abundance, in certain circumstances, as an index of the degree of openness of an environment.

Man's animals have also affected the snail fauna. Sheep and rabbits particularly, in maintaining large areas of limestone grassland in a closely cropped state, have favoured certain xerophile species. A number of writers (Stratton, 1963; Block, 1964; Kerney, 1968b) have pointed out how a relaxation of grazing pasture leading to an increase in tall grasses results in the extermination of various species of *Helicella*, particularly *H. itala*.

Several of man's animals harbour parasites whose intermediate hosts are land snails. While the effect of these is not clear, there is a strong possibility that in some instances they may exert a considerable check on the abundance of certain species (p. 107), a situation which could not have obtained prior to the Neolithic period.

On the credit side, there has been an increase in the variability of the landscape since the monotony of the primeval woodland was first broken up, which has culminated in a mosaic of different habitats— woodland, pasture, cultivated land, chalk downland and settlement (Fig. 145). Miles of hedgerows and field ditches provide bountiful refuges for many species of snail in an environment otherwise impoverished by thousands of years of farming, and now only supporting a few specialized forms. Both Boycott (1934) and Kerney (1966b) suggested that this diversification of the landscape has resulted in a richer and more varied fauna than that of the climax forest of the Atlantic period, at least in overall terms, if not in the individual habitat.

The pattern of land boundaries extant today is largely of medieval and later origin. But the parcelling up of land by ditches, hedges, walls and other means goes back well into the prehistoric period (Fowler, 1971; Fowler and Evans, 1967). In addition, the construction of burial mounds, defensive earthworks and ritual monuments has added a third, vertical dimension to the landscape. The Fussell's Lodge Long Barrow on Salisbury Plain, built some 6000 years ago, constituted in its original form a height difference of 20 ft from the bottom of the ditch to the crest of the mound (Ashbee, 1966); and at Avebury in north Wiltshire the vertical distance from the top of the henge bank to the bottom of the ditch was 55 ft (Smith, I. F., 1965b). Such diversification proved very acceptable to many species of snail, as the rich faunas from numerous ditches show.

We have discussed too, in the section on general principles (p. 29), how man has frequently created a calcareous habitat, advantageous to Mollusca, from one which was originally neutral or acid, the Roman town of Caerwent and the Neolithic flint mines of Grimes Graves (Fig. 8) being classic examples. The liming of soils and quarrying of calcareous rocks, particularly when overlain by acidic drift as on many parts of the Chalk, are further processes which lead to the upgrading of the habitat for Mollusca. In some instances where the drift cover is thin, as at Kilham (p. 277), even ploughing, in converting a brown earth into a rendsina soil, has created a more favourable habitat, at least with regard to its lime content. It is also possible that road dust provides lime for plants and animals in a finely powdered and easily assimilable state (Boycott, 1921d) thus bringing about an enrichment of roadside faunas.

There are two groups of species (p. 201) which are defined by their relationship to man. These are the synanthropic and anthropophobic snails, and it is important to appreciate that prior to the Neolithic period, this differentiation does not apply; both groups were living in the same environment. *Hygromia striolata*, for example, was characteristic of the forest faunas of the Atlantic period, rarely occurring in great abundance, but consistently present nevertheless. After the Neolithic period it became less frequent, and is rare in faunas of secondary forest, a fact well brought out in the snail diagram from the Rifle Butts section at Brook (Kerney *et al.*, 1964). Today, it has made a dramatic comeback, but only as a synanthropic species living in man-made habitats such as gardens, farm yards, refuse heaps and hedgerows, not in woodland. Not only does *H. striolata* tolerate man's interference, it is dependent on him for its very existence; in the north of Britain, where its spread may have been effected by man in the first place, it is possible that *H. striolata* could not exist were human interference to cease (Kerney, 1966b: 9).

At the opposite extreme, there are other species such as *Ena montana* and *Helicodonta obvoluta* which shun human habitations and all places where man's influence is strongly felt. These are the anthropophobes and, naturally, they are rare in Britain today. But *Ena montana* is " ... surprisingly common in Neolithic and Bronze Age ... layers in archaeological sites in the Wiltshire area, in contexts where much human interference was clearly already present ..." (Kerney, 1968a: 286). The same applies to *Helicodonta obvoluta* in West Sussex and Hampshire (Fig. 56). On the Continent both are common hedgerow species. Kerney suggests that such habitat changes are due to a climatic worsening, probably a fall in summer temperature, since the Neolithic and Bronze Age periods.

The existence of these two groups may be a reflection of the same phenomenon, members of both requiring habitats rich in food and with abundant shelter. What factors tip the balance in favour of a species adopting an anthropogenic or synanthropic mode of life is not clear. But the fact remains that changes from one to the other have taken place in both directions in the Post-glacial—*Ena montana* becoming anthropophobic, *Hygromia striolata* synanthropic—and from the environmental point of view this is a clear warning against the rigid interpretation of these species from subfossil assemblages in terms of their present-day habitat preferences.

Man has brought about the introduction of a number of species into Britain in the later prehistoric period, generally in Iron Age times and later. Most are synanthropic as would be expected, but some such as *Helix aspersa* have become naturalized, particularly in the south-west of Britain. The influence of these species on the native fauna is difficult to assess but the introduction and successful establishment of three species of *Helicella*, particularly *H. gigaxi*, in arable habitats once occupied by the native *H. itala*, has probably caused this species to become a lot rarer than would have been the case had these introductions not occurred.

In sum, man has brought about total deforestation of the British Isles, either directly by felling and burning or indirectly by grazing his stock; virtually all extant woodland is secondary, being situated on land which has once been cleared. He has made for increasingly dry and disturbed soils, with agriculture and grazing generally leading to loss of nutrient status, and these processes have led to a fauna in which grassland species and xerophiles predominate. As well as causing changes in the structure of the vegetation, man's animals have introduced internal parasites, the intermediate hosts of some of which are land snails. Adversity has been offset to some extent by the increase in lime content of the soil surface caused by processes such as quarrying, liming and ploughing; the story of man's assault on the landscape is not always one of degradation, at least as far as the snails are concerned. Man has created, too, a more variable environment than that of the primeval forest, a variability which extends not only laterally but vertically as well. The addition of a third dimension to the rolling landscape of the Chalk enabled many species to flourish which would otherwise have died out. Habitats strongly influenced by human habitation, providing shelter and unlimited nutrients, have engendered their own special fauna of synanthropic snails. And finally, the accidental or deliberate introduction of new species of snail by man has probably led, at least in the case of *Helicella itala*, to the restriction of our native species.

On the whole, the influence of man on the snail fauna has been beneficial, even though locally the fauna may have become impoverished. Obligate marsh species have undoubtedly suffered most. And while twenty species have been introduced and become established, man has caused no extinctions.

Changes of ecological tolerance. Comparison of faunas from different times in the past, or of past faunas with those living today, intimates that certain species have changed their ecological requirements. *Pupilla muscorum*, for example, is today an obligatory xerophile, but in the colder periods of the Pleistocene was able to live in marsh habitats as well (p. 150). *Vertigo pygmaea* on the other hand appears in the Late Weichselian only as a marsh species, and did not live in dry situations; today it can tolerate both types of habitat. The synanthropic and anthropophobic groups of snails are further examples of species which have apparently changed their habitat preferences in the past.

In the case of the latter group, we have already suggested that the adoption of a synanthropic mode of life may be due not to a change in ecological requirements, but simply to the way in which these are satisfied; and with regard to their counterparts, the anthropophobes, it seems likely that adverse climatic conditions, particularly low summer temperatures, are the cause of their present-day exacting needs. It is more difficult to justify that the avoidance of marsh habitats by *Pupilla* in the Post-glacial is due solely to a difference in the climatic regime from that which obtained during the colder periods of the Pleistocene, particularly as the change in habitat requirements was accompanied by a change in morphology (p. 150). Nevertheless, in view of these climatic differences and the absence of close present-day parallels to the environment of southern England during the Last Glaciation, it would be unwise to assume unreservedly that a change of habitat requirements has taken place. The differences may be due simply to the way in which these requirements are satisfied.

In other words, two groups of factors appear to be operating, some of immediate relevance to the snail, such as food, vegetation structure, soil type and other factors controlling local humidity; others, more subtle, such as climate, other species of snail, the nutrient status of the soil and the presence of parasites. The latter factors determine the way in which the former needs are met.

For example, sufficient food may be present in a habitat to support a population of given size, but other factors such as temperature, in controlling the activity of the snails, determine whether or not all the potentially available food is reached and used up. Food thus becomes a limiting factor only above (or below) a certain level of temperature. In the same way humidity, as controlled by precipitation, may determine the degree to which places of shelter satisfy the requirements of certain species; we cannot judge the ability of a habitat to support a snail population on its visual physical attributes alone.

A related phenomenon is that habitat development (one might almost call it evolution) imposes changes on the snail fauna which may take some time to beome stabilized, longer in fact that the habitat changes themselves. This applies, or may apply, to our grasslands, a type of habitat unknown in Britain in its present form prior to the Neolithic period, and due almost entirely to the introduction by man of domesticated grazing animals—sheep and cattle. Many of the Neolithic grassland faunas described in this book probably had no more than a few hundred years to become established and adapted to what was then an entirely new environment which they had not previously experienced, whereas those of today must be seen against a time background of up to 6000 years.

To the palaeoecologist, these are matters of supreme importance, for they present threats to the very foundations on which stand the basic principles of environmental reconstruction from fossil faunas. If we do not appreciate the many factors controlling distribution and abundance, the different levels and different ways in which they operate, and the compensation of one by another in various circumstances and through time, then we shall fail on every occasion to make realistic assessments of our ancient faunas in environmental terms. But if we are fully sensible to all this, and if we know, at least from our modern faunas, the range of habitats which a species can occupy, and under what conditions it can do so, then, adding to this the information we can acquire from other sources such as soils, animal bones and archaeology, I believe we have available a technique capable of detecting the finest details of ancient land use, particularly those of most relevance to the prehistorian such as nutrient status, vegetation structure and nitrogen metabolism in the environment.

Evolution. "The outburst of evolutionary development appears to have taken place at an earlier date among land and freshwater Mollusca than it did among the mammals and already to have slowed down to such an extent that little more than subspecific differentiation occurred during the Pleistocene" (Kimball *in* Zeuner, 1959: 329). Sparks, of the British fauna, writes: "It seems practically impossible to detect evolutionary changes in non-marine Mollusca in the Quaternary period" (1969: 397). With the possible exception of the physiological changes just discussed, there is no question of evolution having taken place during the Upper Pleistocene and Post-glacial (Evans, 1969a: 172).

5
Distribution, Habitats and History

This chapter should be used in conjunction with the census of the distribution of British non-marine Mollusca (Ellis, 1951), the census supplement (Kerney, 1966a) and "British Snails" (Ellis, 1969). Kerney's papers on the Late Weichselian and Post-glacial faunas of south-east England should also be consulted (1963; Kerney et al., 1964). The order of species followed is that used by Ellis (1951).

* = Species introduced into Britain, probably in the last 200 years.
† = Species now extinct in Britain.

Family POMATIIDAE

Pomatias elegans (Müller)

South and west Europe, north to Denmark (Kerney, 1968a: Fig. 2). *P. elegans* inhabits "scrub, woods, sandhills, etc., always on highly calcareous soil" (Ellis, 1951), and is our most characteristic calciphile (Fig. 43). It is absent from marshes. In France the snail is classed as a xerophile (Germain, 1930: 39), and in southern Italy in the Matera Gorge it is amazing, in view of its virtual restriction in Britain today to habitats of high humidity, to see this species active in the mid-day sun. In Britain it has been recorded live from sandhills (Boycott, 1921a) and arable land (M. P. Kerney and R. A. D. Cameron, personal communication), and in the past it also appears to have occurred on occasion in open habitats, as at Badbury (buried soil beneath bank, Fig. 134).

In general, however, the species favours shaded and moist habitats, with broken ground and loose soil into which it can burrow, and its presence in abundance generally indicates some form of disturbance of the soil surface, as for example, by forest clearance. At Pink Hill (Fig. 116), Pitstone (Evans, 1966b) (Fig. 53), South Street (buried soil and ditch, Figs 90 and 129), Avebury (Fig. 96) and Brook (Kerney et al., 1964: Fig. 14) it shows a marked increase in the clearance horizon, accompanied by a decrease of woodland species and an increase of open-country species, only declining when the open-country fauna has become fully established (Fig. 53).

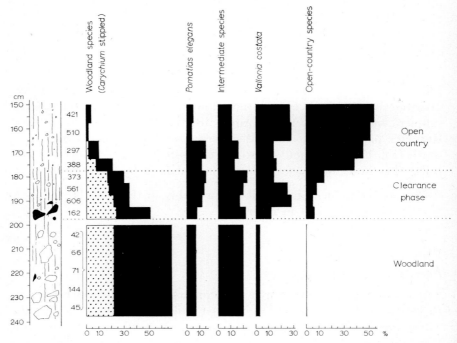

Figure 53. Pitstone, Bucks., south-east section. Clearance horizon. (Redrawn from Evans, 1966b: Fig. 7.)

Kerney (1968a) has shown, by comparing the modern and fossil distribution of *P. elegans*, that there has been a southward and westward contraction of its range (Fig. 43) during the Post-glacial. *P. elegans* is intolerant of winter cold, and it has been suggested that this change in distribution is correlated with the lowering of winter temperatures which has taken place since the Sub-boreal period.

We can thus see the concentration of *P. elegans* in the soil horizons of ditch deposits (the "*Cyclostoma elegans* zone" of the early excava-

Cochlicopa lubricella (Porro)

Widely distributed in Britain. This species is extremely common in archaeological deposits, and is particularly characteristic of grassland faunas on the Chalk (Fig. 144, in which *Cochlicopa* refers mainly to this species). It tends to occur in drier and more exposed habitats than *C. lubrica* (Quick, 1954), but is not restricted to them, sometimes living in heavily shaded woodland and damp grassland. Both species of *Cochlicopa*, in fact, are catholic species, being able to live in a wide range of habitats.

C. lubricella is present in the Late Weichselian. It has been recorded from the Hoxnian (Kerney, 1959) and Eemian (Sparks, 1964b) Interglacials.

Family VERTIGINIDAE

Pyramidula rupestris (Draparnaud)

Central, western and southern Europe (Ellis, 1951). In Britain, it has an essentially western distribution, living on dry, exposed rocks and walls. Apart from a record from a cave site in Ireland (Williams and Williams, 1966) of possible Atlantic or Sub-boreal age, *P. rupestris* is not known with certainty from Pleistocene and Post-glacial deposits. In view of its tolerance of exposed habitats—it occurs for example up to 3050 ft on Skiddaw (Boycott, 1934: 15)—this is odd.

Columella edentula (Draparnaud)

This species extends to 70° N in Scandinavia, and is widespread in Britain (Ellis, 1969: 155). "*C. edentula* occurs in wet woods, fields and marshes which are calcareous and generally 'good' localities for molluscs" (Paul, 1971). It usually occurs in low abundance, particularly in the south and east of England and is of very rare occurrence in archaeological deposits of strictly terrestrial type. It was present, for example, at Ascott-under-Wychwood, Pink Hill and Pitstone (Evans, 1966b) in association with woodland faunas.

It is recorded from the Hoxnian (Zeuner, 1959: 330) and Eemian (Sparks, 1964b) Interglacials.

Columella aspera Waldén

This newly described species, formerly included in *C. edentula*, has a north European distribution (Waldén, 1966); in Britain, "*C. aspera* will probably prove to be more northerly and westerly in distribution than *C. edentula*" (Paul, 1971). Probably due to its very recent recognition, *C. aspera* has not been recorded subfossil, and it may be that some of the records for *C. edentula* refer to this species. However, in view of its preference for ". . . acid woods, marshes and peat bogs, etc.: habitats which are generally 'poor' for molluscs" and non-calcareous, fossil records are likely to be rare.

†*Columella columella* (Benz)

Today, living in Norway, Sweden, Finland and the Alps. In the colder periods of the Pleistocene its range was more extensive, and in Britain it is a characteristic member of the fauna of the Middle Weichselian (Fig. 140) (Large and Sparks, 1961; Sparks *in* Shotton, 1968; Kerney, 1971a). It is present too in the Late Weichselian (Kerney, 1963), but does not appear to have survived into the Post-glacial: (cf. p. 101).

"It is uncertain whether this form is a distinct species, or only a variant, perhaps a subspecies, of the common Holarctic *Columella edentula* . . ." (Kerney, 1963). Certainly the differences between the two are no greater than those between the Pleistocene and Post-glacial forms of *Pupilla muscorum* (Fig. 22).

Truncatellina cylindrica (Férussac)

Europe, north to southern Scandinavia. In Britain, a rare and local xerophile species living in dry exposed places, hillsides, sandhills and maritime turf (Ellis, 1951).

There are a number of isolated Post-glacial records for *T. cylindrica*. In Wiltshire, for example, where today it is extinct, it was once common during the early second millennium BC particularly in the area around Stonehenge, occurring at Durrington Walls (Wainwright and Longworth, 1971: 329), Woodhenge (Evans, unpublished), two sites on Earl's Farm Down (Kerney *in* Christie, 1964; 1967), and on Boscombe Down (Kennard and Woodward *in* Newall, 1931). It is not clear why this species has become extinct in Wiltshire, for suitable habitats are still widespread and have probably been so continuously since the Bronze Age. The Post-glacial thermal decline can hardly be invoked, in view of the northerly distribution of the species in Britain (as far as Edinburgh), but increased rainfall may be responsible. From other Neolithic and Bronze Age sites of similar habitat around the headwaters of the Kennet in north Wiltshire, *T. cylindrica* is absent (e.g. Silbury Hill, Roughridge Hill, Hemp Knoll and Avebury). Nor was it recorded from Bronze Age sites at Stockbridge, Hants.

(Kennard *in* Stone and Hill, 1938) and Arreton Down, Isle of Wight (Sparks *in* Alexander *et al.*, 1960) where in both cases the environment was one of dry open downland. In other words, its present-day disjunct distribution was also a feature of the past.

Truncatellina cylindrica is the only indigenous obligatory xerophile with a widespread distribution in Britain which was not present in the Late Weichselian. Its Post-glacial distribution, unlike that of *Abida secale* (p. 152), is not a relict one. It is not clear when it entered Britain, nor how it managed to become so widely dispersed. It does not occur in shaded places, and its spread into Wiltshire must, one assumes, have taken place after the clearance of woodland by Neolithic man, from open refugia along the coast. What prevented it from ever becoming common is equally puzzling, and particularly so in view of its widespread distribution, apparent climatic tolerance and the fact that where it occurs, it is often abundant.

T. cylindrica occurs in the Hoxnian (Kerney, 1959; 1971b) and Eemian (Sparks, 1964b) Interglacials. It is one of the species cited by Sparks (1964b) as persisting from the Eemian Interglacial into the early stages of the Weichselian, and it has been recorded from Marden in the Vale of Pewsey, Wiltshire, in just such a context (Evans, *in* Wainwright, 1972).

Truncatellina cylindrica britannica Pilsbry

This form is at most a subspecies of *T. cylindrica s.s.* (Ellis, 1969). The ecology of the two is similar, if not identical. *T. c. britannica* is today probably confined to sites along the south coast, particularly in Devon, Dorset and the Isle of Wight. As a fossil, it is recorded from the Neolithic flint mines (probably in a Bronze Age or even later context) at Grimes Graves in Norfolk (Kennard and Woodward *in* Clarke, 1915: 220).

It is recorded from the Hoxnian (Davis, 1953: 357) and Eemian (Zeuner, 1959: 330) Interglacials.

Vertigo pusilla Müller

A rare species, occurring in woodland habitats, "amongst moss, ivy and dead leaves on old walls and dry banks, generally under the shade of trees" (Ellis, 1969). It occasionally turns up in archaeological deposits in the south and east of England (South Street, buried soil and ditch; Ascott-under-Wychwood, subsoil hollow and buried soil; Wayland's Smithy, ditch (Kerney, personal communication) and

Grimes Graves (Kennard and Woodward *in* Clarke, 1915)), but is more common in the earlier part of the Post-glacial, its distribution having been severely restricted since the onset of forest clearance in Neolithic times. It is frequently associated, both today and in the past, with *Vertigo alpestris* (Ellis, 1969).

V. pusilla is absent from the Late Weichselian, but occurs early on in the Post-glacial. It was present in the Hoxnian (Kerney, 1971b) and Eemian (Sparks, 1964b) Interglacials.

Vertigo antivertigo (Draparnaud)

To 63° N in Scandinavia, and ubiquitous in Britain. *V. antivertigo* is a marsh species, common in Late Weichselian and Post-glacial deposits. It occurs in the Cromerian (Sparks *in* Duigan, 1963: 197), Hoxnian (Kerney, 1971b) and Eemian (Sparks, 1964b) Interglacials.

Vertigo substriata (Jeffreys)

A Boreal/Alpine species, extending to 67° N in Norway. It is widespread in Britain, but more frequent in the north than in the south, living in woodland and marsh habitats.

V. substriata is surprisingly rare on archaeological sites, even in woodland faunas of pre-clearance age, and has not been recorded from any of the rich assemblages (Avebury, Ascott-under-Wychwood, Beckhampton Road and Pink Hill) discussed in this book. At Northton in the Isle of Harris (Fig. 106) it is present below 295 cm to the virtual exclusion of *V. pusilla* and *V. angustior*; above this level, when the sand surface becomes more stabilized and the environment shaded, it shows a relative decrease in abundance antipathetic to a rise of the other two species. At Brook in Kent (Kerney *et al.*, 1964) during the Boreal period, *V. substriata* shows two successive decreases in abundance in opposition to a relative rise of *V. pusilla*, again in an environment becoming increasingly shaded and altogether more suitable for molluscan life. These changes suggest that *V. substriata* is more tolerant of open/drier habitats than is *V. pusilla*. In shaded/damp habitats, where *V. pusilla* can thrive, *V. substriata* appears to be prevented from attaining its potential abundance—perhaps by some form of competition.

Vertigo substriata was absent from the Late Weichselian, but present in Britain from an early stage in the Post-glacial. It occurred in the Eemian Interglacial (Sparks, 1964b).

Vertigo pygmaea (Draparnaud)

A Holarctic species with a wide range in Europe, extending to 63° 30′ N in Norway, where it lives mainly in maritime habitats (Kerney, 1963). It is ubiquitous in Britain.

Today, it lives in a variety of habitats—pasture, sand-dune slacks, marshes and rocky places. Essentially it is a species of open country, and in both fossil and modern assemblages is rare or absent in woodland contexts (Boycott, 1934: 14). In southern Sweden, Waldén (1955) maintains that a light shading by trees causes its disappearance. On the other hand, Ellis (1951) records it as occurring in woods, and on the Chiltern scarp at Pulpit Hill (MS I, p. 233) it is present in open woodland. In north-west Germany, Ant (1963) records it from habitats only at the dark end (but not the darkest values at which some species can live) of the lux scale (Fig. 39) in conjunction with species such as *Pomatias elegans, Columella edentula, Discus rotundatus* and various zonitids.

It occurs in dry habitats, and is often common in short-turved, grazed grassland as today at Knap Hill (personal observation) and Windmill Hill (Fig. 82), Wiltshire, and the grassy limestone plateau of the Gower Peninsular, Glamorgan (Quick, 1943); it thrives in grassland kept in an almost lawnlike condition by man, particularly in parks and beauty spots, as at Longleat House, Wiltshire and Watlington Hill, Oxon. (Chappell *et al.*, 1971). It also frequents places of longer grassland such as damp meadows (Swanton, 1912; Boycott, 1934), sometimes in association with *Vallonia excentrica*. Its ecological amplitude in Sweden (Waldén, 1955) extends to moderately wet conditions, and the same can be said of its distribution in Britain (Boycott, 1934). Quick (1943) notes that in West Glamorgan it usually lives in damp places, and records it from sand dune slacks with *Vertigo antivertigo*. Ellis (1941) records it from Norfolk in swamp carr and fen with moisture-loving species and, with the exception of rare *Vallonia costata*, no associated heliophiles.

From the evidence of several modern and fossil soil horizons, in which *Vertigo pygmaea* displays a marked increase as the habitat becomes more stable, the snail would appear to have a preference for places in which there is a complete vegetation cover. This is shown particularly well in modern soils at Overton and Fyfield Down (Fig. 120), Windmill Hill (Fig. 82), Badbury Earthwork (Fig. 134) and Ascott-under-Wychwood (Fig. 137), and in buried soils at South Street (Fig. 90). Against this, *V. pygmaea* is common in the blown sand deposits at Northton (Fig. 106), in habitats where the surface was frequently unstable, while today it is absent from the modern turf—an environment of grazed grassland in which surface stability is virtually total. Admittedly, this site is much further north and west than the others quoted, which may account for differences in habitat preferences (cf. p. 173), but whatever the cause, the variation shown by *V. pygmaea* in its choice of habitats, not only with regard to surface stability but to humidity and shade as well, warns against interpreting one site in the light of others without qualification as to their overall environmental properties.

One of the most characteristic features of populations of *V. pygmaea* both today (Chappell *et al.*, 1971) and in the past is their low abundance, values greater than 6% being sufficiently rare to list:

	%
MS III (Fig. 77)	9
Silbury Hill (Fig. 93)	9
Windmill Hill (Fig. 82)	10
Earl's Farm Down, G.70 (Christie, 1964)	14
Comb's Ditch, Dorset (Evans, unpublished)	15

Clearly it is not the environment which is responsible for maintaining the general level of abundance, but some inherent property of the species itself, for its common associates, *Vallonia costata*, *V. excentrica* and *Pupilla*, can reach much higher values.

During the Late Weichselian, *Vertigo pygmaea* was restricted to marshes (Kerney, 1963) where it frequently abounded with other obligatory hygrophiles such as *V. antivertigo*. Its extension into the drier downland habitats where it now thrives did not take place until the Post-glacial (Kerney *et al.*, 1964).

In the Post-glacial, *V. pygmaea* is generally absent from terrestrial woodland faunas of Atlantic age, e.g. Pitstone (Evans, 1966b), Ascott-under-Wychwood (Fig. 88) and Avebury (Fig. 96), but due to its tolerance of dampness, was frequent in marsh habitats, being present in a number of tufa deposits. It was not eclipsed to the same degree as *Pupilla*, *Helicella itala* and *Abida* by the spread of forests. From Neolithic times onwards wherever man has destroyed the forest cover and created a grassland environment, *V. pygmaea* is present; its response to forest clearance is well shown at sites such as Northton (Fig. 106), South Street (Fig. 90) and Avebury (Fig. 96).

The species is known from the Hoxnian (Kerney, 1971b) and Eemian

(Sparks, 1964b) Interglacials. During the latter, it was less widespread than today, which may relate to the more continental climate of the Eemian relative to that of the Post-glacial.

Vertigo geyeri Lindholm

Formerly included in *Vertigo genesii* Gredler. *V. geyeri* was common in marshy habitats in the Late Weichselian, but less frequent in the Post-glacial (Kerney *et al.*, 1964). Today it lives in a few isolated sites in central and western Ireland on the margins of peat bogs.

†*Vertigo genesii* Lindholm

(= *V. concinna* Scott; *V. parcedentata* Braun). This species is closely related to *V. geyeri*, and a common congener in Late Weichselian and early Post-glacial marsh habitats (e.g. Kerney *et al.*, 1964; Kennard and Musham, 1937). It occurs in Eemian Interglacial deposits (Sparks, 1964b).

Vertigo moulinsiana (Dupuy)

Absent from Scandinavia, and with an essentially southern distribution in Britain. *V. moulinsiana* is a marsh species, today rare and local, but in the earlier part of the Post-glacial more widespread and common particularly in tufa deposits. Recorded from the Hoxnian (Zeuner, 1959: 330) and Eemian (Sparks, 1964b) Interglacials.

Vertigo lilljeborgi Westerlund

A species with an essentially northern and western distribution in Britain and Eire (Kevan and Waterston, 1933; Dance, 1972). As a Post-glacial fossil it is very rare, there being records for Buckinghamshire and Lincolnshire (Ellis, 1951: 186). The Upper Pleistocene record from Barrington refers to *Vertigo genesii* (Sparks, 1952: 173).

Vertigo alpestris Alder

A northern European and Alpine species. *V. alpestris* is a terrestrial species, today confined to rocks and stone walls, often living in conjunction with *V. pusilla* (Dean and Kendall, 1908). It has a local distribution, occurring in North Wales and the Lake District, but, like *V. pusilla* was once more widespread having been recorded from a number of sites of Atlantic age, mainly in woodland contexts, e.g. Ascott-under-Wychwood, Pink Hill, Avebury and Pitstone (Evans,

1966b), and from several tufa deposits, e.g. Millpark in Ireland (Kerney, 1957a) and Leckwith in South Wales (Kennard, MSS).

Pleistocene records are erroneous (Ellis, 1969: 276).

Vertigo angustior Jeffreys

A rare marsh species, but again, like many of the other species of *Vertigo*, more widespread and abundant in the Atlantic period, than today. *V. angustior* is said to be a marsh species, but it may not be confined to such habitats. At Northton (Fig. 106) it occurs subfossil in a strictly terrestrial context, and Quick (1943: 8) records dead shells in sand dune pockets at Oxwich in the Gower.

It occurred in the Hoxnian (Davis, 1953) and Eemian (Sparks, 1964b) Interglacials.

Pupilla muscorum (Linné)

A Holarctic species extending to 70° N in Scandinavia, *Pupilla* is virtually ubiquitous in Britain in suitable habitats, but is rare in Scotland.

P. muscorum is an open-country form, characteristic of grassland habitats, walls and sand dunes; it rarely enters woods or other shaded places. Ant (1963) records it in north-west Germany mainly from the light end of the lux scale (Fig. 39), with a few occurrences at the dark end. Stratton (1964) records it from open woodland on the side of a valley in Derbyshire; and it is present in Wychwood Forest, Oxon. on the loose, rubbly sides of certain coombes, and on the slopes of Pulpit Hill (MS I, Fig. 76), both of which are lightly wooded. H. H. Brindley (*in* Marr and Shipley, 1904: 123) records it from "... woods which border the Roman Road over the Gogmagogs" in Cambridgeshire.

A characteristic habitat of *Pupilla* is earth bare of vegetation. On grassy chalk slopes for example it is often present in patches of broken ground induced by sheep or around rabbit burrows. It is particularly common in sand dunes, and occurs too on stone walls. Clearly *Pupilla* can tolerate not only dry habitats but those in which there is a considerable range of diurnal temperature as well. It does not, however, like intensive agriculture, and today is generally absent from arable land, as for example at South Street (Fig. 129). This has not always been the case, for *Pupilla* is sometimes extraordinarily abundant in hillwash deposits where there was evidently a considerable degree of mechanical disturbance, as at Badbury (Fig. 134; up to 50%), Pink Hill (Fig. 116; 40%) and Pitstone (Evans, 1966b; 20%).

In a series of measurements on the relative humidity of *Pupilla* habitats in north-west Germany (Ant, 1963), two peaks of preference were shown. One, between 40% and 50% R.H. included such species as *Helicella itala*, and the other, at 60% R.H., coincided with the dry end of the range of *Discus rotundatus*, *Vitrea crystallina* and *Clausilia bidentata* (Fig. 38).

In Britain, *Pupilla* has very occasionally been found in marshes (Boycott, 1934: 18).

In grassland, whether short-turfed or long, *Pupilla* is today often very abundant, occasionally comprising over 40% of the total fauna. On some sites, however, it is inexplicably rare or absent. The following list comprises percentages of *Pupilla* in the total fauna (excluding *Cecilioides*) extracted from modern grassland soils (cf. Chappell et al., 1971: Table 4).

SITE	ABUNDANCE (%)	HABITAT
Fyfield Down (Fig. 120)	62	⎫ Short-turfed grassland; exposed site.
Overton Down (Fig. 120)	47	⎭
Badbury (Fig. 134)	45	Sheltered; some scrub.
MS III (Fig. 77b)	15–29	⎫ Long grass; some scrub; hawthorn sere (Fig. 78).
MS II (Fig. 77a)	16–26	⎭
MS V (Fig. 77c)	20	Short grass; juniper sere.
Hemp Knoll (Fig. 132; 15–22·5 cm)	20	Short grass?
Pink Hill (Fig. 116)	17	Long grass; sheltered meadow.
Devil's Kneadingtrough (Kerney, et al., 1964)	12	Long grass; sheltered.
Windmill Hill (Fig. 82)	12	Short grass; exposed.
Coombe Hole (Fig. 80)	7	Long grass; sheltered.
Waden Hill (Evans, 1968a)	1	Long grass; sheltered.
Ascott-under-Wychwood (Fig. 137)	1	Long grass; some scrub.
Northton (Fig. 106)	0	Short grass; exposed.

On these sites there does not appear to be a clear association between either exposure or the type of vegetation on the one hand and the abundance of *Pupilla* on the other though admittedly the number of sites is pitifully small. While clearly able to abound in the most exposed habitats, *Pupilla* is sometimes common in the more sheltered and moister ones.

Nor does a consideration of fossil faunas make an understanding of the ecology of *Pupilla* any easier. There are a number of sites, the pedological and molluscan evidence from which suggests an environment of open, short-turfed grassland. In these, *Pupilla* never attains the higher values found in modern populations, but there is nevertheless considerable variation from site to site (Fig. 144).

SITE	ABUNDANCE (%)	AGE
Earl's Farm Down, G.71	29	Beaker
Earl's Farm Down, G.70 (Christie, 1964)	28	Bronze Age
Durrington Walls, DW I (Wainwright *et al.*, 1971)	17–28	Late Neolithic
Durrington Walls, DW II	9	Late Neolithic
Ascott-under-Wychwood, soil under mound	26	Neolithic
South Street, soil in ditch (Fig. 129; 130–137·5 cm)	22	Bronze Age
Badbury, soil under bank (Fig. 134)	17	Iron Age
Julliberries Grave	14	Neolithic
Arreton Down	9	Bronze Age
Hemp Knoll, soil under mound (Fig. 132)	8	Beaker
Avebury	2–5	Late Neolithic
Roughridge Hill "B"	4	Bronze Age
South Street, soil under mound (0–3 cm)	4	Neolithic
Thickthorn Down	5	Neolithic
Horslip	3	Neolithic
Beckhampton Road	2	Neolithic
Silbury Hill	2	Neolithic
West Kennet Long Barrow	1	Neolithic
Wayland's Smithy II	1	Neolithic
Farncombe Down	0	Bronze Age

With the exception of Ascott-under-Wychwood, all the early/middle third millennium sites are poor in *Pupilla*. This might be thought of as being due to a slow rate of dispersal and establishment once forest clearance had taken place, if it were not for the fact that at South Street (Fig. 90), *Pupilla* is present beneath the turf line of the buried soil in a cultivation horizon (3–17 cm) at 14% abundance, three-and-a-half times its value in the turf line. Quite clearly *Pupilla* on this site was more successful in an environment whose soil surface was exposed than in one with a complete vegetation cover. And yet on the same site in the Bronze Age, in an environment again of open grassland with a complete vegetation cover, *Pupilla* is present at 22% abundance (Fig. 129).

In the area around Stonehenge (Earl's Farm Down and Durrington Walls) *Pupilla* was very abundant in the Late Neolithic/Early Bronze inland cliffs and the steeper chalk and limestone escarpments (Kerney, 1962). Its present disjunct distribution is thus essentially a relict one. Unlike other xerophile and open-country species such as *Pupilla* and *Vallonia, Abida* has not been able to recolonize successfully many of the secondary habitats created by man. As Kerney has pointed out, this is odd in view of its rapid initial spread in the Late Weichselian.

It is unknown in Britain prior to the Late Weichselian.

Family VALLONIIDAE

Acanthinula aculeata (Müller)

Extends to 60° N in Scandinavia. *A. aculeata* is a common woodland species, generally distributed in Britain, though rarely occurring in any great abundance (Boycott, 1934: 12). It is essentially a rupestral species, often living affixed to the under-surface of logs, but may also occur amongst dead leaves. I have, too, found it on quite open grassy glades in Wytham Wood, Berks. (p. 113), though never far from sheltered habitats. In archaeological deposits it is of fairly consistent occurrence, being one of the more frequent "other woodland species".

It was present in Britain by the Boreal period. It occurred in the Hoxnian (Kerney, 1971b) and Eemian (Sparks, 1964b) Interglacials.

Acanthinula lamellata (Jeffreys)

A species of old woodland. It is local in its occurrence and rare in the south, but generally distributed north from Stafford, and in Ireland (Ellis, 1951: 187). Like other species with northern and western distributions—notably *Vertigo substriata, V. alpestris* and *Lauria anglica*— *A. lamellata* was once widespread and common in southern England, having been recorded from numerous tufa deposits (e.g. Davis, 1955). In the present study it has been found at Ascott-under-Wychwood and Coombe Hole in subsoil hollows of probable Atlantic age. It probably entered Britain towards the end of the Boreal period.

A. lamellata occurred in the Eemian Interglacial (Sparks, 1964b).

Vallonia costata (Müller)

V. costata is a Holarctic species, climatically very tolerant and extending beyond 70° N in Scandinavia. It was among the first species to enter Britain at the beginning of the Late Weichselian, and has persisted in a variety of habitats, though in varying degrees of abundance, ever since.

Today, it is without doubt a species of open habitats, rarely entering woods. It may occur in profusion on chalk downland though is frequently absent or rare in what appear to be suitable habitats; it is obviously less uniformly distributed and less consistent in its occurrence now than in the Neolithic and Bronze Age periods. Its absence from many places is clearly due to farming. Thus it is generally absent or rare in the upper levels of ploughwash deposits, as at South Street (Fig. 129) and Pink Hill (Fig. 116), and I have only once—in fields adjacent to Woodhenge—found it living in arable land. In Oxfordshire around Tackley and Ascott-under-Wychwood, and at Haresfield Beacon in Gloucestershire, it does not occur in pasture, while *V. excentrica* is common in such habitats. In both these areas, however, *V. costata* is common on stone walls.

Along the Chiltern scarp between Ivinghoe and Princes Risborough *V. costata* is more common in juniper sere than in hawthorn sere habitats (Figs 76 and 77). In Wychwood Forest, Oxon., it is absent from tall grass in the forest glades where *V. excentrica* is common, but present on drier slopes where the grass cover is less dense. In the area around Avebury, however, the situation is reversed, *V. costata* occurring in abundance in two tall grass habitats today (West Kennet Long Barrow and Waden Hill) while virtually absent from short-turfed grassland on Windmill Hill, Fyfield Down and Overton Down, where *V. excentrica* thrives. But only 7 km distant from these sites, both species are abundant in short-turfed grassland on the escarpment at Knap Hill.

Some modern records for *V. costata* are listed below. Where soil samples have been included, no distinction has been made between shells

SITE	ABUNDANCE	*V. excentrica*
Long grass, with occasional thorn scrub		
St. Catherine's Hill, Winchester	Common	Present
West Kennet Long Barrow, Wilts.	Common	Absent
Waden Hill, Avebury, Wilts. (Evans, 1968a), 0–15 cm	17%	5%
Pitstone, Bucks. (Evans, 1966b)		
N.E. section, 0–10 cm	Absent	36%
S.E. section, 0–7·5 cm	3%	26%
Pink Hill, Bucks. (Fig. 116), 0–10 cm	8%	37%
MS II (Fig. 77), 0–7·5 cm	Absent	7%
MS III (Fig. 77), 0–7·5 cm	1%	41%
Wychwood Forest, Oxon., grass rides	Absent	Present

Short grass, grazed by sheep or rabbits

SITE	ABUNDANCE	V. excentrica
Radbrook Common, Wytham Wood, Berks.	Present	Present
Wychwood Forest, Oxon.	Present	Present
Knap Hill, Wilts.	Common	Present
Beacon Hill, Ellesborough, Bucks. (100 specimens collected)	40%	39%
MS V (Fig. 77), 0–6·5 cm	7%	12%
Windmill Hill (Fig. 82), 0–6·5 cm	Absent	33%
Fyfield Down (Fig. 120), top soil	2%	24%
Overton Down (Fig. 120), 5–12 cm	2%	31%

which have been in the soil for some time and those which derive from living or recently dead animals. However, this is not thought to affect seriously the true values of relative abundance.

From a consideration of these grassland habitats, it is not clear what factors are controlling the distribution and abundance of *V. costata* in them. Certainly there is no clear correlation between abundance and grassland type. Ellis (1951, 1969) and Boycott (1934) suggest that *V. costata* is a species of dry habitats, though not a xerophile in the strict sense: "... the commonest species of its genus in the British Isles, and widely distributed in dry situations, amongst dead leaves and grass, on stone walls, in quarries ... becoming rather local in Scotland" (Ellis); "Characteristically though possibly not exclusively affects dry places" (Boycott). But with regard to chalk and limestone grassland it does not appear to show any tendency to favour the drier habitats, and indeed is absent from some of the driest. We have noted its avoidance of arable habitats, and it is rare or absent from lowland pasture, particularly where cattle are present and the vegetation unkempt; but on the trim, compact sward of downland grazed by sheep it is often abundant, as at Knap Hill and several of the Chiltern sites. The sites on Windmill Hill, Fyfield and Overton Down have not been cultivated since the twelfth century AD, if at all, so that the absence of *V. costata* from them is not due to this factor. We will return to this problem shortly, after a consideration of other types of habitat in which *V. costata* can occur, and after looking at some aspects of its Late Weichselian and Post-glacial history.

V. costata is frequently present on walls. Thus at "The Trout", Godstow, Oxon. it abounds beneath ivy in association with *Lauria cylindracea* and *Ena obscura*; and in the area around Tackley it is confined to stone walls, where it is abundant, while in the same area

V. excentrica is confined to pasture where it too thrives. A similar antipathetic distribution is present around Haresfield Beacon, Gloucestershire. In such situations, *V. costata* appears to be occupying drier habitats than *V. excentrica*.

On the other hand, *V. costata* occasionally lives in damp situations, although the exact moisture regime of some of these is not clear. "The occurrence of *V. costata* in swamp-carr and in fen-carr is unexpected, as this species is normally an inhabitant of dry situations in the open, yet at Wheatfen it may sometimes be found on the same log as *Zonitoides nitidus*, *Agriolimax laevis* and *Carychium minimum*" (Ellis, 1941). I have found it at Little Missenden, Bucks., at the edge of a pond with *Succinea pfeifferi* and *Monacha granulata*. In the vicinity of Stockholm, Waldén (1955) records that it does not like dry localities as much as *V. excentrica*.

V. costata may occasionally be found in woods. In north-west Germany (Ant, 1963) it is tolerant of more shaded conditions than *V. excentrica* (Fig. 39), and in southern Sweden (Waldén, 1955) it is "... not rare in slopes, rivulet valleys, deciduous woods and groves, in shady as well as open places ...". Its typical habitat on the Isle of Wight is said to be "woods and hedges" (Thistleton, 1966). On the Chiltern scarp (Fig. 76; MS I) it occasionally occurs in open beech woodland (Fig. 75).

Finally, it should be pointed out that *V. costata*, although apparently avoiding pasture where there is considerable disturbance by cattle, can occur in close association with man as a synanthropic species. Boycott (1934) notes that it sometimes occurs in gardens, and I have found it in four situations on rubbish tips, under cardboard or on waste ground, in all of which *V. excentrica* was absent. Thus at Barnes Bridge, London, it lives with *Milax* spp., *Oxychilus draparnaldi* and *Laciniaria biplicata*. Verdcourt (1951) found a specimen hibernating in a hollow asparagus stem.

V. costata was a common species in the Late Weichselian, present in the earliest faunas of the period, together with about six other climatically tolerant forms (Kerney, 1963: 232). At this time, it occurred in a variety of habitats, but appears to have been best suited to stable, dry grassland, thriving in the mild Bølling and Allerød Interstadials. During the deterioration of climate in zone III, though persisting, it generally suffered a greater reduction in abundance than did certain xerophile species, notably *Pupilla*, *Abida* and *Helicella*. At Holborough in Kent (Kerney, 1963; section D) *V. costata* comprised

51% of the fauna during zone II in an environment of a permanent and continuous cover of grasses and herbaceous plants. At other sites in Kent described by Kerney on steeper and more exposed slopes, its abundance was about half this value—still, however, a sizeable component of the fauna.

At Pitstone, Bucks. (Evans, 1966b), *V. costata* was less abundant in zone II, comprising from 5% to 11% of the fauna. This site was more low-lying than most of those described by Kerney and was characterized by the presence of *Catinella arenaria*, a species which lives in damper situations than those of chalk hillsides. On the other hand, a damper environment does not necessarily exclude *V. costata* altogether, for at Brook (Kerney *et al.*, 1964) in a fossil marsh soil of Allerød age in which freshwater and obligatory land hygrophile species were present, its abundance ranged from 1% to 16%. The faunas illustrated in Fig. 141 show clearly, however, that *V. costata* is favoured by drier habitats and *V. pulchella* by wetter ones.

During the Boreal and Atlantic periods, *V. costata* persisted in marsh and woodland habitats, at values of abundance ranging from 1% to 12%. Thus it is recorded in a Boreal context from Brook in Kent (Kerney *et al.*, 1964). The environment is interpreted as "wet marsh becoming increasingly shaded", and *V. costata*, present initially at 6% falls to 3% by the end of the period. Other open-country species were absent; here, *V. costata* is undoubtedly present in a shaded and damp environment. It occurs too in a probable Boreal context at three rather more terrestrial sites, Avebury (Fig. 96) (7%), South Street (Fig. 90) (7%) and Ascott-under-Wychwood (Fig. 88) (10%–12%). At all these sites, the environment was a woodland one, but with a somewhat open canopy.

Some records from Atlantic woodland environments, in which there was probably a completely closed canopy, are listed below.

From this consideration of Late Weichselian and early Post-glacial occurrences of *V. costata*, it is clear that the animal can live in several kinds of habitat, but is more abundant in some than in others. As a grassland species it may comprise 50% of the fauna, on chalk hillsides where the vegetation cover is patchy, *ca.* 25%. It can occur in marsh and swamp environments (up to 10%), in open woodland (up to 12%) and in closed-canopy woodland (up to 6%).

When woodland is cleared and open country created, *V. costata* is generally the first of the open-country species to invade the habitat and become abundant. At Pitstone (Fig. 53) it rose rapidly from a value

SITE	*V. costata*	MARSH SPECIES
Brook (Kerney et al., 1964); Pit B, tufa	1%–2%	Up to 26%
Wateringbury (Kerney, 1956); tufa	ca. 7%	ca. 3%
Huntonbridge (Kennard, 1943); beds 2–5, tufa	ca. 11%	ca. 20%
Box (Bury and Kennard, 1940); tufa	ca. 1%	ca. 6%
Norton Common (Kerney, 1955); tufa	ca. 6%	ca. 73%
Millpark (Kerney, 1957a); tufa	Common	Common
Ascott-under-Wychwood (Fig. 88b)	0%–6%	Absent
Coombe Hole (Fig. 80b)	7%	Absent
Pink Hill (Fig. 116), 100–120 cm	2%	Absent
Pitstone (Fig. 53), 200–237·5 cm	3%	Absent

Open-country species are present at 1%–2% at Box, Norton Common and Millpark; otherwise they are absent.

of 3% in closed woodland to about 25% in the clearance phase; *V. excentrica*, *Pupilla*, *Vertigo pygmaea* and *Helicella itala* came in for the first time in the clearance phase, but each at not more than 4% abundance. At Brook (Kerney et al., 1964; Rifle Butts section) during a phase of early partial clearance, *Vallonia costata* rose to 7% and *Abida secale* to 3%, but no other open-country species was present; in a second phase of clearance, *V. costata* was again the first of the open-country species to attain abundance. At Northton (Fig. 106), *V. costata*, with *Pupilla* and *V. pygmaea*, preceded *V. excentrica* in invading an area of machair during a phase of forest clearance. During a subsequent phase of woodland regeneration, *V. costata* maintained its abundance, while *V. excentrica* and *Pupilla* declined to almost nothing.

One factor contributing to the rapid rise to abundance of *V. costata* in cleared habitats is its presence in the fauna of the primeval forest where, unlike other open-country species, it had maintained itself in small numbers throughout the Post-glacial. Another factor is that *V. costata* is perhaps better able to live as a rupestral species than are other open-country forms, and thus act as a pioneer of freshly exposed soil surfaces. For example, in the secondary fill of the ditch of the South Street Long Barrow, where the environment was probably one of patchy vegetation and small, actively eroding areas of bare chalky soil, *V. costata* was particularly abundant (*ca.* 30%) while *V. excentrica* fell from its previous level of about 35% in a grassland environment to less than 10% in the ditch. A similar pattern was recorded at Wayland's Smithy (M. P. Kerney, personal communication).

Together with *V. excentrica*, *V. costata* (each at 20% to 40% abundance) dominated many of the open-country faunas of Neolithic and

Bronze Age Britain, in an environment of short-turfed grassland, probably grazed by sheep (Fig. 144). There is considerable variation from site to site: in some they are present in equal abundance, in others one may exceed the other by more than 50%, but always both are important elements in the fauna. What such variation implies in environmental terms is hard to tell with such a small number of sites and such inadequate modern data, but the sequence in the buried soil at South Street provides a clue. Here, in the cultivation horizon (Fig. 90; 3–17 cm) where there was undoubtedly an unstable soil surface as indicated by the stony nature of the soil and the presence of *Pupilla*, *V. costata* predominates. In the turf line (0–3 cm), however, where the stone-free mull humus horizon and the paucity of *Pupilla* suggest a more stable surface, probably with an unbroken grass sward, *V. costata* declines and *V. excentrica* shows a pronounced increase. Similarly, in five of the six faunas graphed in Fig. 144, where *Pupilla* is abundant, *V. costata* predominates over *V. excentrica* suggesting that dryness and disruption of the soil surface favour the former, and surface stability the latter.

Today, *V. costata* is much less abundant than formerly, and its decline and absence from many grassland habitats may, as already suggested, be due to intensive agriculture. At Northton (Fig. 106) and South Street (Fig. 129), however, its initial decrease in abundance is not connected with cultivation but occurs in grassland soil horizons, so that a factor, or factors, other than ploughing appear to be involved. It is possible, for example, that as the environment became more specialized and unsuitable for molluscan life, and as the Post-glacial thermal decline intensified, closely related species of snail began to compete for food and shelter, the most successful in any given set of circumstances ousting the other. Clearly, as can be seen from a consideration of Neolithic and Bronze Age sites the ecological requirements of *V. costata* and *V. excentrica* in grassland are closely similar.

The situation in the Chilterns, although not clear, leads one to suspect that here the humidity of the environment, as controlled by depth of soil and grass type, is the key factor, *V. costata* occurring in the drier habitats. The anomalous area, and particularly so in the light of the Chiltern habitats, is that around Avebury. Here, the same factor which is said to control the antipathetic distribution of the two species of *Cepaea* (p. 171), namely temperature, may be operating. This is supported by the fact that *V. costata* is the more northerly ranging of the two and thus better able to survive the frost-hollow habitats in lowlying areas and valleys. In upland places and others suitable for *V. excentrica*, it is excluded. It is also relevant that, with one exception, at all the grassland sites investigated, *V. excentrica* is present; *V. costata* on the other hand is often totally absent. How such competition operates, if this indeed is what is taking place, is unknown.

V. costata occurs in the Hoxnian (Kerney, 1971b) and Eemian

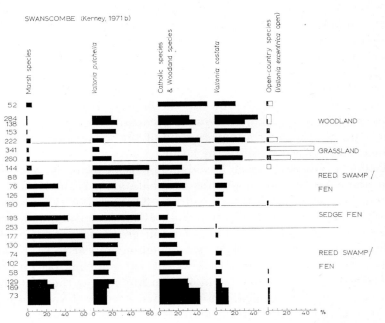

Figure 54. Changes in the abundance of three species of *Vallonia* in the interglacial deposits at Swanscombe, Kent. (Redrawn from Kerney, 1971b: Fig. 3.)

(Sparks, 1964b) Interglacials. In the Hoxnian deposits at Swanscombe (Kerney, 1971b: Fig. 3) the species occurs in a shaded context at the beginning of the sequence in which other open-country species are virtually absent (Fig. 54). Later on during a phase of dry open grassland when *V. excentrica*, *Truncatellina* and *Pupilla* became present it increases in abundance. Finally it declines slightly when the environment reverts to dry woodland. As in the Post-glacial, *V. costata* is showing a high degree of adaptation to a changing environment.

5. DISTRIBUTION, HABITATS AND HISTORY

Vallonia pulchella (Müller)

A Holarctic species of wide climatic tolerance, extending beyond 70° N in Scandinavia. Generally distributed in Britain, *V. pulchella* occurs in wetter habitats than *V. excentrica*, particularly "... at the roots of grass in moist fields and meadows" (Ellis, 1969: 162). In the present work it has been recorded so rarely that the specimens have usually been considered as extreme examples of *V. excentrica*. Nevertheless it does occur in strictly terrestrial (non-marsh) habitats on the Chalk, as at Brook (Kerney *et al.*, 1964) and Wayland's Smithy, Berkshire (Kerney, personal communication).

It is characteristic of the Late Weichselian, first occurring in zone I, and of periglacial environments generally (Sparks, 1952; Burchell and Davis, 1957). But it is absent, as are the other *Vallonia* species, from the colder horizons of the Middle Weichselian (Fig. 140). In the Postglacial it is present in marsh and flood-loam deposits, but appears to dislike heavy shading, being often scarce in tufa.

V. pulchella occurs in Hoxnian (Kerney, 1971b) and Eemian (Sparks, 1964b) Interglacial deposits. According to Sparks, (1953a: 115) "... the earliest record is difficult to determine, because of the difficulty of distinguishing it from *enniensis* when both are worn. It certainly occurs at Swanscombe and probably in earlier deposits".

Vallonia excentrica Sterki

The range of this Holarctic species is not well known, but may extend far north in Scandinavia (to 70° N) (Kerney, 1963). In Britain it is widespread, reaching to the north coast of Scotland and to Ireland.

V. excentrica was not separated from *V. pulchella* until the beginning of this century (Woodward, 1904), and is still considered by some authors to be a variety or subspecies of *V. pulchella*. In Britain, most authors regard the two as separate species (Sparks, 1953a) but throughout the literature there is evident confusion and many records of *V. pulchella* from archaeological deposits probably refer to *V. excentrica*. This cannot, however, be automatically assumed, as *V. pulchella* can on occasion live in dry habitats (see above).

V. excentrica is a snail of open habitats, and characteristic of grassland. In north-west Germany (Ant, 1963) it extends into lighter places than *V. costata*, though at the dark end of the lux scale the tolerance of the two species coincides (Fig. 39). It is virtually unknown from woodland or shaded habitats in Britain either living (Boycott, 1934) or in subfossil contexts, though there are a few records from scrub habitats. It is, however, often found under stones and logs in otherwise open places.

It is difficult to be sure of its moisture preferences due to the confusion with *V. pulchella*; Ant (1963) does not consider the two species separately but notes that *V. pulchella* agg. occurs in habitats of 60–70% relative humidity—damper than those of *V. costata* (Fig. 38) but not as damp as most woodland species. Ellis (1951) considers *V. excentrica* to occur in drier habitats than *V. pulchella* and as a general rule this seems to hold. We can thus arrange the three living British species of *Vallonia* in the following order with regard to their moisture tolerances:

V. pulchella—wettest habitats
V. excentrica
V. costata—driest habitats

V. excentrica is less common, too, in rupestral habitats than *V. costata* or *Pupilla*, and this applies to the ecology of the two species of *Vallonia* in France, as pointed out by Germain (1930). It occasionally occurs on walls (Ellis, 1969) and in rocky places (Kerney, 1969), but my own observations, particularly around the villages of Tackley and Ascott-under-Wychwood in Oxfordshire, suggest that it is confined to grassland habitats, *V. costata* being the predominant form on walls.

But *V. excentrica* occasionally lives in marshy habitats, as pointed out by Boycott (1934: 14); and Ellis (1941) records it in *Glyceria* reed swamp and swamp-carr—in the latter habitat with *V. costata* and *Vertigo pygmaea*.

Vallonia excentrica is common in arable land. In grassland generally, it is present, whether this be lowland pasture, heavily grazed limestone downland, or tall, rather damp and overgrown grass of hawthorn sere type.

Some records for living *V. excentrica* are given opposite.

As a fossil, *V. excentrica* is ubiquitous in archaeological deposits of Post-glacial age representative of terrestrial open-country habitats. In the Late Weichselian it is rare, but present in the later part of the Allerød Interstadial at Oxted, Surrey (Kerney, 1963: 219). It was subsequently present throughout zone III and in the earliest part of the Post-glacial, but probably became widely exterminated by the spread of forest over the British Isles and is absent from woodland faunas of Atlantic age in terrestrial situations. It is, however, recorded

1969), Hoxnian (Kerney, 1971b) and Eemian (Sparks, 1964b) Interglacials.

Theba pisana (Müller)

A rare coastal xerophile occurring locally in Cornwall, South Wales and eastern Ireland. It is unknown from Pleistocene or Post-glacial deposits, and is probably a human introduction of recent origin (Kerney, 1966b).

Helix (Cepaea) hortensis Müller

To 67°N in Scandinavia and generally distributed in Great Britain.

It favours damper places than its congener, *Cepaea nemoralis*, and has a more northerly range. It is frequently associated with *Arianta* but can occupy drier habitats, thus being intermediate in its moisture requirements between this species and *C. nemoralis*. In laboratory studies on *Arianta* and *Cepaea*, Cameron has shown that their geographical and local distributions can be explained in terms of resistance to water loss and behaviour patterns at various temperatures and humidities (Cameron, 1970a; 1970b). "*C. nemoralis* survives better, ceases activity more rapidly, and loses proportionally less weight in low humidity than *C. hortensis*, which in turn shows the same properties with respect to *A. arbustorum*." "*A. arbustorum* is relatively more active than *Cepaea* at low rather than high temperatures, as is *C. hortensis* with respect to *C. nemoralis*."

Today, mixed colonies of the two species of *Cepaea* are uncommon and in some downland areas, for example the Marlborough Downs in north Wiltshire, there is a clear pattern of distribution, *C. hortensis* predominating in the valleys, *C. nemoralis* in the uplands, the changeover often taking place over very short distances (Fig. 57). In the areas where they predominate, both species are found in all types of habitat, but where *C. hortensis* is present in a predominantly *C. nemoralis* area, it is restricted to the vicinity of trees, suggesting that there is some form of competition between the two species, relating to food or moisture, which is relaxed in shaded habitats (Cain and Currey, 1963a). It has been suggested that the absence of *C. nemoralis* from valleys is due to the extreme low temperatures which sometimes prevail in them, brought about by the downward movement of cold air from higher land on still, clear nights, creating "frost hollows". In such places the more cold-tolerant *C. hortensis* is able to survive.

C. hortensis is a common species in the Boreal and Atlantic periods.

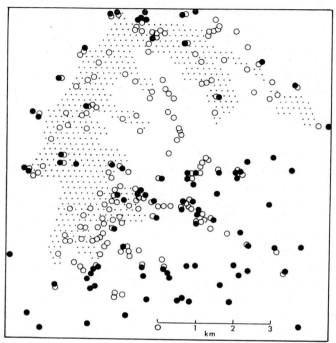

Figure 57. Distribution of *Cepaea nemoralis* (open circles) and *C. hortensis* (black dots) on the Marlborough Downs. Land over 700 ft stippled. (After Cain and Currey, 1963a: Fig. 2.)

It is also frequent in the Sub-boreal in Neolithic and Bronze Age deposits, and often occurs on high plateau areas of downland in association with *Cepaea nemoralis* and *Arianta*, and from which it is now absent (Cain and Currey, 1963a: 32, Evans, 1968a: 298). In fact the two species of *Cepaea* are more often associated than not on Sub-boreal sites (Fig. 58). These differences in past and present distribution may have been brought about by a variety of factors, but the most likely is the progressive clearance of woodland which has taken place from Neolithic times onwards. This led both to the destruction of suitable shaded and damp refugia for *C. hortensis* on the downland plateau, with an intensification of competition between the two species, and to the creation of frost hollows, inimical to *C. nemoralis*, in valleys —a phenomenon which was probably absent prior to forest clearance. Cain and Currey (1963a) aptly point out that frost hollows develop most strongly in the least wooded and most open country, and particularly where expanses of rolling upland are cut by well-marked valleys. Under forested conditions the environment both in upland and valley situations would have been altogether more equable.

In Scotland, beyond the range of *C. nemoralis*, *C. hortensis* occurs widely on fixed-dune pasture and limestone grassland, ". . . taking up habitats which, as defined on vegetation, are occupied in England and southern Scotland by *C. nemoralis*" (Cain et al., 1969). Oldham (1929) and Boycott (1934: 7) suggested that the absence of *C. hortensis* from sand-dune habitats within the geographical range of *C. nemoralis* might be due to its exclusion through competition. But Cain (Cain et al., 1969) pointed out that as the habitats in the far north are more humid and cooler than those further south, they are not strictly equivalent, and the presence of *C. hortensis* on dunes in Scotland is not therefore positive evidence for its competitive exclusion by *C. nemoralis* further south.

At Northton on the Isle of Harris, *C. hortensis* has clearly declined since the Iron Age, a phenomenon which may not altogether be associated with the introduction of *Cochlicella* and *Helicella itala* (Evans, 1971b: Table 3). It was abundant in the prehistoric levels, but today is rare in South Harris, only one living juvenile being found over two successive years' collecting (cf. Jones and Clarke, 1969: Fig. 2). This may be due to the decline of temperature which has taken place since the prehistoric period, for suitable habitats are abundant, and occupied by no other large helicid. It is difficult, however, to see the decline of this species in the downland areas of Britain as being due to the same factor, in view of the supposed climatic tolerance of *Cepaea hortensis* to low temperatures.

C. hortensis, unlike *C. nemoralis*, shows no significant change of banding pattern during the Post-glacial, unbandeds and bandeds occurring throughout (Currey and Cain, 1968; Cain, 1971: 86).

C. hortensis occurs in the Hoxnian (Zeuner, 1959: 331) and Eemian (Sparks, 1964b) Interglacials.

Helix (Cepaea) nemoralis Linné

In Scandinavia to 63° N; in Britain, generally distributed, but absent from the north of Scotland and the Outer Isles (Jones and Clarke, 1969: Fig. 1). This species occurs in a variety of habitats—woods, hedges, marsh and downland; in relation to *C. hortensis*, it occurs in drier and warmer habitats.

C. nemoralis is common in archaeological horizons, probably entering Britain early on in the Boreal, or even Pre-boreal, period. As already discussed (p. 109), it occurs on sites of Mesolithic, Neolithic and Bronze Age date (in conjunction with *C. hortensis* and *Arianta*) in areas from which it is now absent, the contraction in its range taking place at some time between about 2000 and 0 BC (Cain, 1971) (Fig. 58). This phenomenon has been ascribed to the induction of strong local climatic gradients and frost hollows in open downland

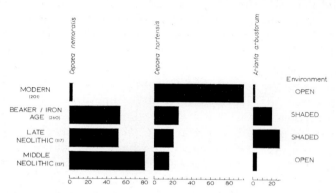

Figure 58. Changes in the abundance of *Cepaea* and *Arianta* from the Neolithic period to the present at South Street, Wilts. (After Cain, 1971.)

areas, stemming, in all probability, from the clearance of forests by prehistoric agriculturalists (Cain and Currey, 1963a).

Most of the changes in morph frequencies in mixed populations of *C. nemoralis* which have taken place since the Atlantic period are consistent with climatic selection (Currey and Cain, 1968; Cain, 1971). These are white lip, colour, and banding. In shells from the Caerwys tufa and Neolithic levels at South Street (Avebury Trusloe of Cain, 1971), white lip comprises 10% and 16% respectively of the totals. It becomes rare by the first millennium BC, and although present in animals from the Avebury area today, is limited and sporadic (Cain, 1971). In Britain and Europe, white lip is associated with high rainfall areas, being particularly common in western Ireland and the Pyrenees, so that its prevalence in the Atlantic and early Sub-boreal periods, may be a reflection of the damper climate at that time.

There has been an increase in brown shells in populations of *C. nemoralis*, from the Atlantic period to the present day. From tufa sites brown shells are virtually absent; at South Street there is an increase from 8% in Neolithic levels to 41% in first/second millennium levels, and browns are still locally common in the district (Cain, 1971). The modern distribution of browns in Europe suggests an association with relatively cool summers. The increase in browns which takes place in the second/first millennium BC may thus be related to the decline in summer temperatures which occurred at this time as determined from pollen analytical work and other biological evidence (Godwin, 1956).

Similarly, the decrease in unbandeds which has taken place since the Atlantic period may be related to the Post-glacial thermal decline. "The distributions of banding morphs in subfossil samples and over western Europe at present were shown by Currey and Cain (1968) to be consistent with each other and with experimental work, and to suggest the prevalence of unbandeds in hotter times and regions, five bandeds in cooler and wetter areas" (Cain, 1971).

Currey and Cain (1968) also showed that the type of geographical distribution known as the "area effect", in which morph frequencies over a large and diverse area are constant in spite of visual selection (Cain and Currey, 1963a, 1963b), were more widespread in pre-Iron Age times than now, and that some area effects may have lessened in intensity.

C. nemoralis is not present in the Late Weichselian, but occurs at an early stage in the Post-glacial—probably Pre-boreal or earliest Boreal. It occurred in the Hoxnian (Davis, 1953) and Eemian (Sparks, 1964b) Interglacials, and in the Lower Pleistocene (Zeuner, 1959: 331; Ellis, 1951: 190).

Helix aspersa Müller

This species is distributed over much of the British Isles, north to the Outer Hebrides and Orkney, but is rare in eastern Scotland on account of its susceptibility to winter frosts. It is a synanthropic species, occurring in gardens, hedgerows and waste ground, but may also be found in wild places, particularly in the west. In Cornwall it is a common sand-dune species, living in association with *Cochlicella acuta* and *Helicella virgata*.

It is fairly certain that *Helix aspersa* is an introduction of the first century AD. It occurs at Owslebury, Hants, and South Cadbury, Somerset, in the second half of the first century AD, although absent from late pre-Roman Iron Age deposits on these sites (Evans, unpublished), at Lullingstone, Kent, in the third century (Kerney *et al.*, 1964: 172) and at Arbury Road, Cambridge and Whitton, Glam. in the fourth century. It is, however, absent from the Welland Valley in the first and second centuries AD (p. 348). It was presumably eaten for food by the Romans, as it still is today in parts of Britain, but the shells seldom occur in the same enormous abundance as do those of oysters.

All pre-Roman subfossil records are either erroneous or doubtful, and the Upper Pleistocene finds of this species must also be rejected (Kerney, 1966b: 11).

Helix pomatia Linné

This species has a markedly southern distribution in England; it is local in its occurrence. It is recorded from Roman levels (Kerney, 1966b) and there is "no conclusive evidence to prove wrong the popular belief that it was deliberately introduced by the Romans for food . . .".

*Hygromia limbata (Draparnaud)

South Devon. A twentieth century introduction (Kerney, 1966b).

*Hygromia cinctella (Draparnaud)

South Devon. A twentieth century introduction (Kerney, 1966b).

Hygromia subrufescens (Miller)

A species of old woodland and wild places generally, occurring mainly in the north and west of Britain (Fig. 37). Due to its extremely fragile shell, this species has never been recorded subfossil, but there seems no reason to doubt that it is other than native.

Hygromia striolata (C. Pfeiffer)

This species is synanthropic (p. 201) occurring in arable land (South Street, Fig. 129; Waden Hill, Evans, 1968a), gardens, hedgerows and waste places generally. In the south of England it is also found in semi-natural woodland, but in the north it is restricted to man-made habitats. As Kerney has pointed out ". . . it may even be that *Hygromia striolata* could not continue to exist in these northern parts of Britain if human interference ceased . . ." (1966b).

In the Boreal, Atlantic and Sub-boreal periods, *H. striolata* occurred in southern Britain in woodland habitats, free from human influence, in degrees of abundance varying from 1% to 20% of the total fauna. It is generally less abundant than *H. hispida* in archaeological deposits of prehistoric age, in Wiltshire being extremely rare during the Neolithic and Bronze Age periods (Evans, 1968a: Fig. 2). Its comeback as a synanthropic species is of Roman or medieval origin.

It occurs in the Upper Pleistocene (Ellis, 1951).

Hygromia hispida (Linné)

A Holarctic species reaching to 66° N in Scandinavia. It is generally distributed in Britain, but rare or absent from much of northern Scotland. It occurs in a wide variety of habitats, shaded and open, moist and dry, but is most abundant in damper places such as meadows and marsh (Boycott, 1934; Ellis, 1969).

A difficulty in considering the ecology of *H. hispida* is that we are not always dealing with a single taxonomic or ecological unit, there being a number of forms, subspecies or even species included within *H. hispida* (Forcart, 1965) which have different ecological tolerances. Perhaps the most prevalent of these is *H. hispida* var. *nana* (Jeffreys) which occurs in drier habitats, particularly chalk downland. In the present work, this has not been separated from the more normal *H. hispida* due to the difficulties of identifying all but adult specimens. In the Late Weichselian, Kerney (1963) has drawn attention to the large form during zone III which sometimes comprised 80% of the assemblages. It was clearly adapted to damp habitats, fluctuating in sympathy with *Succinea pfeifferi* and *Columella columella*, but was rare in assemblages of strongly xerophilous character.

It would be interesting to know more about the history of var. *nana* for it is conceivable that it is of recent origin, a form which has evolved, in fact, as a result of forest clearance and the creation of chalk grassland. Thus Kerney (1963) in studying the occurrence of *H. hispida* in the Late Weichselian in Kent, where the large form was prevalent, found that every site showed its own particular facies of the species, suggesting that once established, local races quickly developed. *H. hispida*, there, appeared readily adaptable to a changing environment. Another point of possible relevance is the paucity of *H. hispida* in grassland environments of Neolithic and Bronze Age date (Fig. 144), when occurrences of over 5% (Silbury Hill, 8%, and Avebury, 19%) were rare, by comparison with its frequent abundance on Iron Age sites and today in similar environments. This suggests that *H. hispida* was ill-adapted in Neolithic and Bronze Age times to these open grassland habitats. That it could flourish where conditions were suitable is shown by its prevalence in ditches as at South Street, 28%

(Fig. 129; 165–215 cm) and Horslip (Connah and McMillan, 1964), where the large form occurred.

H. hispida appears in zone I (Bølling Interstadial) on the south coast at Beachy Head, becomes widespread in southern England in zone II, flourishes in damp habitats during zone III, and occurs consistently in natural and archaeological deposits throughout the Post-glacial. It is a Cromerian (Sparks *in* Duigan, 1963: 197), Hoxnian (Kerney, 1971b) and Eemian (Sparks, 1964b) fossil, and also occurred in the Middle Weichselian and colder periods of the Pleistocene (Sparks, 1957b; Kerney, 1971a).

Hygromia liberta (Westerlund)

Mainly central England and East Anglia. It lives in woods, hedgerows and in long grass along river banks. In the present work it is recorded from Roman ditches at Arbury Road and Maxey (p. 346). It is abundant in Post-glacial deposits at Apethorpe, Northants, possibly in zone VI (Sparks and Lambert, 1961), but is not known from the Late Weichselian. It occurred in the Eemian Interglacial (Sparks, 1964).

Hygromia subvirescens (Bellamy)

South-west England and Pembroke, living on grassy slopes near the sea (Ellis, 1951); a xerophile (Boycott, 1934). This snail occurs in Holocene deposits of calcareous sand at various sites in north Cornwall (Davis, 1956: 100), reputed to be of Bronze Age origin, but which are probably post-Iron Age.

Monacha granulata (Alder)

Said to be endemic of Britain. It is a snail of damp, shaded places often occurring along river banks, particularly in the south and east of England; in the west (Cornwall to Oban) it is often abundant in dry hedgerows and similar habitats (Boycott, 1934: 14). As a fossil it is rare, probably due to its thin and feebly calcified shell. It occurs in blown sand deposits in Cornwall of Post-glacial age (Woodward, 1908), and is recorded from the Middle Pleistocene (Hoxnian?) deposits at Ingress Vale (Davis, 1953).

Monacha cartusiana (Müller)

Western and southern Europe, north to Holland. A rare xerophile species, in Britain restricted to the extreme south and east of England

(Kerney, 1970)—Hampshire, Sussex, Kent and Suffolk. It has been recorded from several Holocene deposits extending its range into Essex (Ellis, 1969; Kerney *et al.*, 1964), but is unknown from the Late Weichselian and the British Pleistocene.

Monacha cantiana (Montagu)

An alien species of western European origin, first occurring in the Post-glacial in Roman times (Kerney *et al.*, 1964); it is unknown from the British Pleistocene. The life-history and ecology of *M. cantiana* have been discussed by Chatfield (1968, 1972) and its distribution by Kerney (1970) (Fig. 45).

Helicella caperata (Montagu)

A grassland and xerophile species (Baker, 1968). *H. caperata* is alien over most, if not all, of its range in Britain (Kerney, 1966b), having been introduced in the last few centuries—probably since the medieval period. It has been recorded from deposits of blown sand in Cornwall of allegedly Iron Age date along with *Helicella virgata*, *Cochlicella acuta* and *Hygromia subvirescens* (Bullen, 1902), but the archaeological and geological evidence for the age of these is unsatisfactory, and they are likely to be comparatively recent in origin (post-Iron Age).

Helicella gigaxi (L. Pfeiffer)

A xerophile species, particularly characteristic of arable habitats. *H. gigaxi* is an alien, occurring in a few post-Roman hillwashes in the south of England (Kerney, 1966b: 10).

Helicella virgata (da Costa)

A xerophile species (Boycott, 1921c), and, like *H. caperata* and *H. gigaxi*, probably alien (Kerney, 1966b). In Cornwall, it has been recorded from Harlyn Bay in deposits of blown sand, allegedly Iron Age in origin (Bullen, 1902, 1912) and from Newquay in deposits considered to be Neolithic or earlier (Kennard and Warren, 1903; Woodward, 1908) (Fig. 105). The evidence for their age, however, is unsatisfactory and until a detailed examination of the environment and archaeology of these deposits is made, the status of *H. virgata* must remain unclear. In south-east England, the animal is present in a possible thirteenth century AD deposit (Kerney, 1966b: 10).

H. virgata was present in Britain during the Eemian Interglacial (Sparks, 1964b).

Helicella neglecta (Draparnaud)

A twentieth century introduction, now apparently extinct (Kerney, 1966b); "extirpated in 1922 by a rapacious collector" (Ellis 1969: 279).

†*Helicella crayfordensis* Jackson

An extinct species occurring in deposits of Cromerian (Zeuner, 1959: 331), Hoxnian (Davis, 1953; Kerney, 1971b) and Eemian (Sparks, 1964b) age.

Helicella itala (Linné)

A western European species, absent from the Scandinavian mainland (Kerney, 1963: Fig. 18). It is present throughout Britain, but being more or less a calcicole (Boycott, 1934: 31) it is absent from the mountainous regions of the West Country, Wales and Scotland, where it has a generally maritime distribution.

In Britain today it inhabits dry grassland areas particularly where the sward is kept short by grazing, and is often plentiful on chalk downland, scarp slopes being favoured. It appears to avoid lowland pasture and arable habitats generally, preferring wilder places where the influence of man is not strongly felt, though it has been reported from ploughed fields in Gloucestershire (Boycott, 1921c). It is also common on sand dunes. It is more local than *H. virgata* and *H. caperata* (Quick, 1943), but generally abundant wherever it occurs (Boycott, 1934).

Ant (1963) has shown that in north-west Germany, *H. itala* occurs only in very dry habitats (R.H. 40–50%) (Fig. 38). It is also an obligatory heliophile, never occurring in shaded places. This seems to be the case in Britain as well: *H. itala* never occurs in marshes, and there is only one record known to me of the animal living in woodland— "woods in Cambridgeshire" (Pennant, 1812). *H. itala* in fact is probably our most characteristic open-country species, having developed to an extreme degree the ability to live in places which are very open and very dry. It is however essentially a grassland species, not flourishing in rupestral habitats. "It seems definitely to prefer the hottest and often steepest part of a hillside" (Boycott, 1934). In the Outer Hebrides it occurs only in the driest habitats on the dunes with *Cochlicella acuta* and *Vitrina pellucida*; it is rarely present in damper situations, and then never abundant. In the far north of Scotland it is characteristic of the most open and short vegetation (Cain et al., 1969), and the same applies to its occurrence in southern Britain. Around Eastbourne, "several large colonies occur on the slopes of the Downs. Preference seems to be shown for a more or less southern aspect" (Shrubsole, 1933). In Bedfordshire it occurs on dry chalk hills in the south, where the soil is of an arid nature; locally it is extremely abundant (Verdcourt, 1945). In Somerset it is found "chiefly on downs and pastures in hilly districts on calcareous soils" (Swanton, 1912). On the south-facing slopes of Knap Hill, Wiltshire, it is abundant in short-turfed grassland. And in west Glamorgan it is present on high limestone ground covered by blown sand (Quick, 1943).

Kerney (1968b) notes that *H. itala* is becoming increasingly local in most inland parts of England, perhaps because of the disappearance of suitable short-turfed grassland caused by the scarcity of rabbits and by the decline of sheep grazing. Block (1964) reports on the destruction of a colony of *H. itala* living on heavily grazed turf around rabbit burrows: "When myxomatosis destroyed the rabbits the grass grew up tall, and in the more humid conditions the *itala* all died out".

H. itala was present in Britain during the Late Weichselian, the earliest record being from Beachy Head in the Bølling Interstadial (Kerney, 1963: Fig. 14). It persisted throughout the period, but was widely exterminated by the spread of forest during the Post-glacial. Unlike *Vallonia* and *Vertigo pygmaea*, *H. itala* had no ability to colonize alternative habitats such as marsh and woodland, and this lack, coupled with the fact that it is probably a calcicole resulted in a less rapid increase at the onset of Neolithic clearance than is shown by certain other grassland species. This is particularly noticeable at Brook (Kerney *et al.*, 1964), but is not so apparent in north Wiltshire where *H. itala* was present in abundance at an early stage (*ca.* 3300 BC at Horslip). In the latter area, which today is so very open, there is always the possibility that *H. itala* persisted in isolated localities throughout the earlier part of the Post-glacial, and Thomas Pennant's record should guard us against the too ready assumption that this snail has always been rigid in its habitat requirements. At Ascott-under-Wychwood in the Oxfordshire Cotswolds, however, *H. itala* was not present in the first clearance fauna on the site (*ca.* 2800 BC) and indeed may never have reached the Jurassic ridge during the Late Weichselian in the first place due to the impenetrable barrier of the Oxford Clay. And at Northton in the Outer Hebrides, the snail is quite clearly an introduction of Iron Age or later origin, due, most likely, to the absence of sand dune habitats along the coast, for it was probably not until the Neolithic period that these began to form (p. 295).

In general, *H. itala* is less abundant than the two species of *Vallonia* in Neolithic and Bronze Age sites on the Chalk, seldom attaining more than 15%. It is noticeably rare in faunas with abundant *Pupilla*, strengthening the ascription of these to habitats with much bare ground—at least in Neolithic and Bronze Age times. Later on, in hillwash and lynchet deposits *H. itala* is often very abundant as at South Street, 28% (Fig. 129; 40–120 cm), Waden Hill, 25% (Evans, 1968b), Overton Down, 38% (Fig. 120; 25–65 cm) and Badbury Earthwork, 30% (Fig. 134; 30–70 cm), where it is clearly present in an arable environment. As has been pointed out, it is rarely found in such situations today, and this change in habitat preference may, therefore, be due to competition from more recently introduced helicellids, particularly *H. gigaxi* which is common in arable fields.

H. itala occurred in the Eemian (Sparks, 1964b) Interglacial.

†*Helicella (Helicopsis) striata* (Müller)

A xerophile species, having an essentially south-eastern European distribution, but extending as isolated colonies to southern Scandinavia (Schlesch, 1951). It occurs in Eemian deposits in the Cambridge gravels in conjunction with many other thermophile species of southern distribution in Europe (Sparks, 1953b; 1964b). In Kent it is found in Late Weichselian and early Holocene deposits (Sparks, 1953b; Davis, 1954).

†*Helicella (Xeroplexa) geyeri* (Soós)

A western European species absent from the Scandinavian mainland (Kerney, 1963). Like *H. striata*, this species is a xerophile. It has been recorded from an interglacial deposit at Little Oakley, Essex (Sparks, 1953b) (Cromerian?) and in the Post-glacial (Davis, 1954). Otherwise it is associated with "cold" Pleistocene faunas (Sparks, 1953b); in Kent it is present in zones II and III of the Late Weichselian, and in Wiltshire was recovered from involutions at Durrington Walls (Wainwright and Longworth, 1971: 330).

*Helicella elegans (Gmelin)

A rare xerophile occurring in a few localities in the extreme south-east of England. Probably introduced in the nineteenth century (Kerney, 1966b).

Cochlicella acuta (Müller)

A xerophile species, commonly occurring on sand dunes and other dry open habitats on the south and west coasts of England, Wales and Scotland (De Leersnyder and Hoestlandt, 1958); in Ireland, as well as being coastal, its distribution extends inland (Ellis, 1951). It occurs in blown sand deposits of late Holocene age in Cornwall at Harlyn Bay (Bullen, 1902) and Newquay (Kennard and Warren, 1903) (p. 292). In the Outer Hebrides it occurs in deposits of sixteenth century AD date at Udal on North Uist (Evans, unpublished) and in the "Iron Age" II horizon at Northton, South Harris (Fig. 106); the date of the latter, however, is uncertain.

Family ENDODONTIDAE

Punctum pygmaeum (Draparnaud)

A Palaearctic species extending to beyond 70° N in Scandinavia. Ubiquitous in Britain. *P. pygmaeum* occupies a wide range of habitats, occurring in hedgerows and woods where it is often particularly abundant amongst dead leaves, in marshes and in dry situations. Generally it is classed with the "other woodland species" in archaeological contexts, rarely being sufficiently abundant to warrant separate presentation. However, its preferences sometimes appear to be for more open and dry habitats such as chalk grassland, in which it occurs with other catholic species—*Cochlicopa, Euconulus, Vitrea, Retinella radiatula* and *Vitrina*. The ecological characteristics of this group will be discussed below (p. 195).

P. pygmaeum is characteristic of Late Weichselian assemblages, appearing in zone I and persisting throughout. In the Post-glacial, it occurs in a variety of sites being common in tufas, hillwash and various archaeological deposits. Formerly, it was present in the Hoxnian (Kerney, 1971b) and Eemian (Sparks, 1964b) Interglacials.

†Discus ruderatus (Férussac)

A widely distributed Holarctic species occurring in Central Europe, Norway to Spain, north Italy and Transcaucasia (Ellis, 1951; Scharff, 1907: Fig. 39). It inhabits woodland and is abundant in damp places.

D. ruderatus does not occur in the Late Weichselian, but is present in a number of early Post-glacial deposits, often in conjunction with *Vertigo geyeri* and *V. genesii*, as at Brook (Kerney et al., 1964),

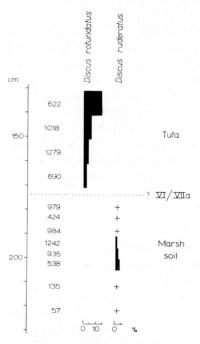

Figure 59. Changes in the abundance of the two species of *Discus* in the early Postglacial at Brook, Kent. (After Kerney *et al.*, 1964: Fig. 18.)

Broughton-Brigg, Lincs. (Kennard and Musham, 1937), Norton Common, Herts. (Kerney, 1955), Southwell, Notts. (Davis and Pitchford, 1958) and Wilstone (Kennard, 1943). Thereafter, it appears to die out, when it becomes replaced by *D. rotundatus*. At Brook, the changeover between the two species is particularly elegant, the more so in that it takes place at a stratigraphical and environmental break thought to correspond to the zone VI/VIIa transition (Fig. 59). Kerney (1968a) has suggested that the behaviour of these two species is controlled by a climatic shift towards greater oceanicity at the beginning of the Atlantic period. Their present-day ranges would certainly support such a hypothesis, *D. ruderatus* having an "Alpine/Scandinavian/east European/Asiatic distribution" and *D. rotundatus* a southern, more maritime distribution.

We must, however, be wary of ascribing to these two species the properties of zone fossils until more sites like Brook have been studied, and C-14 dates obtained from them. Thus at Cherhill, Wiltshire

(p. 304; Fig. 109) *Discus rotundatus* is present in abundance in a Boreal horizon dated by radiocarbon to pre-5300 BC (i.e. prior to the zone VI/VIIa transition), while *D. ruderatus* is absent.

D. ruderatus occurred in the Cromerian (Sparks *in* Duigan, 1963: 197), Hoxnian (Kerney, 1971b) and Eemian (Sparks, 1964b) Interglacials.

Discus rotundatus (Müller)

A Palaearctic species, extending to 63° N in Norway, and south to Spain and Sicily (Scharff, 1907: Fig. 38). It is virtually ubiquitous in Britain, but becomes rare in certain eastern counties, for example in East Anglia, where the relatively continental climate is unfavourable to it.

D. rotundatus is a common woodland species, occurring in leaf litter, under logs and in hedgerows often in enormous numbers—far more than most other species of equivalent size. Its range of habitats, however, is more narrow than that of *Punctum pygmaeum* and *Carychium tridentatum*, for it is rare in marshes and dry, open downland.

Subfossil it is one of the commonest species in Post-glacial deposits, often occurring in sufficient abundance to warrant separate presentation. It is common in archaeological contexts and has been recorded from almost every site discussed in this book. It appears early on in the Post-glacial, possibly as early as the Pre-boreal period. But its spread may have been slow, and the paucity of the animal in the lower horizons of subsoil hollows at Ascott-under-Wychwood (Fig. 88b) and Avebury (Fig. 96) may be of chronological significance (p. 269).

Discus rotundatus is a Hoxnian (Kerney, 1971b) and Eemian (Sparks, 1964b) fossil, in the latter persisting into the cool *Pinus* zone at the end of the Interglacial. It is abundant in a probably Early Weichselian deposit at Pitstone, Bucks. (Evans, unpublished).

Family ARIONIDAE

Geomalacus maculosus Allman

Known only from Kerry and west Cork where it eats lichen (Boycott and Oldham, 1930). There are no subfossil records.

Arion

Of the nine species of *Arion* in Britain, two, *A. silvaticus* and *A. fasciatus* are local in their occurrence; the others are widespread. They generally live in damp habitats. Thus at Halling, Kent, in a Middle Weichselian deposit, the Arionidae fluctuated in sympathy with *Succinea oblonga* and *Agriolimax* (Kerney, 1971a). *Arion* granules are frequent in Pleistocene and Post-glacial deposits, sometimes occurring in enormous numbers. Apart, however, from the possibility of separating *Arion ater* agg. from the smaller species (Fig. 29), they are not specifically determinable (p. 73). The species are:

- *A. intermedius* Normand
- *A. circumscriptus* Johnston
- *A. silvaticus* Lohmander
- *A. fasciatus* Nilsson
- *A. hortensis* Férussac
- *A. subfuscus* (Draparnaud)
- *A. ater* agg. (Linné)
- *A. ater* seg. (Linné)
- *A. lusitanicus* Mabille
- *A. rufus* (Linné)

Family ARIOPHANTIDAE

Euconulus fulvus (Müller)

A Palaearctic species, extending beyond 70° N in Scandinavia, and climatically very tolerant. Ubiquitous in Britain. It occurs in a variety of habitats such as marshes, leaf litter in woodland and occasionally in quite open and dry places; but it is rarely present in abundance. It is generally grouped with the woodland species, but is more sensibly included with the catholic species when these are present in sufficient abundance to plot as a separate entity (p. 195).

E. fulvus is present in zone I of the Late Weichselian, and thereafter occurs sporadically in deposits of natural and archaeological origin throughout the Post-glacial. It is particularly common in marsh and swamp deposits of Atlantic age. In dry chalkland sites of Neolithic and later date, however, it is infrequent. Formerly it was present in the Hoxnian (Kerney, 1971b) and Eemian (Sparks, 1964b) Interglacials.

Euconulus fulvus var. *alderi* Gray

This variety is possibly a distinct species (Ellis, 1969: 281).

Family ZONITIDAE

Vitrea crystallina (Müller)

A Palaearctic species extending to 66° N in Scandinavia. Ubiquitous in Britain. This snail occurs in a variety of habitats, generally of a fairly moist nature such as hedgerows, marshes and woods. It is classed as a woodland species but has affinities with the catholic

As would be expected, it is a common Late Weichselian species, present from zone I onwards. On archaeological sites its affinities are with the catholic species, particularly *Punctum* and *Retinella radiatula*, with which it often behaves in sympathy, as at South Street (Fig. 90) and Ascott-under-Wychwood (Fig. 88). At Northton (Fig. 106) it comprised a sizeable component of the sand dune fauna.

It was present in the Hoxnian (Kerney, 1971b) and Eemian (Sparks, 1964b) Interglacials. Cromerian records probably refer to *Vitrina (Semilimax) semilimax* (Ellis, 1969: 281).

Vitrina major (Férussac)

A species of old woodland and wild places generally, widespread but local in southern England and Wales (Boycott, 1922; 1927). There are no fossil records, but this is probably due to the difficulty of distinguishing it from *V. pellucida*, and there seems no reason to doubt its native status.

**Vitrina (Semilimax) pyrenaica* (Férussac)

Lives in a small area on the boundary of Co. Meath and Co. Louth. Probably a recent introduction (Kerney, 1966b: 11).

†*Vitrina (Semilimax) semilimax* (Férussac)

An Alpine and central European species. Recorded from the Cromerian (Sparks *in* Duigan, 1963: 197) and Hoxnian (Kerney, 1959) Interglacials. Cromerian records of *V. pellucida* probably refer to this species (Ellis, 1969: 281).

Family LIMACIDAE

Milax

The three common species of *Milax* occur mainly in gardens and cultivated land, only invading natural habitats in the south and west of England. They are:

> *M. gagates* (Draparnaud)
> *M. sowerbyi* (Férussac)
> *M. budapestensis* (Hazay)

A fourth species, *M. insularis* (Lessona and Pollonera), occurs at Bexhill, East Sussex (Ellis, 1969: 267).

Milax sp. has been recorded from a possible zone VIII deposit at

Apethorpe, Northants. (Sparks and Lambert, 1961) but is otherwise unknown with certainty as a Post-glacial fossil. It is likely that the present widespread distribution of the three common species has been brought about by man (Kerney, 1966b: 12).

Formerly, *Milax* sp. occurred in the Eemian Interglacial (Sparks, 1964b).

Limax

Limax flavus is a synanthropic species. The other four live mainly in woodland habitats. They are:

> *L. tenellus* Müller
> *L. maximus* Linné
> *L. cinereoniger* Wolf
> *L. flavus* Linné
> *L.* (*Lehmannia*) *marginatus* Müller

Limax sp. is recorded from Pleistocene (Cromerian (Sparks *in* Duigan, 1963, 197); Eemian (Sparks, 1964b)) and Holocene deposits, but is very infrequent by comparison with *Agriolimax*. Records of particular species must be considered as tentative, for the shells are impossible to separate with certainty. Large shells, probably of *Limax* sp., occasionally turn up in archaeological deposits, particularly in woodland contexts, but in the present work these have generally been listed as "Limacidae".

Agriolimax agrestis (Linné)

Occurs in marshes and damp places (Ellis, 1967). *A.* cf. *agrestis* has been recorded from the Post-glacial (Sparks and Lambert, 1961) and the Pleistocene (Sparks *in* Duigan, 1963: 197; Sparks, 1957a; 1964b).

Agriolimax reticulatus (Müller)

A Palaearctic species, ubiquitous in Britain. It lives in a wide variety of habitats (South, 1965), and according to Ellis (1969) is our commonest land mollusc; it is particularly abundant in arable land.

It is probable that the vast majority of shells of *Agriolimax* from archaeological deposits of Post-glacial age in dry land (as opposed to marsh) situations belong to this species. At South Street (Fig. 129) and Pitstone (Evans, 1966b: Fig. 7) the steady increase of Limacidae (cf. *A. reticulatus*) reflects in a spectacular way the beneficial influence of agriculture on these species.

A. cf. *reticulatus* has been recorded from Post-glacial deposits (Sparks, 1955; Sparks and Lambert, 1961) and the Pleistocene (Sparks *in* Duigan, 1963: 197; Sparks, 1957a; 1964b).

Agriolimax caruanae Pollonera

Widespread, mainly in gardens, but in wild places in south-west England (Ellis, 1969: 268). It is recorded as *A.* cf. *caruanae* from Post-glacial (Hayward, 1954; Sparks and Lambert, 1961) and Eemian (Sparks, 1957a; 1964b) deposits.

Agriolimax laevis (Müller)

A Holarctic species, ubiquitous in Britain. It lives in marshes and damp places generally. As *A.* cf. *laevis* it is recorded from the Post-glacial (Sparks and Lambert, 1961) and the Eemian Interglacial (Sparks, 1957a; 1964b).

6
Ecological Groups

In this chapter the main ecological groups and more important individual species of snail occurring in Late Weichselian and Post-glacial deposits are described. Most of the faunas are wholly terrestrial and comprise about sixty species which can be grouped into three categories: woodland or shade-loving species, intermediate or catholic species, and open-country species. The basis for this classification is Boycott's fundamental paper on the ecology of British land snails (1934).

The woodland species constitute the largest group. Their essential requirements are for relatively damp, undisturbed and sheltered habitats such as are provided by leaf litter on a woodland floor or the tangled mass of dead leaves and straw at the base of grasses in ungrazed grassland. Ecologically they are a heterogeneous collection, and few are restricted to woodland *per se*. It is generally useful to split them into four sub-groups: the Zonitidae, *Carychium tridentatum*, *Discus rotundatus* and the other woodland species.

The **Zonitidae** are characterized by a rather thin, fragile shell and flimsy lip, and are thus suited to life amongst leaf litter rather than to rupestral habitats, though they are not infrequent on the undersurface of fallen branches and amongst stones. Six species are common on archaeological sites. They are:

Vitrea crystallina *Retinella radiatula*
V. contracta *R. pura*
Oxychilus cellarius *R. nitidula*

Three other species, *Oxychilus alliarius*, *O. helveticus* and *Zonitoides excavatus*, though native are much less common, particularly on archaeological sites of the later prehistoric periods.

Carychium tridentatum and **Discus rotundatus** are generally plotted as individual graphs as they are often the most abundant of the shade-loving species, outnumbering all others. Their habitats are broadly similar—under logs and amongst leaf litter—though *Carychium* is common in long grass, well down at the bases of the leaves. It can probably occupy a greater range of habitats than *Discus* due to its smaller size (less than 2·0 mm high), though it dislikes any form of disturbance which leads to drying out of the environment. Its disappearance is thus a useful indicator of the onset of cultivation or heavy grazing.

The **other shade-loving species** comprise those which generally occur in such low abundance as to make individual plots meaningless. There are just over twenty species. These are:

Acicula fusca
Azeca goodalli
Columella edentula
Vertigo pusilla
V. substriata
V. alpestris
Lauria cylindracea
L. anglica
Acanthinula aculeata
A. lamellata
Ena montana
E. obscura
Marpessa laminata
Clausilia bidentata
C. rolphi
Balea perversa
Helicodonta obvoluta
Helicigona lapicida
Hygromia striolata
Punctum pygmaeum
Discus ruderatus
Euconulus fulvus
Vitrina pellucida

Many have adopted the rupestral habit, occurring under logs and stones, on walls and even on the trunks of trees.

It is sometimes convenient to separate from the shade-loving species a group of four which are particularly tolerant of a wide range of habitats, especially the more open ones, and having affinities with the catholic species. They are:

Punctum pygmaeum
Euconulus fulvus
Retinella radiatula
Vitrina pellucida

All are characteristic of the Late Weichselian fauna and today extend beyond 70° N in Scandinavia (Kerney, 1963). Boycott (1921b; 1934) considered them to be tolerant of poor habitats and indifferent to lime. All occur in woodland but are frequent in poor habitats characterized either by a scarcity of some environmental component, e.g. food in an actively accumulating sand-dune habitat; or by an unstable surface as on an active dune or bare chalk slope; or by a marked variability of temperature as on closely grazed chalk downland; or of water level as in a reed swamp. Thus they are particularly characteristic of short-turfed grassland and of open habitats where there is much bare ground—for example the secondary fill of ditches (South Street, p. 330 and Ascott-under-Wychwood, p. 345). Along the Chiltern scarp they appear to be more frequent in juniper than in hawthorn sere habitats (p. 89); *Vallonia costata* is a commonly associated species. Ellis (1941) lists *Punctum pygmaeum*, *Euconulus fulvus* and *Retinella radiatula* from a reed swamp in association with a fauna of which only *Carychium tridentatum* and *Agriolimax agrestis* are not obligatory marsh or freshwater species. The two species of *Vitrea* have affinities with this group but have generally been retained with the Zonitidae.

Pomatias elegans (p. 133) is best considered separately, for its habitat preferences are fairly clear cut. In general, it favours shaded places, but is particularly characteristic of habitats in which the soil surface is bare of vegetation and somewhat rubbly. In some instances its increase in a habitat is associated with woodland clearance (Fig. 53).

The **intermediate** or **catholic species** comprise a group of about eight, none of which is particularly diagnostic of either shaded or open habitats. They are:

Cochlicopa lubrica
C. lubricella
Arianta arbustorum
Helix (Cepaea) hortensis
H. (Cepaea) nemoralis
Hygromia hispida
Limacidae (*Limax* and *Agriolimax*)

With the exception of the two species of *Helix*, and of *Limax* spp. they were all present in the Late Weichselian.

Another reason for treating this group separately is that they all present difficulties of identification. Thus the two species of *Cochlicopa* are difficult, if not impossible to separate as juveniles (p. 51) and rarely occur in sufficient abundance as adults for separate plots to be meaningful. The ecological tolerances of the two are somewhat different, *C. lubrica* preferring relatively shaded habitats, *C. lubricella* the more open. But unless we can be certain that only one of the two species is present on a site, we cannot justifiably include them with either the woodland or open-country species.

Likewise with *Arianta* and *Helix* (*Cepaea*) spp. *Arianta* prefers damp places such as hedgerows, woodland and marsh; on the Chalk it generally occurs in valleys amongst lush herbage (Cameron *et al.*, 1971). Today, *Helix nemoralis* and *H. hortensis* do not usually occur together, the former occurring on high downs, the latter in valleys, and sometimes in association with *Arianta*. But in the past, as Kennard frequently pointed out, all three species were commonly associated (Fig. 58).

Hygromia hispida (p. 177) is common on archaeological sites and occurs in a wide range of habitats. On chalk downland it is generally replaced by the dwarf form *H. hispida* var. *nana* which is ecologically distinct from normal *H. hispida* but difficult to distinguish from it in the juvenile state. In central England, *H. liberta* often replaces *H. hispida* and is common in damp vegetation along the banks of rivers and streams. Again however, this form is difficult to separate from true *H. hispida* in all but adult examples.

With the Limacidae, similar problems arise. Species of *Limax* generally occur in woodland habitats, and the larger internal plates of this group are no doubt referable to one or other of the *Limax* species (p. 78). But the smaller plates are difficult to identify with certainty and could belong to any one of nine species of *Limax* or *Agriolimax* which between them occupy a variety of habitats. Often the smaller plates are found in profusion in ploughwash deposits, when one can be fairly sure that one is dealing with *A. reticulatus*, a common species in arable land today (p. 192). In theory too the internal shells of *Milax* are distinguishable from those of *Limax* and *Agriolimax*, though seldom found subfossil (p. 191). But on the whole, the Limacidae, like the other intermediate species, are of limited value in environmental interpretation.

Essentially the identification of the intermediate species depends on their abundance on a site and their degree of preservation, it being sometimes possible, for example, to separate *Arianta* and *Helix* (*Cepaea*) spp. as quite small juveniles. Another factor to be considered is the type of information required. If one is solely interested in working out the environmental history of a site, sufficient data may generally be obtained from species which are easy to identify. If on the other hand one's interest is in the history of a particular species of snail or slug, much greater care is necessary in obtaining accurate identifications of that species.

There are five common **open-country species** which occur on almost every site on the Chalk of south and east England. They are:

Vertigo pygmaea
Pupilla muscorum
Vallonia costata
V. excentrica
Helicella itala

With the exception of *Vallonia costata* which can occur in small numbers in shaded habitats, these species almost never enter woods and their presence on a site is thus good evidence of open ground. The ecological tolerances of these species differ, but it is not altogether clear what factors are controlling their distribution and abundance in any one instance, due largely to the lack of quantitative information from modern populations in known habitats. This is particularly unfortunate for two reasons. First of all these species comprise a sizeable proportion of the land-snail fauna on archaeological sites from the beginning of the Neolithic period onwards; and secondly they frequently show clear cut and opposing patterns of change through a deposit or soil, which are almost certainly controlled by external environmental influences. If we could interpret these changes in a more satisfactory manner than we can at present we might learn a lot more about the land use and environment of early man. I feel quite sure that the key to many of the problems concerning the early agricultural history of the Chalk rests with an understanding of the changes in distribution and abundance of these five species.

Truncatellina cylindrica, *Abida secale* and *Monacha cartusiana* also belong to the open-country group but are local or rare in their occurrence.

In addition, two species, *Cochlicopa lubricella* and *Hygromia hispida* var. *nana*, although not restricted to open-country habitats, frequently comprise a substantial proportion of their fauna. In a similar category are the four catholic species, *Punctum*, *Euconulus*, *Retinella radiatula* and *Vitrina pellucida* already discussed (Chappell *et al.*, 1971).

These then are the three main ecological groups of land snails which are used in the reconstruction of terrestrial environments. Their classification is based largely on their relationships with the moisture régime and light intensity of the environment, though it will be appreciated that other factors may be operating as well (Chapter 4).

On three sites, Northton, Maxey and Arbury Road, two other groups are present: the **marsh species** and the **freshwater slum species**. Their occurrence too is controlled largely by the moisture régime of the habitat, though factors such as nutrient status are involved as well.

Fourteen species are peculiar to marshes, occurring neither in freshwater, nor the drier terrestrial habitats. They are:

Catinella arenaria	*V. geyeri*
Succinea oblonga	*V. genesii*
S. putris	*V. moulinsiana*
S. pfeifferi	*V. lilljeborgi*
S. sarsi	*V. angustior*
Columella columella	*Zonitoides nitidus*
Vertigo antivertigo	*Agriolimax laevis*

In addition there are four species which, although technically belonging to the freshwater slum group, are characteristic of marsh faunas as well, viz. *Lymnaea truncatula, Pisidium casertanum, P. personatum* and *P. obtusale. Lymnaea glabra* and *L. palustris*, though strictly speaking freshwater species, also occur in marshes. At the other extreme, we can recognize five terrestrial species which more often than otherwise occur in marsh habitats, viz. *Carychium minimum, Lauria anglica, Vallonia pulchella, Monacha granulata* and *Agriolimax agrestis*; and there are many species which frequently live in marshes but whose typical habitats are more terrestrial, namely *Carychium tridentatum, Cochlicopa lubrica, Columella edentula, Vertigo substriata, V. pygmaea, Arianta arbustorum, Helix nemoralis, Hygromia hispida, Punctum pygmaeum, Euconulus fulvus, Vitrea crystallina, Retinella radiatula, Vitrina pellucida* and *Agriolimax reticulatus*. This list includes a number of species which are characteristic of the Late Weichselian fauna, and which we have already classified as catholic, suggesting that we are dealing here with species which can tolerate poor environments generally but have no special affinities with the marsh habitat in particular.

Marsh habitats may thus contain species belonging to four ecological units:

> Amphibious species such as *Lymnaea truncatula*
> Obligatory marsh species
> Species characteristic of marshes but not confined to them

Terrestrial species having no special affinities with the marsh habitat

Within the obligatory marsh species we may recognize varying degrees of preference for wetness and shade. Thus *Agriolimax laevis, Zonitoides nitidus, Vertigo antivertigo* and *Succinea pfeifferi* live in wetter places than others (Boycott, 1934), and may be associated with *Lymnaea truncatula* and *Pisidium* spp.; and, with the exception of *Agriolimax laevis*, all prefer open habitats, the species of *Succinea* being positively phototropic (Boycott, 1934).

The slum species are essentially a freshwater group with a preference for, or tolerance of, poor water conditions such as obtain in small bodies of water subject to drying, stagnation and considerable temperature variations (Sparks, 1961: 76). They are mentioned here as they frequently occur on archaeological sites on river gravel as at Arbury Road and Maxey (pp. 350 and 346). The main species are *Lymnaea truncatula, Aplexa hypnorum, Planorbis leucostoma, Sphaerium lacustre, Pisidium casertanum, P. personatum* and *P. obtusale*.

There are many other ways in which we could construct an ecological classification of land snails, depending on the kind of information required. For example, temperature is a limiting factor in the distribution of certain species, lime in the distribution of others (Chapter 4). But for the present purpose, the categories described above are the most useful.

We can, however, recognize four other groups, which, although rarely playing an important role in environmental reconstruction, are of sufficiently frequent occurrence to warrant consideration. They are the alien species, the burrowing species, the anthropophobic species and the synanthropic species.

The alien species, which have been introduced into Britain by man, some accidentally, others deliberately, at various times since the Bronze Age are:

Testacella spp.
Fruticicola fruticum
Theba pisana
*Helix aspersa
H. pomatia
Hygromia limbata
H. cinctella
*Monacha cantiana

**Helicella caperata*
**H. gigaxi*
**H. virgata*
H. neglecta
H. elegans
**Cochlicella acuta*
Oxychilus draparnaldi
Vitrina pyrenaica

Those most commonly occurring in modern soil profiles or on archaeological sites of late prehistoric, Roman or medieval age are marked with an asterisk; the details of their ecology and period of introduction into Britain are discussed in Chapter 5. In the histograms they are conveniently placed to the right of the open-country species. With the exception of *Helix pomatia* which has occasionally been recorded from Roman and later sites (e.g. Cain *in* Brodribb *et al.*, 1968–1972), the others have never been recorded with certainty in fossil contexts from Post-glacial archaeological sites, and are unlikely to turn up in the future.

Burrowing species include *Pomatias elegans* and *Cecilioides acicula*. *P. elegans* is largely subterranean, frequenting loose, friable earth into which it can burrow—hence its abundance in scrub habitats. It does not appear to burrow in excess of a few centimetres below the surface, however, and in subfossil assemblages is rarely out of context with the associated fauna. *Cecilioides acicula* on the other hand extends to depths of over 2 m, and many of the shells extracted from ancient deposits are quite fresh in appearance, frequently containing the living animal (p. 168; Fig. 55). Many snail species probably burrow some way into the soil, particularly when the surface becomes excessively dry or frozen. Slugs have a diurnal rhythm, concealing themselves in the soil during the day and moving around above ground at night. Biennial species may burrow to hibernate, *Helix aspersa* and *H. pomatia* for example spending over half the year in hibernation. In the case of these species it is important that fossil records are based on well-fossilized shell fragments and juveniles; adult individuals alone may be of modern origin.

Anthropophobic species are those which shun the presence of man and are confined to wild habitats approaching the conditions of primeval woodland. They include *Lauria anglica*, *Ena montana*, *Helicodonta obvoluta* and *Limax tenellus*, though in the past there is some evidence—for example the persistent occurrence of *Ena montana* on archaeological sites—to suppose that they may not have been so restricted (Kerney, 1968a). At the other extreme are the **synanthropic species** which live almost exclusively in man-made habitats such as hedgerows, gardens and refuse heaps, rarely occurring in the wild. Examples are *Hygromia striolata*, *Helix aspersa* and *Limax flavus*. These species are not simply tolerating man's influence: they depend on him for their very existence. But as with the anthropophobes there is evidence that synanthropic snails can occur in wild conditions if other environmental factors such as overall climate, are congenial. Thus *Hygromia striolata* during the Atlantic and Sub-boreal periods was a species of woodland; and today, *Helix aspersa* is known from semi-natural habitats in the extreme south-west of Britain. Thus although from man's point of view the distinction between anthropophoby and synanthropy appears clear cut in any one place or at any one time, this does not apply over the country as a whole nor did it apply in the past. It is important that these changes in habitat preferences be appreciated for they have an important bearing on the interpretation of subfossil assemblages, especially from archaeological sites where man was rife (p. 129).

These groups can now be summarized as follows:

Freshwater slum species
Marsh species
Terrestrial species

"Woodland" or shade-loving species

Zonitidae
Carychium tridentatum
Discus rotundatus
Other shade-loving species (including some catholic species such as *Retinella radiatula*)

Pomatias elegans
Intermediate or catholic species

Cochlicopa
Arianta, Helix (Cepaea) spp.
Hygromia hispida
Limacidae

Open-country species

Vertigo pygmaea
Pupilla muscorum
Vallonia costata
V. excentrica
Helicella itala

Cutting across this grouping we have:

Alien species

7
Soils

From the genetic point of view three main groups of calcareous soils and sediments can be recognized. The first comprises soils which form *in situ* by weathering of the parent material—soils in the strict sense. The second group comprises colluvial deposits or sediments, which form by accumulation of material by sedimentation. Included here are deposits such as hillwash which form in terrestrial situations by down-slope movement (sub-aerial processes), water-lain sediments and wind-lain or aeolian deposits such as loess. Intermediate between these two is a third group comprising colluvial deposits which are weathered as they accumulate; these are termed colluvial soils.

In practice the distinction between these groups is often arbitrary and difficult to define. For example, a colluvial deposit or soil may be derived from soil material originally formed by *in situ* weathering; or a soil *sensu stricto* may be formed in colluvial material. Almost inevitably, a colluvial deposit will be subjected to chemical weathering and converted into a soil, once accumulation has ceased.

The multiple origin of many soils and sediments has an important bearing on the interpretation of their contained shell assemblages, and from this point of view it is absolutely essential that these origins are clearly understood.

In this and the next chapter we will consider soils formed *in situ*. Colluvial soils and sediments are dealt with in Chapters 9 and 10.

The Incorporation of Shells into Calcareous Soils

The type of soil which most commonly forms on chalk and other highly calcareous parent materials is a *rendsina*, a soil whose main characteristics are a high pH (7·5–8·0) and high calcium carbonate

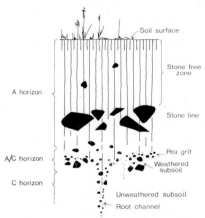

Figure 60. Rendsina soil profile under long-standing grassland. (Depth from surface to base of root channel is 50 cm)

content throughout the profile. Under woodland or permanent grassland, a rendsina shows some or all of the following features (Figs 60 and 61):

(1) A stone-free, mull humus horizon up to 20 cm or exceptionally 25–30 cm thick. This is the A horizon or turf line, and its stone-free nature is due to the sorting action of earthworms.

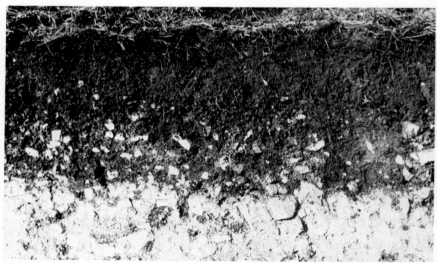

Figure 61. Rendsina soil on Upper Chalk, under long-standing grassland. (Depth of stone-free zone, 20 cm)

(2) A layer of flints or limestone lumps at the base of the turf line. When exposed in plan by stripping off the turf this often appears as a substantial and unbroken layer, which if of flint may be mistaken for an artificial "floor" or cobbling.
(3) A transition zone of brashy rubble and humus penetrating the subsoil as tapering pockets—natural fissures in the rock or cracks induced and widened by root action. This is the A/C horizon.
(4) A zone of split pea or pea grit. This forms in the A/C horizon and comprises a mass of small stones in the size range 2·0–5·0 mm, brought down through the soil by earthworms to line their aestivation/hibernation chambers (Atkinson, 1957).
(5) The unweathered subsoil—chalk, limestone or some form of highly calcareous drift; e.g. coombe rock. Rendsina soils also form on wind-blown shell sand, fine marls, tufa, and other highly calcareous sediments of secondary origin.

The formation of a rendsina soil takes place mainly through the dissolution of calcium carbonate by carbonic acid (carbon dioxide dissolved in rainwater) and organic acids released from dead organic material in the soil. The acid-insoluble fraction of the parent material provides the mineral component—largely silt and clay (Avery, 1964: 91, 157; Perrin, 1956)—while calcium carbonate in solution combines with degraded organic material to form a dark-coloured, base-saturated humus which is strongly bound to the clay fraction forming granular aggregates, or soil crumbs. Such base-rich soils are, if sufficiently moisture-retentive, very favourable to earthworm activity, and where these animals are abundant the A horizon of a rendsina may consist almost entirely of worm castings (Avery, 1964; Cornwall, 1958: 51). Worm-action is responsible for the generally stone-free nature of the mull-humus horizon under woodland or semi-natural grassland. The activities of worms in bringing about the burial of stones and other objects lying on the soil surface were originally discussed and recorded by Darwin (1881) and recently, with particular reference to archaeological sites, by Atkinson (1957) and Cornwall (1958). Two species, *Allolobophora longa* and *A. nocturna*, eject their castings onto the surface of the soil. By this process, aided by the collapse of worm burrows within the soil from which the castings have come, all objects too large to pass through a worm's gut (greater than 2·0 mm; p. 212) are gradually worked down through the soil and buried.

There is little doubt that worm action is also the main agency responsible for incorporating snail shells into a soil, though other processes may be involved to a lesser degree. For example, moles, ants and other burrowing animals may be of importance locally. Shells may fall down cracks in the soil opened up in conditions of exceptional dryness, while some species, during periods of adverse weather, burrow into the soil where they may become entombed. Shells may become incorporated into the replacement material of decayed tree roots or uprooted trees (p. 219), for the hollows created by these provide congenial habitats for many species. But all these processes are patchy in their occurrence and uncertain in their effect by comparison with the overall and relentless activities of worms.

As shells are wormed-down through the soil, some are destroyed by leaching and perhaps by the worms themselves (but cf. p. 212). But a number survive to reach the A/C horizon where they may remain for some time, though not indefinitely, for once its periostracum has been lost, a shell is susceptible to the same weathering processes as any other fragment of calcium carbonate in the soil and is eventually destroyed. In theory, this implies that a gradient of shell numbers will become established in a soil, with the greatest number at the surface and the least at the base. (This I have called a positive gradient.) The exact form of the gradient will depend on factors such as the size of the worm population and the rate and intensity of shell destruction by leaching. The decrease in the rate of down-worming, which necessarily occurs with depth, is perhaps an opposing influence, resulting in a build-up of shells at the base of the A horizon in extreme cases. (The latter situation I have called a negative gradient.) In practice, out of nine sites, only three show a strong positive gradient in the worm-sorted zone. Of the others, four show a weakly positive or negative gradient, or no gradient at all; and two, a strong negative gradient.

Even allowing for the fact that shells can become re-incorporated from the lower into the upper levels of a soil by earthworm mixing, it is clear that the shells at the top are generally younger than those at the base; i.e. there is an age gradient in the soil. Thus any changes in the snail fauna which occur during life are preserved for a time in the soil, and if during this time the soil is buried, resulting in the inactivation of the earthworm population, no further movement of shells occurs, and a stratified sequence reflecting changes which have taken place at the surface is preserved.

Stone-free A horizons of modern and buried soils from which two or more samples have been analysed for shells

SITE	GRADIENT OF SHELL NUMBERS	FAUNAL CHANGE
MS IV (Fig. 76b)	Strong positive	Present
West Kennet (Fig. 91b)	Strong positive	Present
Durrington Walls, DW I (Wainwright et al., 1971: 330)	Strong positive	Present
Windmill Hill, (Fig. 82; 0–12·5 cm)	Strong negative	Absent
Avebury (Fig. 96)	Strong negative	Present
MS I (Fig. 76a)	Weak/Absent	Present
MS II (Fig. 77a)	Weak/Absent	Absent
MS III (Fig. 77b)	Weak/Absent	Present
Pink Hill, (Fig. 116; 0–20 cm)	Weak/Absent	Absent

There appears to be little correlation between the presence of a gradient of numbers in a soil and the preservation of a pattern of faunal change. A strong negative gradient as occurred at Avebury and Windmill Hill is possibly due to incipient decalcification of the upper few centimetres of the soil. A strong positive gradient is probably due to the rapid incorporation of shells into the soil by worm action. Where no gradient is present, the two processes probably balance each other in maintaining a constant level of abundance in the soil; alternatively there may be rapid earthworm mixing of shells within the A horizon.

Only at Pink Hill and MS II, where there is no gradient either of abundance or faunal change, can we envisage complete earthworm mixing of the A horizon—two out of nine sites—and even with these, other factors may be responsible for the absence of gradients as we have discussed. From the environmental point of view this has an important bearing on the analysis of chalk soils. Previously it has been held that the A horizon of a rendsina is completely mixed by earthworms (e.g. Dimbleby *in* Smith, I. F., 1965b: 36). That this is not the case, and that a sequence of environmental change can be preserved in the A horizon, implies that we cannot treat such horizons as uniform entities when sampling for molluscan analysis. When possible, they must be subdivided, even though no stratification is visibly apparent in the profile.

So far only the worm-sorted mull-humus horizon has been considered. Below this level, shells are generally fewer in number, and their mode of incorporation is more problematical. Again, however, worms play an important part. Frequently there is an abundance of shells in the A/C horizon below the stone-free zone. Some of these undoubtably constitute material moved down passively through the profile by the processes of worm-sorting which lead to the formation of the stone-free zone. But others are more actively incorporated, and are often to be seen in soil sections lining earthworm tunnels and their terminal aestivation chambers. These are commonly sorted in the 2·0–5·0 mm range, and are shells which have been seized by worms and carried down well below the zone of passive earthworm sorting (Fig. 80). This is the pea-grit zone. In the modern soil at Windmill Hill (Fig. 82; 0–38 cm), where the distinction between the turf-line and A/C horizon was particularly well-marked, the shells were counted on a size basis. They were found to conform to the distribution of chalk fragments as estimated visually and which was in fact the basis for the division between the A and A/C horizons. Thus active incorporation of shells into the A/C horizon is probably taking place, and may also distort the ecological picture by sorting the larger from the smaller species.

A number of the processes which we have been discussing may give rise to a stratified sequence of shells which is not related to, and indeed may be a distortion of, the pattern of ecological change. Passive and active earthworm sorting are two of these. Thus analysis of worm casts collected from the surface of a grass sward over a chalk soil showed that shells and shell fragments up to 2·0 mm in diameter, but no larger, can pass unharmed through a worm's gut. In a sample of 320 g of earthworm casts, eleven apical fragments were counted. Active earthworm sorting may take place as in the case of the modern soils at Windmill Hill (Fig. 82) and Coombe Hole (Fig. 80). Both these processes lead to the segregation of shells less than 2·0 mm from those larger than this in the A and A/C horizons respectively, but only exceptionally, as at Windmill Hill, does this distort the ecological picture, since most of the shells in a soil are in any case less than 2·0 mm in diameter.

Artificial (non-ecological) stratification may also be caused by the differential preservation of shells of certain species whose apices are tougher and more resistant to destruction than most. Some may even accrete calcium carbonate and thus acquire still greater resistant properties (p. 23). *Pomatias elegans, Cepaea, Arianta*, the Clausiliidae and the Limacidae are the more important groups whose apices become concentrated towards the base of the soil. The phenomenon has also been recognized in *Cochlicopa*, *Acanthinula* and *Vitrina*. In *Acanthinula aculeata*, small granules on the inside of the shell are a characteristic feature of well-worn examples (Fig. 23). In profiles MS I and MS IV, where these worn and pitted apices were counted, their concentration at the base of the profiles is apparent, amounting to about 20% of the fauna (Fig. 76). The abundance of woodland species in the lower levels of the soil at West Kennet (Fig. 91b) was also in part due to the presence of worn apices of *Clausilia* and *Pomatias elegans*. A similar phenomenon has been noticed in ploughwash horizons, for example at Badbury Earthwork (Fig. 134a) and Pink Hill (Fig. 116), where *Clausilia* and *Pomatias* are represented solely by worn apices.

Serious problems of interpretation arise when the parent material of a soil contains shells. Such is the case when a soil forms in any of the deposits classed as colluvial (Chapters 9 and 10)—the most frequent being periglacial sediments such as loess or brickearth, and the secondary infill of ditches. The shells in these sediments are weathered and destroyed as is any other lump of limestone in the parent material; they are thus absent from the upper, mature horizons of a soil. But at the junction with the parent material (A/C horizon), where active weathering of fresh subsoil takes place, there is always the chance of shells becoming incorporated into the lowest levels of the soil, particularly where decalcification is incomplete and the transition between the two horizons gradual. The dangers of snail analysis giving misleading results in this kind of situation are manifest. Thus while the fauna of the unweathered parent material on the one hand and that of the mature soil on the other may be distinct, each a reflection of a specific former environment, the fauna in the transition zone will be a mixed one containing elements from both horizons—an artificial assemblage brought about by the processes of *in situ* soil formation from a subsoil already containing shells, and not one reflecting a former environment at all. For example, this arises when a Post-glacial soil forms in a deposit of periglacial origin. The latter usually contains an open-country fauna, while the overlying soil may contain one of woodland type. Analysis of a series of samples through the deposit and soil would suggest an open landscape gradually being invaded by forest. While this may have taken place at some stage between these two environmental extremes, what is actually being recorded is not an environmental but a stratigraphical transition caused by partial weathering of colluvial deposits in the A/C horizon of the soil.

In some ways the problem is an academic one in that it is generally a simple matter to check the presence of shells in the parent material and then, in the subsequent sampling, to avoid the transition zone. But there is always the feeling that in so doing, critical information about the environment during the early stages of soil formation is being missed. This problem has not been satisfactorily solved (p. 222).

Variations on the Basic Rendsina Profile

There are a number of variations on the basic rendsina profile which can be readily recognized in the field. These, and their shell faunas, can be used to give some indication of the environmental history of the soil.

Plough soils. First of all, the effect of ploughing: ploughing destroys the stone-free zone and, if deep enough, the stone line also, producing a uniformly stony soil. On land which has been heavily ploughed for some years, the A/C horizon is eventually destroyed, when the junction with the subsoil becomes sharp and level. This is well brought out in Fig. 62 which shows a modern and Neolithic plough soil in the same section. At the base of a plough soil, excavation may reveal the marks left by ploughing, scored into the subsoil surface (Fig. 63) (Fowler and Evans, 1967). Ploughing and cultivation, if continued for many years, result in the loss of organic matter and breakdown of the soil structure which, if the soil is on a slope, may lead to hillwashing, or, if on level ground, to wind erosion; both these processes are discussed in connection with the formation of ploughwash (p. 282).

Cultivation or disturbance on a less drastic scale may result in a crude stratification within the soil, as was recognized at South Street (Figs 62, 89), due to the lateral transport of stones in the profile.

A plough soil which has been abandoned soon becomes grassed over. Earthworms become active and a thin stone-free horizon quickly forms. This is apparent in the ancient soil beneath the long barrow mound at South Street (Fig. 62), in the ancient turves at Silbury Hill (Fig. 92) and in the modern profile MS V, where a split-pea horizon has also formed at the base of the soil (Fig. 77). On the other hand a plough soil which is being actively cultivated shows no such horizon as the photographs of the modern soils at South Street and Beckhampton Road show (Figs 62 and 84).

Figure 62. South Street Long Barrow. Section through mound showing, successively, the modern plough soil, the barrow mound, the Neolithic plough soil and a periglacial involution. (Depth of section from the modern surface to the base of the Neolithic soil, 60 cm)

Figure 63. Cherhill, Wilts. Ploughmarks associated with a system of ridge-and-furrow preserved beneath a thin layer of ploughwash. (Scale 1·0 m)

Chalk- and limestone-heath soils. Under certain conditions, the A horizon may become decalcified, even though calcareous material may be no more than 20 cm below the surface. This is the case on the harder limestones, such as the Carboniferous, which are less subject to fragmentation than is the Chalk, and particularly so when under permanent grassland with no disturbance of the parent material, e.g. by tillage or tree-root penetration (Fig. 64). Such soils are devoid of shells.

But even on the Chalk, non-calcareous soils may occur, and in many instances it is almost certain that this is due to the presence of a layer of siliceous drift (probably of loessic origin) so thin as to have been entirely incorporated into the A horizon, and not apparent by macroscopic inspection of the profile in the field (Perrin, 1956). The vegetation of such soils is, by virtue of its curious mixture of calcicole and calcifuge species, known as "chalk heath". These soils, too, are devoid of shells.

Sols lessivés. A *sol lessivé* is a variant of the brown earth, often developed on calcareous drift such as coombe rock (p. 289), in which the entire profile, down to the parent material, is non-calcareous. There

Figure 64. Butcombe, Somerset. Modern soil on Carboniferous Limestone. Although limestone is close to the surface, the mull humus horizon of the soil is decalcified and devoid of shells.

is a pronounced clay-illuviation horizon, consisting of particles which have been washed downwards through the profile (Bt horizon), whose junction with the subsoil is usually sharp (Figs 65 and 69). Often the base of this type of soil comprises a series of irregular, clay-filled pockets which taper down into the subsoil, and which are probably the casts of former tree roots (Fig. 97). These are of widespread occurrence on the Chalk, and on archaeological sites are sometimes confused with stakeholes or other artificial features, particularly when coinciding with pockets of periglacial drift (Figs 69 and 97). Like chalk-heath soils, *sols lessivés* are totally devoid of shells. The type has been introduced here because it is extremely common on calcareous parent materials, often occurring, as at Beckhampton Road (p. 248), in close juxtaposition to a rendsina profile. The formation of *sols lessivés* is further discussed on pages 277 and 287.

Figure 65. Chalfont St. Giles, Bucks. *Sol lessivé* profile showing a strongly developed Bt horizon, overlain by hillwash. (Depth of section, 3·0 m)

Figure 66. Pitstone, Bucks. Section through a subsoil hollow overlain by ploughwash. (Evans, 1966b: North-east section.)

Subsoil hollows. At the base of some soils, hollows, filled with humic material, are sometimes present. These are irregular in plan and section, and frequently undercut the unweathered subsoil (Figs 66, 87 and 95). They vary in width and depth from a few centimetres to 2 m or occasionally more. They may be broad and shallow, narrow and tapering or include a combination of these forms. Unlike normal rootholes at the base of a soil, their fill is *discontinuous* with the overlying A horizon, showing a more or less sharp break; i.e. they are not part of the active weathering profile. This is an important criterion in their recognition. Examples occurred at Coombe Hole (Fig. 80), Ascott-under-Wychwood (Fig. 87), South Street and Avebury (Fig. 95), and they have been termed subsoil hollows.

A characteristic feature is that their fill is stratified, and this is manifest both in its gross morphology and its snail fauna. For example, at Ascott-under-Wychwood and Avebury, the lower part of the fill comprised a compact, pale stony loam while above was a more strongly humic deposit. The fauna from these two levels differed significantly in both cases; and in both cases the differences showed parallel trends, suggesting a transition from open-canopy woodland to densely shaded forest. At Ascott, there was further stratification in that the upper levels of the fill of the subsoil hollow were decalcified and devoid of shells. These facts suggest that infilling took place gradually, though against this it should be noted that the fill often contains a mass of small grit perhaps indicating that physical weathering played a substantial part. Another characteristic of their fill is the persistent presence of charcoal fragments. This feature was recorded by Iverson (1964) from subsoil hollows at the base of acid soils in Jutland.

What these features are is not entirely clear, although it seems certain that they were once hollows which have become infilled by the normal processes of physical weathering, rather than features formed by chemical weathering *in situ*, i.e. they are not solution hollows. Perhaps the most reasonable hypothesis is that they were formed originally by tree roots and that their fill is the cast of a former tree-root system. We perhaps tend to forget the vast size and ramifications of roots, and two exposed examples are shown in Figs 67 and 68.

Alternatively, the hollows, though not the fill, may be of periglacial origin, for we can hardly envisage the land surface at the end of the Late Weichselian as being entirely flat. Johnson (1959)

Figure 67. North-east Yorkshire. Exposed tree-root system.

Figure 68. Wychwood Forest, Oxon. Exposed tree-root system.

discusses the distinction between rootholes and periglacial features, such as ice-wedge casts, in Sweden.

Periglacial features. Today, many calcareous soils are developed on solid rock. But soils buried at an earlier period in the Post-glacial and which have not been as severely weathered by the processes of ploughing, hillwashing and solution (p. 32) are almost invariably developed either on some form of drift deposit of periglacial origin, or on a geological solid which shows traces of cryoturbation (frost-heaving) or solifluxion (Evans, 1968b). This may take the form of

Figure 69. Hungerford, Berks. Periglacial involutions overlain by the Bt horizon of a *sol lessivé*.

involutions (Fig. 69) as at South Street (Fig. 62), Avebury (Fig. 94) and Marden (Fig. 97), frost-heaving structures as at Ascott-under-Wychwood or discrete layers of sediment as at Beckhampton Road. With the exception of Ascott, there was at each of these sites an associated land-snail fauna of periglacial type.

Periglacial structures can provide a useful guide to the depth of mechanical disturbance in the overlying soil through the activities of man. Thus at Ascott-under-Wychwood small groups of pitched stones extended intact almost to the soil surface indicating the complete absence of ploughing; and at Beckhampton Road (Fig. 84) cones of chalk marl reached close to the surface. At South Street on the other hand, involutions had been truncated at the *base* of the soil by Neolithic ploughing (Fig. 62), a conclusion which was borne out by the presence of criss-cross ploughmarks on the subsoil surface.

The relevance of periglacial deposits as a whole to archaeology and the stratification of soils has been discussed elsewhere (Evans, 1968b). In connection with the analysis of shells from calcareous soils, the main point is that periglacial deposits frequently constitute the parent material from which these soils developed, and it is well to recognize them when they occur for, as has indeed been found, they may contain shells. With unweathered periglacial deposits—pure white marl or golden brickearth—there is no problem for their fauna is generally uncontaminated and readily recognizable on account of its characteristic and restricted facies (p. 355). But in the transition zone with the overlying soil (A/C horizon), as discussed above (p. 214), it is not always so easy to distinguish between shells which derive from the periglacial sediments and those which belong to the soil profile and are of Post-glacial origin.

Fossil features. Burial of a soil causes compression, but it is uncertain to what extent this takes place. Estimates of up to 60% have been considered likely for the A horizon of a rendsina (Evans *in* Wainwright and Longworth, 1971: 336). Wastage of organic matter is also an important factor, perhaps more so than compression, in reducing the thickness of a buried soil.

Another feature of fossil rendsina soils is the movement of iron and

Figure 70. Pitstone, Bucks. Section through Late Weichselian deposits. Note the pronounced zone of leaching immediately below the Allerød soil, due to the removal of iron, probably after burial. (Scale, 30 cm)

the formation of thin iron "pan" zones as at Avebury (Fig. 95) and in the Allerød soil at Pitstone (Evans, 1966b) (Fig. 70). The significance of this phenomenon is unknown, but may be connected with areas of high organic content. Burial kills the organic matter in a soil and on its surface, and its oxidation by bacteria quickly uses up all available oxygen in the environment. Reducing conditions are thus created and under such conditions iron becomes mobile (Cornwall, 1958: 193). At Avebury, Ascott-under-Wychwood and Pitstone there is considerable variation in the degree of iron movement from place to place along the profile, and it would be useful to know if this reflects variation in some aspect of the environment, such as vegetation, prior to burial. Are we, for example, dealing with different types of grasses—tussock-forming and turf-forming—or patches of vegetation on a generally bare soil surface, or even, in the case of Avebury, with cowpats?

Fossil Rendsina Soils

Fossil rendsina soils of various ages have been recorded from many situations in southern Britain, and have given us much of our information about the terrestrial environments of the Late Weichselian and Post-glacial periods. There are four main groups, defined on their age and mode of burial: Late Weichselian soils buried beneath solifluxion débris of zone III age; early Post-glacial soils buried beneath calcareous swamp deposits, largely of Atlantic age; soils buried later in the Post-glacial by hillwash and blown sand, generally of Sub-boreal or Sub-atlantic age; and a fourth group comprising soils buried beneath archaeological monuments of Neolithic and later date, or sealed in ditches and other man-made structures.

During the Late Weichselian (p. 352) there were two periods when the climate was sufficiently mild for soil to form. One, the earlier, is known as the Bølling Interstadial and the other as the Allerød (Kerney, 1963). The Bølling soil is generally poorly developed and little more than a thin organic streak; it has been recognized in chalk muds and loess deposits from several sites in Kent (Kerney, 1963; 1965). The Allerød soil on the other hand is altogether more substantial and has been widely recognized in southern England from sites in Somerset (ApSimon et al., 1961), Berkshire (Paterson, 1971), Buckinghamshire (Evans, 1966b), Surrey, Sussex, Kent (Kerney, 1963; 1965; Kerney et al., 1964) and possibly Cambridgeshire (Sparks,

1952). In all cases, these soils are preserved beneath solifluxion débris of zone III age (Fig. 71).

Generally the Allerød soil is a rendsina, with a skeletal character and little sign of segregation of coarser fragments towards the bottom of the profile. There is no sign of decalcification, the soils probably being maintained in a flushed condition by minor frost-heaving and hillwashing. Often the Allerød soil shows a degree of stratification in

Figure 71. Pitstone, Bucks. Section through dry valley deposits of Post-glacial and Late Weichselian age. (Scale, 1·0 m)

its molluscan fauna and lithology indicating that the soil may be in part colluvial. Only at one site, Brean Down in Somerset, where the aspect was southwesterly, was there evidence of iron mobilization and incipient lessivation.

At Pitstone there is evidence within the Allerød period of two phases of soil formation, separated by a minor phase of solifluxion (Evans, 1966b); and Paterson (1971) suggests a similar possibility for an Allerød profile at Cholsey in Berkshire. Several pollen diagrams from

England suggest likewise that the Allerød Interstadial was a dual phenomenon (Smith, A. G., 1965: Fig. 20; Pennington, 1970).

Another characteristic of the Allerød soil is that, except where disturbed by later frost heaving, the base is generally level. This is in contrast to Post-glacial soils in which irregularities due to tree-root penetration of the subsoil are often prominent.

Soils of Boreal age are known from a few sites where they have been buried beneath tufa deposits. At Brook (Kerney *et al.*, 1964) and Cherhill (p. 304; Figs 72 and 109) the soil is calcareous throughout. Molluscan analysis of the Brook soil indicated a marked stratification,

Figure 72. Cherhill, Wilts. Section through Post-glacial deposits showing, successively, modern soil and ploughwash, tufa (pale horizon) and buried soil of Boreal age. (Scale in 5- and 25-cm intervals.)

possibly due to the gradual build up of the soil by the addition of increments to its surface. As with certain Allerød profiles, a soil apparently formed *in situ* by weathering of the subsoil, is, at least in part, of colluvial origin. Given time and the necessary conditions of stability, a soil profile of brown-earth type was able to form on calcareous parent material during the Boreal period. One such is present at Nash Point in South Wales (p. 305) developed on highly calcareous Liassic Limestone, and buried by a tufa deposit. As at Brean Down, this is a site of pronounced south-westerly aspect, a fact which may be responsible for the mature character of the soil (cf. Dalrymple, 1955).

Soils of Atlantic age are preserved within deposits of tufa, where they reflect standstill phases in the accumulation of the deposit. At Cherhill, Huntonbridge and Nash Point (p. 304) where such soils have been recorded, they are highly calcareous throughout. But again it is likely these are largely colluvial. At Ascott-under-Wychwood, for example, there is a decalcified soil profile of Atlantic age, formed *in situ* by weathering of underlying calcareous sediment (Fig. 87). And at Willerby Wold (Cornwall *in* Manby, 1963: 201), Kilham (Fig. 99) and Marden (Fig. 97) the soil profile prior to Neolithic disturbance was probably also of brown-earth type, although in the latter two sites there was a substantial non-calcareous component in the parent material.

It is in fact difficult to be certain about the type of soil prevalent on the Chalk under climax forest of the Atlantic period, mainly because of its destruction by the processes of Neolithic forest clearance and agriculture. The effect of forest clearance alone may have brought about lessivation in brown-earth or rendsina soils by severing the replenishment of bases to the soil in the leaf fall (p. 277), and cultivation of a thin brown earth or *sol lessivé* on chalk will rapidly convert them into a rendsina by creating and maintaining a high calcium carbonate content in the soil. Indeed, on some sites there is probably a delicate balance between the existence of a brown earth, rendsina or *sol lessivé* which is dependent on factors such as the concentration of siliceous matter in the parent material, drainage, aspect and land use.

But at several sites in Wiltshire where the lower part of the soil appears to be undisturbed and where Mollusca indicate a woodland environment, the profile is calcareous; the sites are Windmill Hill, Beckhampton Road, South Street and Avebury (Chapter 8). Nevertheless there is still the possibility that the original upper part of these profiles was decalcified. The only undisturbed soil which we have is part of Ascott-under-Wychwood (p. 255) where there was a brown earth. Being on limestone this soil is not strictly comparable with those developed on chalk but it is noteworthy that snails were present at its base in the transition zone with the subsoil, so that it must have passed through a rendsina stage in its development. On balance I am inclined to the view that the natural soil on the Chalk in southern Britain during the Atlantic period was a rendsina and that although decalcification may have occurred locally it was not a widespread phenomenon.

is to be allowed for the formation of the turf line. In the case of the other three sites, however, periods of over half a millennium are involved, and these in a depth of turf no greater than 7·0 cm. Clearly, both from the archaeological and environmental point of view, the excavation of buried turf lines must be done with very great care indeed.

In a smilar way, archaeological material can be used to date the upper and lower limits of a turf line, or buried soil as a whole. At Marden and Durrington Walls the radiocarbon evidence was neatly supported by the presence of different ceramic styles in the turf line and ditch bottom.

Thus the time span of a shell succession in a soil can be estimated by stratigraphical or archaeological means, or by radiocarbon assay, and occasionally by a combination of the three. Sometimes the snail fauna itself may be a guide. Where a soil is sealed between two deposits or horizons of known age, its active life can be estimated. But the shell sequence within the soil probably reflects only the last few hundred years of this time and not its total history

8
Representative Soil Profiles

In this chapter, the snail analyses of six modern soils and eleven of Neolithic age are described, and their interpretation in environmental terms discussed. A number of the sites have been published previously, and these are described briefly. Others, presented here for the first time, are discussed more fully.

The object of this section is twofold. First, some of the problems and technicalities of the snail analysis of calcareous soils, discussed in general terms in the last chapter, are here illustrated with reference to specific sites. And second, the Neolithic sites are later considered in relation to the problem of the early ecological history of the chalklands as a whole (Chapter 11). In this way it is intended to emphasize the relevance of buried soils in the study of ancient environments on both a local (site) and regional basis.

Modern Soils

Five modern soil profiles (MS I–MS V) were investigated from sites along the Chiltern escarpment between Wendover and Princes Risborough in Buckinghamshire (Fig. 117). The shells were divided into three components—live, normal subfossil and weathered—and the percentages of each plotted on the far right of the histograms. The live component comprises snails which were alive or shells whose periostracum was sufficiently well-preserved as to indicate the very recent death of the animal—almost certainly not more than twelve months prior to sampling. Normal subfossil shells are those which have clearly been in the soil for some time but are yet in a reasonable state of preservation. The weathered component consists of the worn apices of resistant shells such as *Pomatias elegans* which have probably been in the soil for a length of time far in excess of the normal subfossil component.

The sixth soil, in Coombe Hole below Ivinghoe Beacon, is included as an example of the problems created by earthworm aestivation chambers.

PROFILE MS I

Location: South-west side of Pulpit Hill, Great Kimble. SP 832047 (Fig. 117).

Situation and geology: The site is on a steep slope (*ca.* 30°) at the head of a dry valley. The geological solid is Upper Chalk.

Vegetation: Open beech woodland of juniper sere type (p. 88). Leaf litter is present only where it has banked up against tree trunks; otherwise the soil surface is bare or covered with fragments of chalk, flint nodules and snail shells. Notwithstanding this, the top of the soil is firmly bound by tree roots, and there is no evidence that erosion is taking place (Fig. 75).

The soil profile was as follows (Fig. 76a):

Depth below surface (cm)	
0–10	Dark-grey stone-free loam (10 YR 3/1). Worm-sorted-mull humus horizon.
10–30	Dark-grey/brown (10 YR 4/2) chalky loam with chalk lumps throughout; becoming increasingly chalky and paler with depth.
30	Zone of pea grit.
30+	Shattered chalk rock.

The pattern of faunal change (Fig. 76a), suggests that the environment was once more open than at present. In fact, only the fauna from the soil surface itself was of shade-loving type; below this open-country species were predominant.

Today, a number of open-country species are living on the site; these are *Pupilla*, *Vertigo pygmaea*, *Abida secale* and *Helicella itala*. Their occurrence in a woodland habitat is probably due to a number of factors such as the proximity of open ground, the poor moisture-

Figure 75. Beech woodland of juniper sere type, close to profile MS I.

The soil profile was as follows (Figs 77 and 78):

Depth below surface (cm)	
0–23	Stone-free, dark (10 YR 3/1) mull-humus horizon.
23–30	Grey/brown (10 YR 5/2) brashy transition zone with the subsoil, which it penetrates as rootholes to 45 cm.
30–35	Pea-grit zone.
35+	Middle Chalk.

The molluscan fauna shows few changes of significance through the profile (Fig. 77a), probably indicating the present environment of scrub and grassland to have been stable for some time. The abundance of shade-loving species is probably caused as much by the rich herb vegetation and water-retaining capacity of the soil—the turf line is 25 cm thick—as by the presence of scrub. The complementary distribution of *Helicella itala* and *H.* cf. *virgata* through the soil profile has already been commented upon in Chapter 7 (p. 229).

PROFILE MS III

Location: Bacombe Hill, Wendover. SP 863073 (Fig. 117).
Situation and geology: On a gentle slope (*ca.* 5°) on the lower part of the scarp. The geological solid is Upper Chalk.
Vegetation: Tall grasses and thorn scrub of hawthorn sere type.

The soil profile was as follows (Fig. 77b).

Depth below surface (cm)	
0–15	Dark-grey (10 YR 3/1) stone-free loam. Mull-humus horizon, earthworm sorted.
15–38	Chalk rubble, and grey/brown (10 YR 5/2) loam, becoming increasingly rubbly with depth. Between 30 and 38 cm, there is a prominent pea-grit zone.
38+	Upper Chalk.

This profile is similar to Profile MS II, differing in the less well developed mull-humus horizon—15 cm as against 25 cm.

The fauna is of open-country type throughout, though there is perhaps a case to be made for a rudimentary sequence of change, open-country species, particularly *Vallonia excentrica*, being more abundant in the turf line than below. The complementary distribution of *Hellicella itala* and *H.* cf. *virgata* is again marked, as in Profile MS II. Also noteworthy, is the virtual absence of *Vallonia costata* in these hawthorn sere habitats (MS III, MS II and MS IV) (p. 154).

PROFILE MS V

Location: North-west side of Pulpit Hill, Little Kimble. SP 829052 (Fig. 117).

Situation and geology: Gently sloping, almost level ground at the foot of the main escarpment. The geological solid is Middle Chalk.

Vegetation: Short grassland of *Festuca* type. Juniper sere habitat (Fig. 79).

The soil profile was as follows (Fig. 77):

Depth below surface (cm)	
0–6·5	Grey (10 YR 5/1) chalky loam, more or less stone-free. Mull-humus horizon.
6·5–17·5	Grey (10 YR 6/1) chalky loam with chalk lumps throughout. Former plough soil.
17·5–20	Pea-grit zone. Sharp, horizontal junction with the layer below.
20 +	Middle Chalk.

The pale colour of this profile, the sharp junction with the subsoil, its slight depth and poorly developed mull-humus horizon suggest it to be a plough soil of recent origin which has been grassed-over for the past ten years or so.

The snail fauna (Fig. 77c) is of open-country type throughout but shows a stratification in keeping with the soil profile. Thus in the lower levels, shade-loving species are virtually absent (3%) while *Pupilla* and *Helicella itala* predominate. Above, "shade-loving" species—largely *Retinella radiatula* and *Punctum pygmaeum* (cf. p. 195) —increase to 11% and *Vallonia costata* and *Vertigo pygmaea* appear. These changes equate with an increase in the stability of the environment as suggested by the soil profile evidence.

COOMBE HOLE

Location: The Chiltern scarp below Ivinghoe Beacon. SP 966174 (Fig. 117).

8. REPRESENTATIVE SOIL PROFILES

Figure 79. Juniper scrub on the Chiltern escarpment, on the west side of the Risborough gap.

Situation and geology: The profile is in the side of a steep coombe known as Coombe Hole. The geological solid is Lower Chalk, though overlain by a thin layer of coombe rock derived from the Middle Chalk.

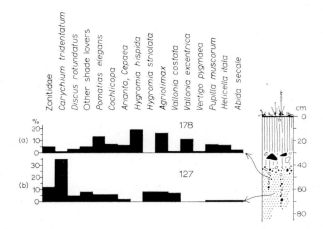

Figure 80. Coombe Hole, modern soil profile. (a) Earthworm burrows. (b) Subsoil hollow.

Vegetation: Tall grasses with some scrub. The profile examined was in the narrow head of the coombe which has almost certainly never been cultivated.

The soil profile was as follows (Fig. 80):

Depth below surface (cm)	
0–30	Dark, stone-free mull-humus horizon.
30–40	Flint and chalk lumps in a loamy matrix; the product of passive worm-sorting.
40–45	Pea-grit zone. Earthworm burrows lined with pea grit penetrate to 75 cm (active earthworm sorting).
45–85	Compact, grey and somewhat gritty calcareous loam. Subsoil hollow, possibly a tree-root cast.
85+	Coombe rock.

Two samples from this profile, one from the pea grit and the burrows below, and the other from the fill of the subsoil hollow (45–85 cm), were analysed (Fig. 80). The fauna from the pea-grit zone is similar to that from Profiles MS II and MS III, and is representative of similar habitat conditions—grassland with some scrub and a thick mull-humus horizon. Neither shade-loving nor open-country species predominate. But the fauna of the subsoil hollow is dominated by shade-loving species (68%) (*Hygromia striolata* is here included as one of these; cf. p. 177). Of the open-country species only *Vallonia costata* is at all well represented (7%) and as discussed above (p. 157) this snail can occur in low abundance in woodland habitats. Also, the state of preservation of the single example of *Pupilla* and the one, non-apical, fragment of *Helicella itala* suggests them to be intrusive. The presence of *Vertigo pusilla* and *Acanthinula lamellata*, both extinct in the area today, suggests that this deposit may be of some considerable age, possibly Atlantic.

The clear-cut distinction between these two faunas illustrates in a dramatic way how active-earthworm transport of shells into the subsoil may contaminate earlier deposits. The age difference between the faunas may be as much as 7000 years.

Neolithic Soils

WINDMILL HILL (Evans, 1966a)

Location: Windmill Hill, 2·2 km north-west of Avebury, Wilts. SU 087715 (Fig. 143).

8. REPRESENTATIVE SOIL PROFILES

Situation and geology: The enclosure is situated on the summit of Windmill Hill, about 30 m above the surrounding plain. The geological solid is Middle Chalk, and is virtually devoid of drift.

Archaeology: A Neolithic causewayed enclosure (Smith, I. F., 1965b), comprising three roughly concentric banks, each with its own external quarry ditch. Charcoal from the buried soil beneath the outer bank yielded a radiocarbon date of 2950 ± 150 BC (BM–73) for the initial occupation of the site; charcoal from the primary fill of the ditch gave

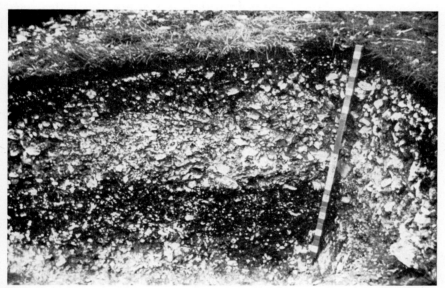

Figure 81. Windmill Hill causewayed enclosure. Section through the bank showing, successively, the modern grassland soil, the bank material and the buried soil. Note the absence from the buried soil of a worm-sorted zone, so striking a feature of the modern profile. Scale in inches. (Photo: G. W. Dimbleby.)

a date of 2570 ± 150 BC (BM–74) which is probably the date of construction of the bank, and hence of the burial of the Neolithic soil.

Vegetation: The south-west section of the enclosure has been levelled by ploughing and is being cultivated at the present day. The remainder, however, is under grass, and it was in this part that the present section was excavated.

By permission of the National Trust and the Department of the Environment, a soil pit, 1·2 m square, was dug in Outer Bank V, 10 m north of the south end of the bank. This revealed the following profile (Figs 81 and 82):

Depth below surface (cm)	
0–12·5	Dark-grey (10 YR 3/1) stone-free loam; mull-humus horizon or turf line.
12·5–38	Sub-angular to rounded chalk lumps in a humic, loamy matrix; shells were visibly abundant, and there was a great quantity of pea grit in this zone. A/C horizon of modern soil.
38–70	Coarse, angular chalk débris, well-compacted in a fine pale-grey matrix. Unweathered bank material.
70–95	Dark-grey (10 YR 4/1) loam with numerous chalk lumps throughout. Buried Neolithic soil. It was from this horizon that the radiocarbon date BM–73 was obtained.
95+	Shattered chalk.

A series of samples from this section was analysed for shells, and the results presented as a histogram of relative abundance (Fig. 82). As there were so many large shells in the pea-grit zone of the modern soil (12·5–38 cm) in contrast to their paucity in the turf line, their sizes were measured in three groups and plotted on the extreme right of the histogram.

In the fauna from the buried soil (70–95 cm), shade-loving species predominate. Open-country species amount to only 21% of which *Vallonia costata* comprises two thirds. This suggests an environment of open woodland, curious in view of the archaeological evidence for occupation of this site prior to the construction of the enclosure. 800 m away, the unequivocable open-country fauna throughout the buried soil beneath the Horslip Long Barrow (p. 261), indicates that forest clearance had taken place in the vicinity, and, as indicated by radiocarbon assay, by 3200 BC.

There are four possible explanations for this anomaly.

(1) The site was occupied but not cleared of woodland. The causewayed enclosure was built in woodland.
(2) The site was cleared, occupied and then totally abandoned, woodland being allowed to regenerate. The enclosure was built in secondary woodland.

8. REPRESENTATIVE SOIL PROFILES

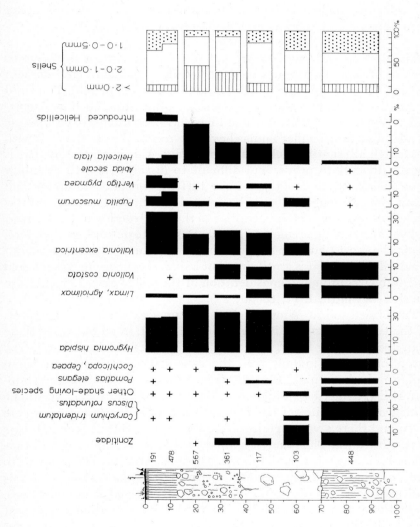

Figure 82. Windmill Hill. Modern soil, bank and buried soil.

(3) The site was cleared and occupied, and the fauna is a synanthropic one whose composition is controlled more by the presence of man directly than through his effect on the forest (p. 201).

(4) The site was cleared and occupied, but before the enclosure was built, the soil was de-turfed. This would account for the absence of an open-country fauna and the stony nature of the soil.

The first two alternatives are not satisfactory since a mull-humus horizon of some sort would be expected in a woodland environment just as in a grassland one (cf. Profiles MS I and IV, p. 233). The third alternative is unlikely, largely because at other sites, such as Ascott-under-Wychwood (p. 251), where there has been intensive occupation on a land surface, similar faunas have not arisen. We are thus left with the fourth alternative, deliberate de-turfing. This is a practice that is always difficult to prove, and difficult to distinguish with certainty from incidental destruction of the turf by constant trampling and occupation, although it should be pointed out that at Ascott there was intensive occupation of the site prior to the construction of the barrow, and yet the turf line remained intact. Deliberate de-turfing, therefore, seems the most plausible explanation of the fauna and soil profile at Windmill Hill, and it is gratifying that a similar conclusion was reached by Dimbleby through a consideration of the pollen and charcoal from the site (*in* Smith, I. F., 1965b: 34).

The buried soil at Windmill Hill is anomalous in another respect. Of the twenty or so Neolithic sites investigated, in all but three there is an open-country fauna in the surface layers of the buried soil. The exceptions are Windmill Hill, the causewayed enclosure on Knap Hill (Sparks *in* Connah, 1965:19) (Fig. 143) and the barrow on Whiteleaf Hill (Kennard *in* Childe and Smith, 1954:230) (Fig. 117). The evidence from Whiteleaf Hill is unsatisfactory in that no numbers are given and there are no details of the soil profile. But at Knap Hill we have more data. The fauna is similar to that at Windmill Hill in that shade-loving species predominate. Open-country species amount to 15%, of which *Vallonia costata* constitutes about half. Pollen analysis indicated hazel woodland (G. W. Dimbleby, personal communication). Here there was not the intensive pre-bank occupation present at Windmill Hill, nor was the surface of the soil stripped. Photographs of the buried soil in the excavation report and others shown to me by Professor G. W. Dimbleby indicate that there was a distinct mull-humus horizon intact on burial. It is perhaps fortuitous that the two causewayed enclosures should prove anomalous, but it is just as likely that their siting was in some way special. In any future excavation of a causewayed enclosure, a rigorous study of the buried soil should be made, which might provide some clue to the purpose of these enigmatic monuments.

Rather surprisingly, the bank material at Windmill Hill contains shells (38–70 cm). The assemblage is largely an open-country one, and possibly derives from the turf horizon of the ancient soil which would naturally be the first material to be incorporated into the bank. This would imply that on the site of the bank de-turfing took place before its construction while on the site of the ditch, from which the bank material derived, no such process occurred—a somewhat unlikely state of affairs. Another possibility, which was pointed out by Atkinson (1957) in the case of Stonehenge, is that the ancient turf line was *in situ* when the bank was built but has since been incorporated into the bank by worm-action. However, it is unlikely that this has happened here, in view of the depth of overlying bank material (60–70 cm). At Stonehenge, Atkinson cites 30 cm as the critical thickness above which destruction of the buried soil does not occur, and in a number of cases I have seen buried soil profiles well-preserved under only 10 cm of deposit. But against this, it should be pointed out that in the Overton Down Experimental Earthwork (Jewell and Dimbleby, 1966:335) earthworms are working upwards from the buried turf line as shown by the redistribution of *Lycopodium* spores, even though sealed by over 2 m of chalk, and there is no obvious reason why the same should not apply to snail shells.

The modern soil at Windmill Hill shows some interesting features. In the turf line (0–12·5 cm), the fauna is a sparse one of open-country type, with *Vallonia excentrica* and *Hygromia hispida* predominant. Below, in the pea-grit zone, the fauna is similarly one of open country but shows certain significant differences from that of today. *Helicella itala* is well-represented and *Vallonia costata* is present, but *V. excentrica* is at only half its present-day abundance. A variety of factors may be responsible for these differences. In the first place, the pea-grit zone was probably once unweathered bank material and as such may have had its own faunal element; this may account for the presence of *V. costata* and *Helicella itala*. Secondly, within the pea-grit zone there may be an element of a fauna which inhabited the bank prior to the present one, i.e. we may be observing a straightforward pattern of ecological change, and the absence of *H. caperata* and *H.*

virgata would support this. Lastly, the action of worms in causing a size gradient of shells between the turf line and the pea-grit zone may have been responsible for a certain amount of separation of a homogeneous fauna into artificial elements. For example, the abundance of *Hygromia hispida* and *Helicella itala* in the upper levels of the pea-grit (12·5–25 cm) may be due to this factor (p. 212).

BECKHAMPTON ROAD

Location: 2·5 km south-west of Beckhampton, Wilts. SU 066677 (Fig. 143).

Situation and geology: The barrow is inconspicuously sited in the bottom of a broad dipslope dry valley. The geological solid is Lower

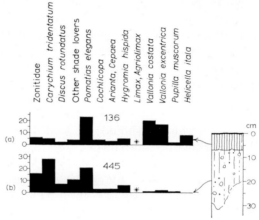

Figure 83. Beckhampton Road, buried soil. (a) Sample i, turf line. (b) Sample iv, A/C horizon. *=present but not counted.

Chalk, but this is overlain by coombe rock and a variety of other drift materials of periglacial origin (Evans, 1968b), mainly chalk marl and a quartzose brickearth.

Archaeology: The site is a Neolithic long barrow (Smith and Evans, 1968). Radiocarbon assay of charcoal from the turf line of the buried soil gave a date of 3250 ± 160 BC (NPL–138); and of antler from the mound, a date of 2517 ± 90 BC (BM–506b) probably dating the burial of the soil.

The soil beneath the barrow showed two kinds of profile which were related to the nature of the subsoil. Where this comprised a highly calcareous marl or gravel, a rendsina profile was present. On the less calcareous quartzose brickearth, there was a soil of brown-earth (*sol lessivé*) type, more or less decalcified and devoid of shells.

On the chalk marl, the profile consisted of an upper, stone-free horizon or turf line of dark-grey (10 YR 4/2) chalky loam, *ca.* 5–7 cm thick, and a lower, more chalky horizon which in places penetrated the subsoil to depths of 30 cm or more (Figs 84 and 85).

The marl yielded a fauna of typical Late Weichselian facies in which *Pupilla muscorum* and *Vallonia pulchella* were the dominant elements (Evans, 1968b:15; Fig. 5).

Figure 84. Beckhampton Road Long Barrow. Transverse section showing, successively, the modern plough soil, the barrow mound and the buried soil. (Scale, 1 ft.)

Four spot samples were taken at various places from the buried soil and analysed for land snails (Fig. 83). Three of the four assemblages (ii, iii and iv) represent a shaded environment, with *Pomatias elegans*, *Carychium tridentatum*, *Discus rotundatus* and various zonitids predominant. The other assemblage (i) reflects an open-country landscape with *Helicella itala*, *Vallonia costata*, *V. excentrica* and *Pomatias elegans* the important elements.

As this profile was the first to be investigated, before it had been realized that a stratified sequence of shells could be preserved in a soil, the exact provenance of the samples in relation to the two horizons of the profile was not recorded. However, in the light of further work on buried soils of this type (and in particular, the closely similar profile at Ascott-under-Wychwood), a reasonable interpretation of these assemblages might be made as follows.

Assemblage (i) comes from the mull-humus horizon (turf line), while assemblages (ii), (iii) and (iv) are from the underlying, more chalky part of the profile. Such a situation is in accord with the soil profile evidence and broadly similar to that at Ascott (p. 252). The main bulk of the soil (below the turf line) has not been mechanically

Figure 85. Beckhampton Road Long Barrow. Detail of buried soil. (Scale, 1 ft.)

disturbed by ploughing or other forms of tillage, for it shows no incipient layering such as was present at South Street (Fig. 62), and unweathered cones of subsoil occur high in the profile, reaching upwards to the base of the turf line (Fig. 84). One would thus expect the fauna of this horizon to reflect an undisturbed habitat.

If this interpretation is correct, the following sequence of events can be suggested. There was originally an open, tundra landscape during which time a variety of drift deposits of periglacial origin were laid down (Evans, 1968b). Later, during the Post-glacial, the site became forested and soil formation took place. A snail fauna of woodland type superseded the original sub-arctic one. Then the forest was cleared, probably by man, and radiocarbon assay of charcoal from the turf line of the buried soil suggests a date of 3250 ± 160 BC for this episode. From then on until the construction of the barrow some 700 years later, the site was kept open. During this time, a turf line developed at the surface of the soil, and the site appears not to have been tilled. An open-country land-snail fauna prevailed.

ASCOTT-UNDER-WYCHWOOD

Location: 1·2 km south of Ascott-under-Wychwood, Oxon. SP 299175 (Fig. 142).
Situation and geology: On gently sloping (less than 2°) ground above a spring, on the side of a tributary valley of the River Evenlode. The geological solid is Inferior Oolite limestone, but with extensive areas of calcareous loamy material; cryoturbation (frost-heaving) structures are numerous.
Archaeology: A Neolithic long barrow dated by radiocarbon assay of charcoal on the surface of the buried soil to 2785 ± 70 BC (BM–492).

The pre-barrow soil profile and the land snail fauna from this site have been discussed in detail elsewhere (Evans, 1971b); there will also be a full discussion in the final excavation report. Here only a summary of the main results is given.

The buried soil stood out clearly as a dark humic horizon, 20 to 25 cm thick (Fig. 86); its profile was as follows:

Depth below buried soil surface (cm)	
0–5	Relatively stone-free mull-humus horizon or turf line.
5–*ca.*25	Dark-brown, calcareous loam with numerous limestone fragments.
*ca.*25–*ca.*50	Localized pockets of humic material, possibly the casts of decayed tree roots. Subsoil hollows.
*ca.*25/*ca.*50+	Shattered limestone rubble.

A series of samples from this profile was analysed for land snails, and the results plotted as a histogram of relative abundance (Fig. 88a). The fauna in the subsoil hollow (26–46 cm) represents light woodland;

shade-loving species predominate, but *Vallonia costata* is present at 6% abundance, perhaps indicating some openness in an otherwise shaded habitat. Above, in the main body of the profile (6–26 cm) the fauna is similar, but the decline of *V. costata* and the richness of the fauna in species suggest a closed woodland cover.

Figure 86. Ascott-under-Wychwood Long Barrow. Transverse section showing, successively, the modern grassland soil, the barrow mound, and the buried soil. (Scale, 1·0 m) The extreme thinness of the turf line of the buried soil by comparison with that of the modern soil is probably due to wastage after burial.

In contrast, the turf line contains an open-country fauna, suggesting forest clearance at this level, and the creation of a grassland environment. The large quantity of pot sherds and flint-knapping débris, spreads of charcoal on and in the turf line, and several pits, all suggest that the change in the environment from closed woodland to open country was a direct result of the artificial destruction of an area of forest by Neolithic man. The formation of the stone-free turf was probably brought about by earthworm sorting under conditions of stable grassland.

8. REPRESENTATIVE SOIL PROFILES

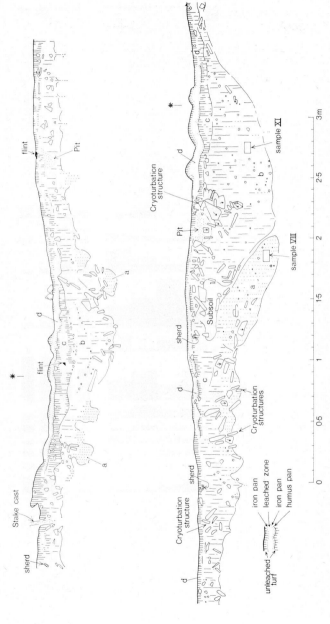

Figure 87. Ascott-under-Wychwood. Sections through the buried soil and subsoil hollow taken at right angles to each other. a = pale-grey, gritty loam; b = dark-brown, gritty loam; c = non-calcareous, orange/brown-mottled loam; d = dark-brown to black, calcareous loam (turf line). * = common point.

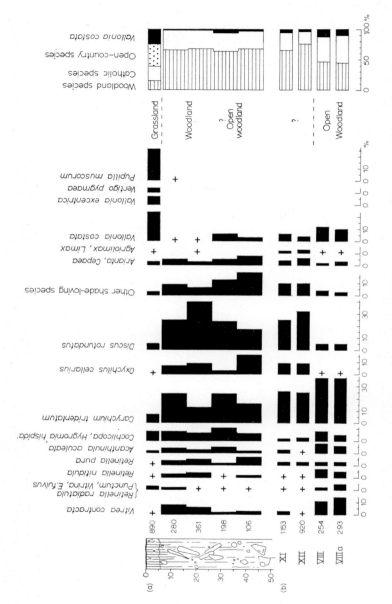

Figure 88. Ascott-under-Wychwood. (a) Buried soil. (b) Subsoil hollow. (Redrawn from Evans, 1971b: Fig. 5.)

8. REPRESENTATIVE SOIL PROFILES

After forest clearance had taken place, there was no further disturbance of the soil. Unlike the situation at South Street, clearance was not followed by ploughing; the main body of the profile remained undisturbed. This is indicated not only by the richness of the fauna and the absence from it of open-country species, but by the fact that structures of periglacial origin—small groups of steeply pitched stones—are preserved intact in the profile (Fig. 87). In this respect, the profile is very similar to that at Beckhampton Road (Fig. 84).

There were two phases of Neolithic occupation. The earlier is represented by pits dug into the soil profile and subsoil, sealed by the turf line, the second by occupation débris on and within the turf line. Radiocarbon dates for the two phases are:

Pit 7, charcoal: 2943 ± 70 BC (BM–491b)
Charcoal *on* turf line: 2785 ± 70 BC (BM–492)

A second profile through the soil was investigated in the area of a complex subsoil hollow (Fig. 87). The hollow consisted basically of two parts separated by a wedge of subsoil. The smaller part, which was the earlier to form, contained a pale, compact fill similar to the material in the subsoil hollow at Coombe Hole (Fig. 80) and in the lower part of the subsoil hollow at Avebury (Fig. 95). Two samples were analysed for snails (Fig. 87: VIII and VIIIa). The fauna (Fig. 88b) suggests an open woodland environment, being dominated by shade-loving species, but with *Vallonia costata* at 12% and 10% abundance. The similarity of this fauna to its analogue from the base of the subsoil hollow at Avebury (Av III 3), in the paucity of *Oxychilus cellarius* and *Discus rotundatus* and abundance of *Vitrea contracta* is striking.

In contrast, the fill of the main part of the hollow was more humic and less gritty. Two samples were analysed (Fig. 87: XI and XII). The fauna (Fig. 88b) shows some interesting differences from that of VIII and VIIIa, which suggest that the arboreal canopy had become denser by this stage. *Vallonia costata* falls to 6%—half its former abundance—while *Oxychilus cellarius* and *Discus rotundatus* attain a level of abundance more usual for woodland faunas.

Above the fill of the subsoil hollow was a layer of dark-brown stone-free loam, decalcified and devoid of shells. This is the B horizon of a brown-earth soil which developed from the underlying rendsiniform material in a fully forested environment undisturbed by man. Overlying this horizon is the turf line seen elsewhere on the site, which is calcareous and has a stone line at its base. The superposition of calcareous material on a decalcified soil indicates the former to have been laterally transported by some form of subaerial process, probably associated with human activity. One possibility is that the soil surface was tilled. Another, that the surface was incidentally disturbed during the process of clearance, when the branches and trunks of trees were dragged across the site.

The sequence of events represented by this profile is as follows:

(1) A phase of open forest (sample VIII and VIIIa).
(2) A phase of closed forest (XI and XII), in the later stages of which a brown-earth soil formed.
(3) A phase of Neolithic settlement during which the soil surface was disrupted, limestone rubble was laterally transported onto the brown-earth soil, and a turf line formed.

There remains to be explained the phase of open forest at the beginning of this sequence. There are two possibilities. The first is that this is an early stage in the natural development of the Post-glacial forest, perhaps in the Boreal period before the closed structure of the climax forest of the late Boreal and Atlantic periods had been attained. The low values for *Discus rotundatus* and *Oxychilus cellarius*, species which do not become widespread in Britain until the end of the Boreal, are certainly suggestive of an early Post-glacial date—particularly when compared with their high values in the later fill of the subsoil hollow.

The second possibility is that there was a *local* opening of the forest canopy, caused by natural or artificial factors. Neolithic material, unless in pits, does not occur below the turf line, and it is virtually impossible that this early phase of open forest is connected with Neolithic man. Indeed it is a characteristic of this site, as of Beckhampton Road, that disturbance of the soil profile by Neolithic man was confined to its surface. Mesolithic artefacts are not, however, thus restricted, and a number of microliths were found in the fill of the subsoil hollow mainly below the brown-earth soil horizon. It would be rash to assume on this evidence alone, that Mesolithic man was responsible for opening the forest canopy—if indeed it was previously closed. But in the present climate of opinion regarding the effect of hunter-gatherer communities on the environment (Smith, A. G., 1970), reference to the possibility of some sort of connection between the phase of open woodland and the contemporary occupation of the site by Mesolithic man cannot be omitted.

8. REPRESENTATIVE SOIL PROFILES

SOUTH STREET

Location: 1·4 km south-west of Avebury, Wilts. SU 091693 (Fig. 143).
Situation and geology: The site is on flat, lowlying land, but above the floodplain of the River Winterbourne. The geological solid is Middle Chalk, overlain by 1·0 m of periglacial drift (Evans, 1968b).
Archaeology: A Neolithic long barrow, much denuded by ploughing (Smith and Evans, 1968). Charcoal from the buried soil surface yielded a radiocarbon date of 2810 ± 130 BC (BM–356) (Evans and Burleigh, 1969), which is probably the date of burial of the soil.

The buried soil beneath the long barrow has already been discussed, along with that at Ascott-under-Wychwood (Evans, 1971b). There will also be a detailed account in the final excavation report. Here only a summary of the main points is necessary.

The relationship between the buried soil and the barrow mound and ditches is shown in Fig. 89a; and a detailed drawing of part of the profile in Fig. 89b.

The parent material comprises a series of involutions of periglacial origin, filled with a fine buff-coloured chalky deposit, in a matrix of coombe rock (Evans, 1968b). Two samples were examined for shells (Fig. 90: Involutions). They contain a characteristic fauna of late-glacial facies in which *Pupilla muscorum* predominates. An open-country tundra environment is indicated.

Penetrating the subsoil, from the base of the soil, are a series of "subsoil hollows" some of which go down to a depth of 60 cm. All are of later origin than the involutions, which they cut through or overlie. Their fill is humic and compact, and characterized by large quantities of small chalky débris, suggesting physical weathering to have been an important factor in their infilling.

Analysis of the snail fauna from one of these yielded a fully temperate assemblage in which shade-loving species predominate, though with a substantial open-country element as well. For a variety of reasons it is felt that this fauna represents an environment of open woodland at some time in the early part of the Post-glacial. For example, it is known from Avebury and Beckhampton Road that a fauna virtually devoid of open-country species once existed on the Chalk in this area, and there is no apparent reason why this should not have been the case here. The presence of open-country species thus implies either a cover of light woodland or that the deposits are

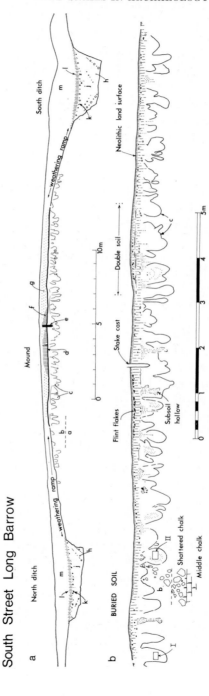

Figure 89. South Street. (a) Transverse section through barrow mound and quarry ditches. (b) Buried soil beneath barrow. a = Middle Chalk; b = coombe rock; c = involutions; d = turf line of Neolithic soil; e = stake cast; f = turf stack; g = chalk rubble; h = primary fill; j = secondary fill; k = Beaker clearance horizon; l = buried soil in ditch; m = ploughwash (tertiary fill).

A monolith from the buried soil beneath the mound showed the following stratigraphy (Fig. 91a):

Depth below buried
soil surface (cm)

0–10   Dark-grey (7·5 YR 3/1) calcareous loam, with a few chalk fragments. Mull-humus horizon.

10–22   Grey (10 YR 6/1) humic loam with numerous chalk fragments; becoming increasingly stony with depth.

22 +   Chalk rock.

The land-snail fauna from the soil (Fig. 91a) is essentially of open-country type, with *Vallonia costata*, *V. excentrica* and *Helicella itala* predominant. Shade-loving species comprise only 7% in the mull-humus horizon, suggesting an environment of dry grassland more or less free of scrub.

Some of the examples of shade-loving species and all those of *Pomatias elegans* are worn and pitted, and are probably the remnants of a former period when the environment was less open. There is a decrease in shade-loving species and an increase in open-country species from the bottom to the top of the profile, which is probably a crude reflection of the detailed ecological changes which were detected at South Street—a shaded habitat becoming open through forest clearance and tillage.

The fauna from the lower buried soil in the ditch of this barrow, dated by pottery to the Beaker period, was analysed by Connah and McMillan (1964). It was of mixed character, open-country species comprising 42% and shade-loving species 32%. A vegetational background of open downland with occasional patches of scrub was suggested.

WEST KENNET

Location: About 0·8 km south of West Kennett, Wilts. SU 105677 (Fig. 143).

Situation and Geology: The site is a north-facing spur, overlooking the Kennet valley. The geological solid is Upper Chalk, overlain by a thin cover of periglacial drift consisting of fine quartzose sandy material filling pockets and channels in the uneven surface of the chalk (Evans, 1968b).

Archaeology: The site is a chambered Neolithic long barrow, thought to have been constructed in the earlier half of the third millennium BC (Piggott, 1962).

By permission of the Department of the Environment, a soil pit, 1·2 m square, was dug into the barrow mound, 33 m west of the façade and 6 m N of the long axis. The buried soil, overlain by compact chalky rubble of the mound, was at 85 cm below the modern surface. It showed the following stratigraphy (Fig. 91b):

Depth below buried soil surface (cm)	
0–15	Dark-brown calcareous loam (10 YR 4/3), with small chalk fragments and a few flints. Mull-humus horizon.
15–25	Pale-brown loam (10 YR 6/4) becoming increasingly chalky with depth.
25+	Coarse chalky débris in a matrix of finer quartzose material. Subsoil.

The snail fauna (Fig. 91b) shows a similar pattern to that revealed at Horslip. In the lower levels of the soil (7·5–22·5 cm) there is a mixed fauna of shade-loving (*ca.* 30%) and open-country (*ca.* 25%) species; *Pomatias elegans* is abundant. Above (0–7·5 cm), a fauna of more open character is present, with *Vallonia costata*, *V. excentrica* and *Helicella itala* predominant. This sequence suggests the development of an environment of dry, open grassland from one previously shaded.

In this particular instance, the main component of the shade-loving species is *Clausilia bidentata* and almost all the examples of this and of *Pomatias elegans* are eroded and pitted apical fragments which appear to have suffered a degree of weathering greater than that of the other shells in the assemblage. Clearly, they have been in the soil far longer, and are probably the resistant apices of the shells of a previous woodland fauna, the more fragile shells of which have been destroyed —for example, those of various zonitids. This does not alter the general interpretation of the pattern of ecological change, but implies that the assemblage in the lower levels of the soil is not strictly representative of the original woodland fauna.

WAYLAND'S SMITHY II

Location: 1·6 km east of Ashbury, Berkshire. SU 281854 (Fig. 142).

Situation and geology: The site is just south of the crest of the Berkshire Downs, adjacent to the prehistoric Berkshire Ridgeway. The geological solid is Upper Chalk.

Archaeology: A chambered Neolithic long barrow (Atkinson, 1965). Charcoal from the surface of the buried soil gave a radiocarbon date of 2820 ± 130 BC (I–1468).

The buried soil beneath the mound and the land-snail fauna were examined by Dr. M. P. Kerney who has kindly allowed his unpublished results to be included here.

The soil was a strongly humic rendsina, about 12 cm thick—rather slight by comparison with other soils of this type. Fragments of chalk were present throughout suggesting that mechanical disturbance, possibly by ploughing, took place prior to its burial.

The land-snail fauna (Fig. 91c) represents open grassland. There is some hint that the environment was becoming more open, there being a stronger woodland element in the lower part of the profile—as at Horslip and West Kennet.

SILBURY HILL

Location: 1·4 km south of Avebury, Wilts. SU 100685 (Fig. 143).

Situation and geology: The site is on a spur of Middle Chalk which juts out into the flood plain of the Winterbourne. There is a layer of clay-with-flints up to 1·0 m thick overlying the chalk.

Archaeology: A round barrow of exceptional size, constituting a height of 30 m from the surface of the buried soil at the centre to the summit. Vegetation from the surface of turves in the turf stack yielded a radiocarbon date of 2145 ± 95 BC (I–4136) (Atkinson, 1967, 1968, 1969, 1970). A series of five C-14 dates was obtained from organic matter from the turf stack by the Radio Carbon Laboratory of the Smithsonian Institute. These, kindly made available by Professor R. J. C. Atkinson, are as follows:

SI–910-A	2725 ± 110 BC
SI–910-B	2365 ± 110 BC
SI–910-C	2620 ± 120 BC
SI–910-C-H	2515 ± 130 BC
SI–910-D	2580 ± 110 BC

The buried soil beneath the mound, being on clay-with-flints was of brown-earth type, and devoid of snails. The core of the primary mound however was made up of stacked turves which were calcareous and derived from a chalky parent material, probably a gravel of periglacial origin similar to that at Beckhampton Road (p. 248). A typical turf showed the following stratigraphy (Fig. 92):

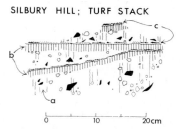

Figure 92. Silbury Hill, turf stack. Section through three turves. a = stony loam; b = turf line; c = vegetation.

Depth below surface
of turf (cm)

0–2   Stone-free mull-humus horizon. Worm-sorted zone.

2–ca. 15   Calcareous stony loam with numerous flints and chalk fragments.

ca. 15+   As above, but becoming increasingly chalky with depth. Chalk fragments sub-angular and sub-rounded.

Like those from South Street and Wayland's Smithy, this profile is probably an ancient plough soil which had lain under grass for a few years prior to burial, during which time a thin turf line formed. The enormous weight of the mound created anaerobic conditions within the turf stack, thus preserving vegetation on the surface of the turves. This comprised largely mosses and grass.

Four samples were analysed for land snails (Fig. 93). In all these, open-country species were predominant, particularly *Vallonia excentrica* which in two cases comprised 40% of the total. An environment of very dry, open grassland is indicated, and this is confirmed by the vegetation of the turf stack.

Because of the anaerobic environment of this material, the periostracum, or proteinaceous coat, of many of the shells was preserved.

8. REPRESENTATIVE SOIL PROFILES

In one sample, Si H 7, which was particularly rich in shells, those in which the periostracum was visible were separated and shown black on the histogram. As already discussed (p. 22), such shells probably belong to animals alive or dead for less than a year at the time of construction of the mound. The other shells which were without their periostracum are shown as an open histogram. The main difference between the two groups is seen in *Helicella itala* where there is a great

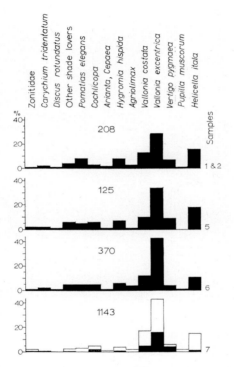

Figure 93. Silbury Hill, turf stack. Open histogram in sample 7 represents shells without a periostracum.

preponderance of dead shells. *H. itala* is a species preferring very short grass, and in prehistoric times it was possibly a species of arable habitats (p. 182). The implied decline in the number of this species at Silbury in the final years before burial may therefore be a reflection either of the change from arable to grassland indicated by the soil profile evidence, or an increase in the height of the grasses, due perhaps to a relaxation of grazing pressure.

AVEBURY

Location: The henge encloses a large part of the village of Avebury in north Wiltshire. The section through the bank described here was at the back of the village school. SU 101698 (Fig. 143).

Situation and geology: The henge is on level ground. The geological solid is Middle Chalk, but this is overlain by 1·4 m of periglacial drift (Evans, 1969c).

Figure 94. Avebury. Section through the bank and buried soil, with periglacial involutions beneath. (Scale in 50-cm intervals.)

Archaeology: The site is a Class II henge, similar to those of Marden and Durrington Walls, and was probably constructed around 2000 BC (Smith, I. F., 1965b). The excavation in 1969 was done by Mrs F. de M. Vatcher for the Department of the Environment.

Beneath the bank, which was preserved to a height of about 2·0 m, was a well-marked buried soil horizon (Figs. 94 and 95). At one point along the section a pit was dug down to the Middle Chalk to study the drift deposits. The profile was as follows:

8. REPRESENTATIVE SOIL PROFILES

Depth below surface
of buried soil (cm)

0–6	Stone-free, dark-brown calcareous loam. Turf line or mull-humus horizon (Fig. 95, j). This shows, in places, zones of leaching (p. 222).
6–22	Stony brown loam with numerous chalk lumps throughout (h).
22–*ca.* 55	Buff chalky loam; coarse chalk lumps festooned. Upper involutions (e).
ca. 55–115	Sandy loam in elongate pockets in a matrix of coombe rock. Lower involutions (c and d).
115–140	Shattered chalk (b).
140+	Middle Chalk (a).

At one point in the section (Fig. 95) there was a subsoil hollow, similar to, but less complex than that at Ascott-under-Wychwood, with a fill of pale gritty loam in the bottom and darker material towards the surface. Its colour was distinctly grey in contrast to the brown of the soil profile above.

A series of samples was analysed for land snails as follows:

Lower involutions: one sample
Upper involutions: one sample
Subsoil hollow, lower levels: one sample—Av III 3
Subsoil hollow, upper levels: two samples—Av III 1 and Av III 2
Soil profile: seven samples

The results have been presented as a histogram of relative abundance (Fig. 96).

The lower involutions were devoid of shells. A sparse fauna from one of the upper involutions was dominated by *Pupilla muscorum*, and is probably of Late Weichselian origin, comparable in age to the fauna from the involutions at South Street (p. 257).

In the lower fill of the subsoil hollow (Av III 3), the fauna represents open woodland; shade-loving species predominate, but open-country species comprise 20% of the fauna. This fauna is similar to the fauna from the fill of the earliest part of the subsoil hollow at Ascott-under-Wychwood (VIII and VIIIa), in the low abundance of *Discus rotundatus*, and the absence of *Oxychilus cellarius*, both species which are well-represented in the upper fill of the subsoil hollow. This

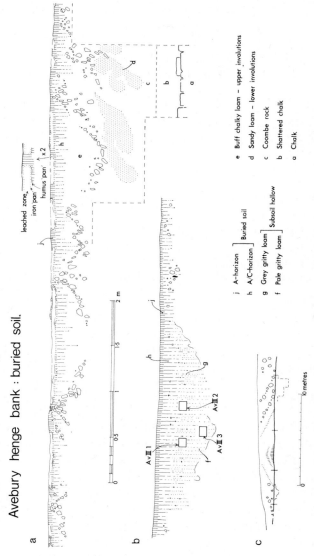

Figure 95. Avebury, buried soil. (a) Soil profile. (b) Subsoil hollow. (c) Skeleton section of bank.

8. REPRESENTATIVE SOIL PROFILES

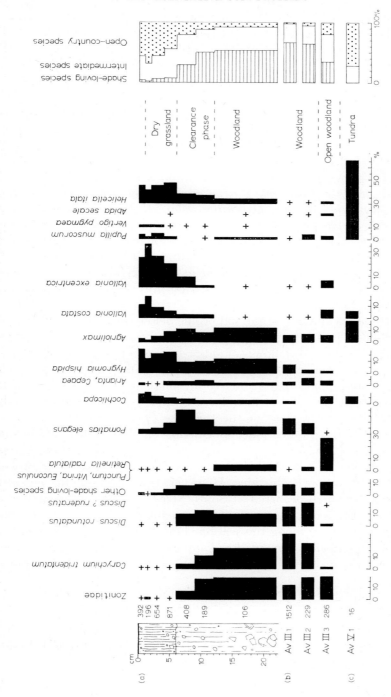

Figure 96. Avebury, buried soil. (a) Soil profile. (b) Subsoil hollow. (c) Involution.

parallelism strengthens the impression that these changes have a chronological basis, and that the faunas in the lower part of the subsoil hollows at both Ascott and Avebury belong to the early part of the Boreal period. *Pomatias elegans*, another species to arrive late in the Boreal period, is also conspicuously rare. Of considerable interest is a possible fragment of *Discus ruderatus* in the Avebury sample, a species which probably became extinct in Britain in the Atlantic period (p. 184).

In the upper fill of the subsoil hollow there is a rich fauna which probably reflects an environment of heavily shaded forest. Open-country species are reduced to a maximum of 6%—including *Vallonia costata*—and the *Punctum*, *Vitrina* group becomes rare. On the other hand, *Pomatias elegans*, *Oxychilus cellarius*, *Discus rotundatus* and *Carychium tridentatum* become abundant, and the presence of all but *Carychium* is probably of chronological significance. The rare species, *Vertigo alpestris*, is present in this fauna and the occurrence of *V. pusilla*, *Columella edentula* and *Lauria cylindracea* is also noteworthy although these latter three do persist on into the Neolithic and later prehistoric periods in Wiltshire.

In the stony horizon of the soil profile below the turf line (6–22 cm), the fauna is basically similar to that just described from the upper levels of the subsoil hollow—of shade-loving type, reflecting a forest environment. The small number of open-country species is probably due to contamination from above although it is not possible to be certain about this. There may have been a little disturbance of the profile between the base of the turf line (6 cm) and 12 cm, as was suggested to have taken place at Durrington Walls (Wainwright and Longworth, 1971: 329), and the strong increase of *Pomatias elegans*, which thrives best in shaded places, but where the ground is broken and rubbly, supports this idea. However, there was not the vigorous disturbance which occurred at South Street through deep ploughing since structures of periglacial origin, reaching to the base of the turf line, are preserved along the section (Fig. 95). A similar argument, it will be remembered, was used to support the essentially undisturbed nature of the profiles below the turf line at Ascott and Beckhampton Road.

The fauna in the turf line is undoubtably of open-country type and probably reflects short-turfed grassland. At 6 cm shade-loving species become virtually absent and open-country species predominate. This change almost certainly represents clearance of the forest by man, although there is no direct evidence that he was involved. Apart from a few flint flakes, no artefactual material was recovered from the soil profile. And if man were involved it is uncertain how clearance took place, or for what purpose. There is a possibility that the soil was ploughed at some stage in its history, but only to the base of what was eventually to become the turf line—no deeper as already explained. Thus in one part of the section grooves filled with humic material and arranged in a criss-cross pattern were present, strongly reminiscent of the ploughmarks at South Street.

At the surface of the turf line (0–1 cm) there was a thin layer of soil which contained rather more flecks of chalk than the material below, and was slightly paler. It was thought, when sampling, that this layer had perhaps been dumped or trampled into the surface during the earliest stages in the construction of the henge, and this is supported by the snail analysis which demonstrates a neat reversal in the pattern of faunal change from the trends in the turf line as a whole. The natural surface of the soil, therefore, is at 1·0 cm.

Through the turf line there is a progressive impoverishment of the fauna in which the eventual predominant elements are five: *Cochlicopa, Hygromia hispida, Vallonia costata, V. excentrica* and *Helicella itala*. Particularly noticeable is the virtual absence of *Pupilla muscorum*. The fauna is similar to many others in the area of Neolithic age—Silbury Hill, South Street and West Kennet. As in all these sites, there is no suggestion that regeneration of woody vegetation took place once the site had been cleared. The habitat was probably being grazed by sheep or cattle.

A final point of interest is the marked decrease in the absolute number of snails through the turf line. This may be due to the fact that the soil was possibly becoming decalcified in the upper part of the turf line, a phenomenon which occurs today in some long-standing grassland soils on the Chalk. The strongly developed zones of iron leaching in the turf may be in part due to its incipient acidity, although the leaching process itself is probably a post-burial phenomenon. Whether the decrease in the number of shells is due to their more rapid destruction in the soil or to the habitat becoming progressively unsuitable for snail life as a whole, is uncertain, but I suspect the latter in view of the fact that all the shells in the upper part of the turf line (1–4 cm) are well preserved.

A summary of the sequence of events at Avebury prior to the construction of the henge can now be given:

(1) Formation of two series of involutions, the lower possibly of Middle Weichselian and the upper of Late Weichselian origin. Tundra environment.
(2) Open forest, probably of Boreal age.
(3) Dense forest of Atlantic age.
(4) Possible disturbance of the soil profile and some clearance as suggested by the increase of open-country species and *Pomatias elegans* (6–12 cm).
(5) Forest clearance, probably by Neolithic man. Possible ploughing of the upper levels of the soil.
(6) Grassland environment, probably of long duration. Kept open by grazing animals.
(7) Construction of henge around 2000 BC.

MARDEN

Location: 0·9 km north-east of Marden, Wilts. SU 091584 (Fig. 142).
Situation and geology: The site is in the Vale of Pewsey on a slight bluff above the River Avon. The geological solid is Upper Greensand, but this is overlain by about 2·0 m of calcareous drift of periglacial origin (Figs 97 and 98).
Archaeology: A Late Neolithic henge monument, dated by radiocarbon assay of charcoal from the ditch bottom to 1989 ± 50 BC (BM–557) (Wainwright, 1972).

The soil profile beneath the henge bank has been discussed in some detail in the excavation report, and only a brief mention of it is needed here. The profile is as follows (Figs 97 and 98):

Depth below buried soil surface (cm)	
0–*ca.* 30	Non-calcareous sandy loam. A, or Eb, horizon. Between 7 and 10 cm is a layer of flints, probably worked down to this level by earthworms.
ca. 30–40/100	Stiff, dark-brown clay and flints. Sharp, irregular boundary with the subsoil. Bt horizon.
40/100 +	Calcareous drift. C horizon.

This profile is a *sol lessivé*, a variant of the brown earth in which clay

8. REPRESENTATIVE SOIL PROFILES 275

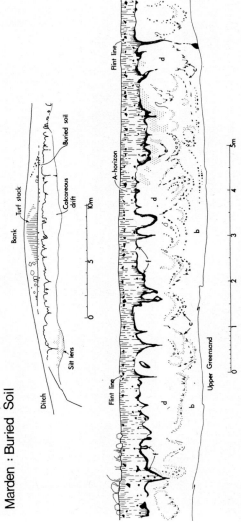

Figure 97. Marden, bank and buried soil in transverse section. b = calcareous gravel; c = lower involutions; d = coombe rock; e = upper involutions; f = Bt horizon of buried soil.

Figure 98. Marden. Transverse section through the bank showing, successively, the modern soil, the bank (turf core and stone capping) the buried *sol lessivé* with its pale A and dark Bt horizons, and the underlying periglacial gravel. (Scale, 2·0 m)

particles have been moved down through the soil, by percolating water, from the A horizon and concentrated at a lower level as the Bt horizon (p. 216).

Although totally decalcified and devoid of shells, this profile is of direct relevance for two reasons. In the first place it is a good example of a *sol lessivé*, a soil type frequently occurring on chalky drift in situations away from the chalk escarpment, and in dip-slope valleys (Figs 65 and 69). A similar soil type was present at Beckhampton Road where the parent material was a barely calcareous quartzose brickearth (p. 248). Decalcification and a low base status are prerequisites for the onset of lessivation, but it is unlikely that such conditions obtained under a vegetation cover of forest, when replenishment of bases lost through leaching was probably taking place in the annual leaf fall. Can we, then, equate the process of lessivation with the removal of the forest cover by man, in the same way that Dimbleby (1962) has suggested for the process of podsolization of heathland soils? There is no doubt that a forest cover once existed at Marden—not only is this a reasonable assumption on the basis of our knowledge of Post-glacial vegetational history, but some, if not all of the tapering pockets at the base of the soil are probably the casts of ancient tree roots.

The second point of interest concerns the flint line (7–10 cm). At this level were found sherds of middle Neolithic pottery and discrete patches of charcoal. C-14 assay of the charcoal, carefully collected from this level, yielded a date of 2654 ± 59 BC (BM–560), which is about 700 years earlier than the construction of the henge. During the intervening period, it is unlikely that the soil was tilled, otherwise discrete patches of charcoal of middle Neolithic origin would not have been preserved so close to the soil surface. The presence of a worm-sorted zone is itself sufficient evidence for a period of soil and surface stability for at least a few decades at the end of its active life.

Thus as at Durrington Walls (Wainwright and Longworth, 1971: 329) and Beckhampton Road, so too at Marden, there is evidence for an episode of forest clearance associated with early or middle Neolithic occupation which was followed by a period, probably to be measured in centuries, when the soil was not tilled.

KILHAM

Location: 12 km west of Bridlington and 9 km north-east of Driffield, Yorkshire (East Riding). TA 056674 (Fig. 142).

Situation and geology: The barrow is situated in the Yorkshire Wolds, on very gently sloping ground overlooking a dip-slope dry valley. The geological solid is Upper Chalk, overlain by a thin cover of periglacial drift.

Archaeology: A Neolithic long barrow, dated by radiocarbon assay of burnt structural elements to 2880 ± 125 BC (BM-293) which is approximately the date of burial of the Neolithic soil. At an earlier stage, there was an episode of Mesolithic occupation comprising a scatter of flint implements and waste, pits and a hearth site (Manby, 1971).

The soil profile beneath the mound is as follows (Fig. 99):

Depth below buried soil surface (cm)	
0–9	Brown (10 YR 5/6–4/4) loam with a very few chalk lumps; non-calcareous. A, or Eb, horizon. At 9·0 cm there is a thin iron "pan".
9–20	Dark-brown (7·5 YR 4/4) fine silty clay loam; more clayey than the A horizon. Bt horizon.
20–48	Shattered chalk in a matrix of pale-brown (10 YR 6/3) chalky loam; becoming paler and more rubbly with depth. C1 horizon.
48–75	Coarse, angular and loose chalk rubble. C2 horizon.
75–86	Compact chalk rubble in a loamy matrix. C3 horizon.
86+	Chalk rock *in situ*. At the junction of the C3 horizon and the Chalk, is a thin layer of light green (5 Y 6/2) clay.

This profile is a further example of a soil of *sol lessivé* type, developed on highly calcareous parent material of periglacial origin. As at Marden, snails were absent. Pollen, however, was present, albeit in small quantities, and the results of analysis are to be published in the final excavation report. The main conclusions are as follows.

Fossil spores were present in the buried soil, suggesting that the material in which the soil developed is derived from a Tertiary deposit either by hillwashing or aeolian action, for such spores are not generally found in the Chalk. In other words, as at Marden, the parent material of the Neolithic (or Post-glacial) soil comprises a non-calcareous, foreign element and is not simply the insoluble residue of the Chalk.

The pollen record indicates two phases of agriculture, separated by a period when the soil was undisturbed, and during which the *sol lessivé* profile probably formed. As was suggested to have been the case at Marden, lessivation appears to have followed as a direct result of forest clearance and agriculture. The second phase of cultivation,

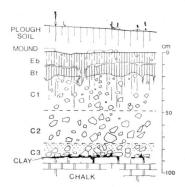

Figure 99. Kilham. Buried soil profile.

which took place shortly prior to the construction of the barrow, failed to disrupt the soil profile sufficiently to convert it into a rendsina. This is in contrast to the present soil on the site which *is* a rendsina, which has been brought about by modern agricultural methods— penetration of the chalky parent material by deep ploughing with a corresponding increase in the base status of the soil.

The origin of *sols lessivés* as a phenomenon of pedogenesis and its causal relationship to prehistoric farming would be well worth full investigation by means of pollen and soil analysis.

9
Colluvial Deposits and Colluvial Soils

Colluvial deposits and soils build up by the addition of increments to their surface. Meaningful interpretation of shell assemblages can only be made with a knowledge of their origin, which itself is a function of the way in which the deposits form.

Three factors have to be taken into account. They are: (a) the rate of accumulation; (b) the rate of *in situ* weathering (soil formation); and (c) the origin of the colluvial material. The first two are interdependent to a certain extent. The rate of accumulation is generally inversely proportional to the number of shells in a deposit, due partly to the deleterious effect of an unstable surface on most species of snail, and partly to the dilution of shells by sediment. During phases of surface stability when erosion and accumulation cease, vegetation colonizes the surface and soil formation takes place. The number of snails increases in response to the more favourable environmental conditions.

In colluvial deposits, the shells may originate from a variety of micro-habitats depending on the derivation of the deposit; they will, at any rate, be more representative of an environment as a whole than the shells from an *in situ* soil. However, because of the rapid build-up of many colluvia, such as blown sand and solifluxion deposits, there is more likelihood of populations of a single generation of shells being preserved than there is in a soil horizon. In other words, colluvial deposits enable us to study the temporal aspects of land-snail populations, while *in situ* soil horizons allow us to study their spatial distribution in a habitat.

During long periods of *in situ* soil formation, weathering of the underlying colluvial material may take place to some depth. Not only does the appearance of the colluvial material become altered, but shells become incorporated into it from the soil surface above, by worm action and the other processes already described (Chapter 7). In this way, the soil horizon comes to contain shells of two different origins and probably from two different environments as well.

The origin of a deposit is relevant, for it may contain shells reworked from earlier colluvial material or soils. Thus in the primary fill of a ditch, shells are generally present from two sources, viz. those living in the ditch as the deposits accumulate, and those which derive from the soil horizon through which the ditch was cut. Sometimes it is possible to isolate the latter when they occur in discrete blocks of turf which have fallen into the ditch, and in these cases the two faunas—the one from the ditch fill and the other from the turf—have been found to be quite different (e.g. South Street, p. 331; Badbury Earthwork, p. 339). More difficult to detect is the reworking of ancient deposits of blown sand or hillwash. Often one must accept the fact that in a soil horizon shells may derive from these three distinct sources—the reworking of ancient deposits, the incorporation of contemporary shells in the colluvial débris, and the downward movement of shells from the soil surface—without being able to do very much about it.

There are various kinds of colluvial deposit which may contain snail shells. They are:

Slopewash deposits (hillwash)
  Ploughwash
  Scree
  Solifluxion débris

Deposits of aeolian origin
  Loess
  Blown sand

Tufa and travertine

Freshwater deposits
  Lake sediments
  River gravel and alluvium

Slopewashes are essentially subaerial in origin. They depend for their formation on two factors, an unstable land surface bare of vegetation (Figs 100, 101 and 102), and a slope; the latter need not be more than 2° in the case of ploughwash and solifluxion débris, but

Figure 100. Chalk escarpment looking south over the Vale of Pewsey, Wilts. Although the slope is steep, the complete grass cover prevents erosion. (Compare Fig. 101.)

with scree, the angle of rest is much higher, that of loose chalk rubble for example being 35°.

Ploughwash

Colluvial material formed as a result of ploughing is called ploughwash. The material generally comprises a pale-brown to brown calcareous loam, is lightly humic, contains numerous small subangular chalk pellets and frequently shows traces of bedding. Ploughwash deposits occur in three main types of situation—valleys, ditches and lynchets—and are characteristically, though by no means exclusively, associated with calcareous soils.

The process of ploughwash formation has not been closely investigated but there is no reason to suppose it to be at all complex. A rendsina soil under woodland or old grassland is strongly humic, maintains a good crumb structure and, in the absence of drift-derived material contains few non-calcareous particles (other than flint nodules) larger than silt size. The soil is freely drained. Even on a steep slope under woodland, where the soil surface may be bare, there

9. COLLUVIAL DEPOSITS AND COLLUVIAL SOILS

Figure 101. Italy. Erosion of calcareous marl due to the removal of the vegetation cover, probably by overgrazing.

Figure 102. Southern Italy, near Matera. Limestone hillsides almost bare of soil. Formerly, this landscape was wooded, and with a thick soil cover, its present state probably having been caused by cultivation and overgrazing.

is often no run-off or erosion during heavy rain. This is particularly apparent on parts of the Chiltern scarp around Princes Risborough (Fig. 75) where ground flora under beechwood canopy is absent, and probably to be attributed to the well-developed crumb structure of the soil—itself a function of high humus and calcium content—maintaining excellent drainage properties.

During the growing season, a tree loses several gallons of water a day, brought up by the roots from the subsoil, through transpiration at the leaf surface. Forest clearance thus reduces the rate of removal of water from the soil and subsoil, and the soil becomes waterlogged more rapidly in wet weather. Cultivation of a rendsina soil by ploughing reduces the humus content, partly by oxidation (cultivated soils being better aerated than those under grass) and partly by the removal of potential humus as crop. Loss of humus and the effect of mechanical disturbance are processes which result in loss of structure, so that the soil becomes compacted and less permeable to water. Under these conditions, during heavy rain, the rate of precipitation exceeds that of drainage, with the result that run-off and erosion take place. In sections of ploughwash, stringers of small chalk lumps are a characteristic feature. These are probably coarse particles sorted by water action in small erosion gullys, as can indeed be seen today in a ploughed field after heavy rain even on quite gentle slopes. The phenomenon of *glazing* may also be relevant in the formation of ploughwash deposits, as discussed by Russell (1961: 48). Soil crumbs in a water-unstable condition collapse under heavy rain and ". . . are reduced to particles so fine that they coat the surface of the soil with a layer that glazes it . . . The glazing of the surface retards the soaking of rainwater into the soil, and the break down of the crumbs in the body of the soil puts it into a poached and sticky condition . . ."

Erosion is probably aided by frost action. Cultivated soils are frozen to greater depths than those whose surface is protected by a cover of vegetation. On thawing out, the surface layers slip downhill over the still frozen levels below—solifluxion, in fact, on a minor scale (p. 289). When dry, soils which have suffered a severe loss of structure through cultivation and frost action are susceptible to the forces of wind erosion, which act irrespective of slope. Thus deposits of ploughwash may be formed by means of a variety of erosional processes—frost action, water and wind erosion, and cultivation.

Ploughwash deposits often contain snail shells, and faunas have been described from several sites (Kennard, 1923; Davis, 1953;

Kerney and Carreck, 1955; Sparks and Lewis, 1957; Kerney et al., 1964; Evans, 1966b) in chalk dry valleys. Often the faunas are rich in species, containing a number such as *Acicula fusca* and *Arianta arbustorum* which require damp and shaded habitats and which are not found today on the adjacent parched hillsides. The origin of the deposits is generally attributed to the destruction of forests which once mantled the downs, the loss of cover permitting erosion of humus and subsoil which accumulated lower down on the hillside as a chalky sludge (Davis, 1953). The faunas do, however, almost invariably contain open-country species, notably *Pupilla muscorum*, *Helicella itala* and *Vallonia excentrica*, which could not have lived in closed forest. These species probably invaded and colonized cleared areas of ground prior to hillwashing, indicating a time lag between clearance and erosion, though Kennard (e.g. *in* Sparks and Lewis, 1957: 32) was inclined to the view that the assemblages probably lived in damp scrub with a coarse herbage, not in woodland or downland.

We can get a clearer picture of the environmental changes which took place by examining the buried soil horizons which frequently occur at the base of ploughwash deposits (Fig. 73). The evidence from Brook (Kerney et al., 1964), Pitstone (Evans, 1966b), Pink Hill (p. 314) and the ditch at South Street (p. 331) suggests that forest clearance and cultivation are not immediately followed by hillwashing, and that there is an interval between the inception of the two processes during which a snail fauna of open-country type was able to colonize the habitat. At Brook, "... the first signs [of clearance] can be detected within the soil itself, for in its upper part *Vallonia* reappears and rises to about 5% of the total fauna. Probably this means that clearance pre-dated by a short while the onset of renewed downwashing, allowing a sufficient interval for the molluscan fauna present previously in the soil to be partially modified by earthworm and root infiltration from its surface" (Kerney et al., 1964: 166). A similar pattern was apparent at Pitstone (Fig. 53) and Pink Hill (Fig. 116). At South Street (Fig. 128), clearance in the ditch during the Beaker period was associated with cross-ploughing and vigorous disturbance of the soil but it was not until the twelfth century AD, some three millennia later, that ploughwashing began.

Indeed on many sites it is suspected on archaeological and general grounds that there is a gap of centuries or millennia between the initial clearance of an area and the onset of hillwashing. It is probable that much of the Chalk was totally and permanently cleared of forest at an early stage in the Neolithic period (p. 363), yet many of the deep deposits of hillwash originated not earlier than the Iron Age. At Brook and Pitstone an early Iron Age date was suggested on the basis of contained pot sherds; at Pink Hill (p. 314), Waden Hill (Evans, 1968a) and Chinnor (Evans, 1971a: Plate 1) an Iron Age or Roman origin is probable; at South Street (p. 331) the deposit is medieval while at Cherhill (Fig. 63) a ploughwash of eighteenth century AD age was present.

Avery (1964: 90) has shown that ". . . when old grassland soils [of rendsina type] are ploughed they retain their dark colour and mellow consistency for a few seasons, but with continued cultivation the organic matter is rapidly reduced, and the surface horizon, which is greyish brown when moist, appears almost white when dry". Thus before a soil becomes susceptible to hillwashing it must suffer a period of structural degradation, and the interval which has been detected in ancient ploughwashes between clearance and cultivation on the one hand and hillwashing on the other may relate to this factor. But in view of the archaeological evidence discussed above it is more likely that cultural differences are responsible, ploughing in the Iron Age period, or even by the Middle Bronze Age (Fowler, 1971) being more intensive and continuous than previously. The increase in precipitation which took place at the beginning of the Sub-atlantic period (500/600 BC) may also have been a contributing factor in the formation of ploughwash deposits as suggested by Kerney (Kerney et al., 1964).

A final point bearing on the formation of ploughwash is that in many instances accumulation appears to have been a rapid process and one not generally taking place today, as is indicated by a decisive break at the junction with the modern soil in the composition of the molluscan fauna and cultural material (e.g. Fyfield Lynchet, Fig. 120).

It has been claimed that there is little evidence for soil erosion on the Chalk during the Post-glacial (Godwin, 1967: 66), but the numerous hillwash deposits which have been recorded would seem to belie this supposition. Against this, however, it must be appreciated that ploughwash deposits are often local in their occurrence, existing solely within the confines of narrow dry valleys, though this is by no means always so, and it is often difficult to estimate with accuracy the area and depth of soil depletion on the surrounding slopes which they represent. Furthermore, ploughwash deposits are characteristic of chalk and limestone escarpments rather than the dip-slope hinterland from which latter they are often absent. This may be due to the gentler gradient and higher non-calcareous drift component of dip-slope soils. These contain a higher proportion of quartz sand than do soils derived from pure chalk, so that loss of structure through intensive cultivation does not necessarily lead to impeded drainage. More often, soils of dip-slope situations are decalcified brown earths of *sol lessivé* type (e.g. Fig. 65) (Avery, 1964).

We can thus define a basic distinction between the soils of scarp areas which are essentially calcareous and in which hillwashing is the predominant type of weathering in the Post-glacial, and those of the dip slopes which are non-calcareous brown earths whose formation has been brought about by leaching and lessivation. The former degrade by sheet and gully erosion to form thick deposits of ploughwash, a process which can be attributed more or less solely to cultivation. Whether we can similarly equate the process of lessivation to human activities such as forest clearance and tillage is a different matter altogether and one which can only be solved by pollen and soil analysis (p. 277).

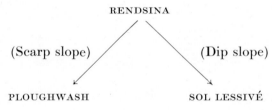

Scree and Limestone Rubble

Scree is coarse angular rock débris, derived by frost shattering and insolation (thermoclastic scree) from a rock face, and which accumulates at its foot as a slope of loosely-packed fragments. Because of the soft nature of the rock and the ease with which its surface weathers and becomes stabilized by vegetation, inland cliffs, and hence scree slopes, are rare on the Chalk. In Yorkshire where the Chalk is generally harder and more brittle than that of southern England, grassed-over screes, perhaps of early Post-glacial age, sometimes occur at the foot of the steeper slopes. Otherwise, deposits of scree are confined to man-made situations such as quarries, ditches and mine-shafts. Thus the primary fill of ditches cut in chalk comprises loosely packed chalk rubble with finer sheets of débris (Jewell and Dimbleby, 1966). Land snails are generally absent from such deposits owing to their dry, unstable surface and lack of vegetation.

On the harder limestones, however, cliffs and hence screes are frequent. Partly active, partly stabilized, and always highly calcareous, they provide a variety of micro-habitats and may support a rich molluscan fauna (Fig. 103). Rupestral species such as *Pyramidula*

Figure 103. Burrington Coombe, Somerset. A limestone hillside, haunt of *Abida secale*.

rupestris, *Abida secale* and *Lauria cylindracea* are characteristic. Few archaeological sites on limestone have been investigated but from some of those which have (Fig. 114), the impression gained is that the fauna of scree deposits is rich in shade-loving species. The same applies to the fauna of stone cairns and of tumbled wall débris. Particularly abundant in such habitats are *Oxychilus cellarius*, *Vitrea contracta* and *Discus rotundatus* (cf. p. 309).

The most plausible interpretation of this type of fauna is that it reflects the local environment existing *within* the scree, for limestone rubble, unlike chalk, weathers but slowly, and retains almost indefinitely a ramifying network of crevices in which humidity is high and temperature relatively uniform. Such conditions are ideal for many species of land snail and result in a rich fauna of "woodland" or "shade-loving" type, even though the over-all environment be open. This is particularly noticeable in the fill of one of the rampart ditches at South Cadbury (Evans, unpublished) where there is an increase in the shade-loving component of the faunas, coincident with an increase in the amount of coarse limestone rubble. On the Chalk, in contrast, rubble horizons become weathered and grassed over rapidly and are generally coincident with an increase in the open-country component of the fauna.

Solifluxion Débris

Two main types of solifluxion deposits occur on the Chalk—coombe rock and chalk meltwater débris. Both are the product of periglacial weathering, and their formation has been discussed elsewhere (Kerney, 1963; Evans, 1968b). Basically they form by the downslope movement of frost-shattered chalk waste, and unlike scree, water plays a substantial part in their formation. Coombe rock comprises coarse, angular/sub-angular chalk lumps in a fine calcareous matrix. Its formation involves the mass movement of semifrozen rock waste over a permanently frozen subsoil. It is characteristic of severe periglacial conditions and generally devoid of snails.

Chalk meltwater deposits are finer and form through the release of water from snow-fields and frozen ground during periods of spring and summer thaw. Unlike coombe rock they are frequently bedded, and characterize milder periods of periglacial weathering. During the summer months their surface became colonized by herbaceous vegetation, and they contain a characteristic molluscan fauna in which open-ground species generally predominate. Some are of Late Weichselian origin (zones I and III) (Kerney, 1963); others may be much older (Sparks, 1952; 1957b).

Soil horizons sometimes occur stratified between sheets of solifluxion débris and have been described from several sites in southern England (Fig. 71), (Sparks, 1952; Kerney, 1963; Evans, 1966b; Paterson, 1971). None has yielded archaeological material; this is particularly unfortunate since they are classic examples of fossilized Pleistocene land surfaces.

The sarsen stone "rock streams" which occur in some of the north Wiltshire dry valleys were transported from their source on the chalk plateaux to their present position by solifluxion (Williams, 1968; Small et al., 1970).

Solifluxed loess is often known as brickearth and may contain land snails; its formation is discussed below.

Loess

Loess is an aeolian deposit consisting largely of silt (grain size 0·002–0·06 mm) and in temperate regions of the world is usually of periglacial origin. Its formation and distribution are discussed by Zeuner (1955, 1957: Fig. 40; 1959) and West (1968). Since loess is a wind-lain deposit it is not restricted to valleys, frequently occurring in plateau situations as well. On the Continent, thick deposits of loess cover large tracts of land from the Rhine valley, across Central Europe to Russia, and were widely settled by Danubian farmers in the fifth millennium BC. But in Britain, loess is rare as discrete deposits of any thickness apart from occurrences in Kent and the south-eastern part of East Anglia. Loess deposits in the Isle of Thanet have been described by Kerney (1965) and ascribed to the later part of the Middle Weichselian (*ca.* 30 000–*ca.* 14 000 BP).

However, certain soils over a wide area of southern Britain, some of them on the Chalk, contain a loessic element, as demonstrated by particle-size and heavy-mineral analysis (Wooldridge and Linton, 1933; Perrin, 1956), which is possibly of periglacial origin. And cryoturbation structures or involutions filled with loess-like material on the surface of the Chalk, have been recorded from beneath Postglacial soils, and are widespread (Williams, 1964; Evans, 1968b); these are certainly of periglacial origin. They contain a characteristic land-snail fauna generally dominated by the Holarctic species *Pupilla muscorum*; the few other species in the assemblages suggest a contemporary landscape, dry and open.

Loess forms primarily in dry conditions. After its deposition, however, it may be secondarily altered by cryoturbation and incorporated into involutions, or be subjected to solifluxion when it is sludged downhill and redeposited, intermixed with chalky débris—in which state it is known as brickearth. A land-snail fauna from a brickearth in Kent of Last Glaciation age is typical of loess deposits (Kerney, 1971a). "It is characterized by the abundance of a small number of

9. COLLUVIAL DEPOSITS AND COLLUVIAL SOILS

A number of ecological groups can be recognized (Fig. 106) which are essentially equivalent to those which occur on inland sites on the Chalk. Thus the hygrophile and shade-loving species include those which prefer relatively moist and shaded habitats. None, apart from *Lauria cylindracea* which has been plotted separately, occurs in any great abundance, and they range from those such as *Lymnaea truncatula* and *Carychium minimum* which are obligatory hygrophiles living in marsh habitats, through others such as *Oxychilus alliarius* and *Carychium tridentatum* which favour relatively undisturbed but more terrestrial situations, to those such as *Clausilia bidentata*, *Vertigo substriata* and *Lauria cylindracea* which are largely rupestral species, living on stone walls or under logs and stones. *Cochlicopa* and *Vitrina pellucida* constitute intermediate or catholic species, less fastidious in their requirements than the former and able to live in moist or dry situations as available. *Vitrina* for example can exist equally abundantly in marsh environments as in the drier (though not the driest) sand-dune habitats. The open-country species comprise three subgroups. *Vallonia costata* shows affinities with the hygrophile and shade-loving species; *V. excentrica*, *Vertigo pygmaea* and *Pupilla muscorum* comprise the bulk of the true xerophiles throughout the deposit; and *Helicella itala* and *Cochlicella acuta* are clearly introductions of recent origin.

The land-snail sequence shows that the site at Northton was originally forested, and that during the Neolithic and Beaker I occupation, deforestation took place, a process possibly attributable to man. Later on, during the Beaker II occupation, there was a phase of woodland regeneration, and stabilization of the machair surface ensued. A second phase of deforestation, accompanied by sand accumulation, then took place, and during the Iron Age the deposit built up to its present height.

The introduction successively of *Helicella itala* and *Cochlicella acuta* is particularly dramatic. Today, these two species predominate over all others in the driest areas of the active dunes. When they first colonized the Outer Hebrides is not, however, clear due to the uncertain age of the "Iron Age" II horizon. Both species were present at Udal in North Uist by the sixteenth century AD. If the date of their introduction could be more accurately fixed they might be used as zone fossils in sites whose age was otherwise difficult to assess, for once introduced into the Hebrides, their spread was no doubt rapid. *Helicella itala*, of course, is one of our Late Weichselian species; its late introduction into the Hebrides is probably due to the northerly situation of these islands and to the fact that migration routes via suitably calcareous habitats have not always been available. It is cited by Boycott (1934) as an obligatory calcicole.

There is evidence that sand accumulation in these west-coastal areas did not begin until the end of the Atlantic period. At Northton, deposition cannot be earlier than about 3000 BC as dated archaeologically (the radiocarbon date of 4100 ± 140 BC for the Neolithic II occupation horizon is unsatisfactory, and probably based on peat charcoal; Simpson, 1966). In Benbecula (Ritchie, 1966) radiocarbon assay of wood from an intertidal organic horizon, overlain by what is alleged to be blown sand, gave a date of *ca.* 3750 BC; and archaeological evidence from sites in Ireland (Kennard and Woodward, 1917), along the South Wales coast (Higgins, 1933) and the north coast of Cornwall (Arkell, 1943) suggests likewise that sand deposition does not pre-date the Neolithic period.

The cause of the onset of sand deposition in the first place is uncertain, but as the material derives from marine deposits which must be dried out before becoming susceptible to wind transport, a fall of sea level, or some form of change in coastal configuration is almost certainly involved. It is of course tempting to invoke the supposed climatic shift to drier conditions at the onset of the Sub-boreal as being in some way responsible, but on the other hand, an increase in the frequency and strength of westerly winds might just as easily have caused the sea to throw up great banks of sand—storm deposits which were quickly dried out and driven shorewards by the wind. Indeed one may speculate as to whether the destruction of woodland at sites such as Northton was not an altogether natural process brought about by the overwhelming action of the sand.

A further point, perhaps connected with the inception of sand accumulation, but whose significance is as yet obscure, is that the earliest deposits of sand at Northton and other prehistoric sites in the Outer Hebrides are non-calcareous. That this is not due to *in situ* decalcification (unless by laterally percolating ground water at a later date) is demonstrated at Northton by the absence of soil formation and the apparent rapidity with which the layer of non-calcareous sand was overlain by shell sand, a process which took place between the two Neolithic occupation horizons. At Morar on the west coast of Scotland (personal observation) and on the Lancashire coast (Hall and Folland, 1970), the calcium carbonate content of the sand decreases with increasing age, the oldest formations being totally decalcified and giving rise to podsols (Fig. 107). Again, if this is not due entirely to the length of time for which the deposits have been subjected to chemical weathering, we may have instances here of the same phenomenon—an initial deposition of non-calcareous sand followed by the deposition of shell sand.

Finally we may point out the value of studying the faunal composition of shell middens—the food débris of early man—which

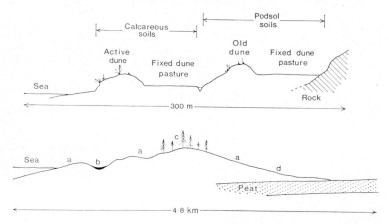

Figure 107. Sections through deposits of blown sand. Upper: Morar, Inverness-shire. Lower: Lancashire coast (After Hall and Folland, 1970: Fig. 20). a = raw sand and pararendsinas (calcareous); b = ground-water and peaty-gley soils; c = micropodsols (acidic); d = ground-water gley soils.

frequently occur in deposits of shell sand (Fig. 108). These may not only yield information about man's economy and subsistence (Evans, 1969b; Shackleton, 1969) but provide us with a valuable insight into the contemporary marine environment from which the shells, and hence the shell sand, derive. At Northton (Evans, 1971b) it was possible to suggest a correlation of sea-level changes based on the shellfish content of the midden with phases of machair build-up, based on the terrestrial snail fauna.

Tufa and Travertine

Tufa is a calcareous deposit formed by precipitation from water heavily charged with lime, and is generally devoid of clastic chalk and limestone débris. It occurs in various forms. Compact and rock-like

Figure 108. Ensay, Outer Hebrides. Shell midden. (Scale, 1·0 m)

occupation débris—bone, flint and charcoal—of basically Maglemosian type with a small proportion of microlithic forms of Sauveterrian affinities (Evans and Smith, 1967). At Prestatyn (Clark, 1938: 330) a Mesolithic industry in keeping with an early Atlantic context was stratified beneath the tufa, the occupation débris being on sand and clay. This industry too showed Sauveterrian affinities and it is possible that we are dealing here with similar industries to those described by Clark (1955) from Peacock's Farm in the Cambridgeshire Fens which also belong to the Boreal/Atlantic transition. Prestatyn, Blashenwell, Cherhill and Box are the only tufa sites which have yielded Mesolithic artefacts. In view of the well-defined stratigraphical and chronological context of the occupation horizon at Cherhill which could reasonably be expected to occur elsewhere, it is to be urged that other sites of this kind be investigated and molluscan analysis applied to their environmental background.

Many tufas (e.g. Brook, Cherhill, Norton Common and Wilstone) now lie a few feet above the level of the active flood plain, comprising slight terraces. Being on the whole friable, calcareous and well-drained they provide good agricultural land. The Romano-British villa at Box is situated on the tufa deposit there, and at Caerwys, the tufa has been extensively quarried for agricultural purposes since 1939. Tufa has been used as a building material, as at the Roman villa of Whitton in Glamorgan and several Roman sites in south-east England (Fig. 111) (Williams, 1971).

The list of tufa sites which follows includes all those from which stratified Mesolithic artefacts have been recorded, and some of the more important deposits with rich molluscan faunas. It is essential to emphasize that tufa occurs, not necessarily as isolated patches, but as continuous layers—ribbons or terraces situated parallel to streams, or in some cases covering whole areas of formerly marshy ground. On the coast of Glamorgan for example, tufa is present in all the valleys cut into Liassic limestone. The location map (Fig. 112) is misleading in this respect and only indicates the sites where particular outcrops have been recorded.

BLASHENWELL, Dorset (Clark, 1938; Bury, 1950; Barker and Mackey, 1961: 40). C-14 date from near the middle of the deposit of 4490 ± 150 BC (BM–89). "... worked flints and charcoal scattered throughout the deposit, although noticeably more abundant in the middle." Industry of Sauveterrian type.

BOX, Wilts. (Bury and Kennard, 1940; Hurst, 1968). "A thin scatter of worked flints was present in disturbed layers throughout the site." One microlith was recorded. "One unretouched blade appeared to underlie the deposit of tufa . . ."

BRIGG, Lincs. (Kennard and Musham, 1937; Smith, A. G., 1958a; 1958b). Tufa overlain by wood and wood peat, dated by pollen analysis to late zone VIIb.

BROOK, Kent (Kerney et al., 1964).

CAERWYS, Flint (McMillan, 1947; Jackson, 1922, 1956).

CHERHILL, Wilts. (Evans and Smith, 1967). Mesolithic occupation horizon of bone, flint and charcoal sealed beneath a tufa deposit and resting on a marsh soil. Industry basically a Maglemosian one with a small proportion of microlithic forms of Sauveterrian affinities. C-14 date of industry 5280 ± 140 BC (BM–447).

GERRARD'S CROSS, Bucks. (Howe and Skeats, 1903–4; Barfield, L., unpublished). Mesolithic occupation site sealed by a deposit of hillwash in which were numerous fragments of derived tufa. Howe and Skeats recorded an *in situ* travertine on this spot (TQ 014880).

HITCHIN, Herts. (Kerney, 1959). An interglacial tufa of probable Hoxnian date.

HOLBOROUGH, Kent (Kerney, 1963).

HUNTONBRIDGE, Herts. (Kennard, 1943). Tufa exposed during road works in 1965. At least two well-defined buried soils were present in the body of the deposit, reflecting periods when the climate became sufficiently dry to prevent the build up of tufa. *Vertigo moulinsiana* was present in this site (Evans, unpublished).

LETCHWORTH, Herts. (Kerney, 1955). Tufa deposit 10 ft above present stream level. Contained *Vertigo genesii* and *Discus ruderatus*.

LLANCARFAN, Glamorgan. Neatly dressed blocks of tufa were used in the construction of buildings in the Roman villa at Whitton, the tufa coming from a nearby site at Llancarfan (M. G. Jarrett, personal communication).

MILLPARK AND GLOSTER, Co. Offaly (Kerney, 1957).

NASH POINT, Glamorgan. (SS 915685) (Dance, S. P. and Evans, J. G., unpublished). Deposit situated in a valley and overlain by a humic hillwash rich in woodland snails. Below the tufa is a brown-earth soil resting on solifluxion débris of shattered Liassic limestone. The tufa itself contains, among others, *Discus ruderatus*, *Vertigo alpestris*, *V. angustior*, *V. pusilla* and *Lauria anglica*. This site and Llancarfan are two of a number of similar sites in the Vale of Glamorgan others being at Leckwith Hill and Barry Island (Strahan and Cantrill, 1902), and the two valleys immediately to the west of Nash Point, Cwm Nash and Cwm Mawr.

NORTH FERRIBY, Yorks. (Wright and Wright, 1947).

PITSTONE, Bucks. (SP 933145) (Evans, unpublished). An extremely hard travertine is present beneath solifluxion deposits of Last Glaciation age. The fauna has not been examined in detail due to difficulties of extraction, but is not noticeably thermophilous in character.

PRESTATYN, Flint (Clark, 1938; 1939; McMillan, 1947). Microlithic chipping floor. It is uncertain how much, if any, tufa had already formed prior to the Mesolithic occupation, but tufa continued to form after its abandonment and sealed it to a depth of 2 ft. Most of the occupation débris was on sand and clay beneath the tufa. The industry was Sauveterrian.

SHAKENOAK, Oxon. (Arkell, 1947; Brodribb *et al.*, 1968: 11, 36; 1972).

SOUTHWELL, Notts. (Davis and Pitchford, 1958). Fauna contains *Discus ruderatus* and *Vertigo genesii*.

TAKELEY, Essex (Kennard *in* Warren, 1945).

TOTLAND, Isle of Wight (Davis, A. G., 1955).

WATERINGBURY, Kent (Kerney, 1956).

WILSTONE, Herts. (Kennard, 1943). Contains *Discus ruderatus* and *Vertigo genesii*. The deposit overlies a pronounced buried soil horizon of brown-earth type.

WOODSTOCK, Oxon. (Arkell, 1947).

Freshwater Deposits

The freshwater deposits of lakes and rivers have been studied largely by Sparks (1961) who has discussed the interpretation of their shell

306   LAND SNAILS IN ARCHAEOLOGY

assemblages in climatic and local environmental terms. Much of the work on these deposits has been done on Pleistocene sites (Sparks, 1957a; Sparks and West, 1959, 1970), Post-glacial deposits having been less frequently investigated (Sparks and Lambert, 1961; Sparks, 1962).

In some instances, archaeological sites in low-lying situations are

Figure 113. Moor Park, Herts. Two buried soil horizons stratified within calcareous freshwater deposits. (Scale, 6 ft)

buried by freshwater deposits (Fig. 113) as in the case of various Iron Age and Roman sites in the Somerset Levels, and the Mesolithic site at Thatcham (Churchill, 1962). The molluscan faunas of these deposits can give some indication of the environment under which burial took place, as, for example, by a sudden flood or by gradual accumulation in a swamp.

Rather more relevant to the terrestrial environment, and to man, is the fact that freshwater sediments frequently contain the shells of land snails, either derived from eroding terrestrial deposits, or, as is more usual, incorporated directly by becoming washed into a river in times of flood. By treating them as an entity separate from the freshwater shells, information can be obtained about changes in the terrestrial environment of the flood plain. This was done in the analysis of the Hoxnian Interglacial deposits at Swanscombe (Fig. 54) (Kerney, 1971b) in which a shaded fen and reed swamp environment gave way to one of grassland, as indicated by an increase in *Vallonia* and other open-country species.

10
Caves, Dry Valleys, Lynchets and Ditches

From the point of view of terrestrial archaeological sites, four types of catchment situation are important in preserving environmental data. They are caves, dry valleys, lynchets and ditches, each of which may contain one or more of the colluvial sediments discussed above in Chapter 9.

Caves

Land snails commonly occur in limestone cave deposits (e.g. Kennard and Woodward, 1917), but there have been no detailed studies of these either from archaeological contexts or generally in Britain. Cave deposits comprise a wide variety of types which may be classified in various ways (e.g. Tratman *et al.*, 1971; Schmid, 1969). Thus there is a basic division between freshwater and terrestrial sediments, and within each of these, autochthonous and allochthonous components may occur. In the Ballymihil Cave, Co. Clare, (Williams and Williams, 1966) where the deposits were terrestrial, secondary calcite (travertine) and stony débris from the ceiling and walls (thermoclastic scree) constituted autochthonous material, while débris such as soil, clay and skeletal remains brought in from outside were considered to be of allochthonous origin.

The fauna from the Ballymihil Cave was of "woodland" type and tentatively assigned to the Atlantic or Sub-boreal periods of the Postglacial. *Oxychilus cellarius* (25%), *Vitrea contracta* (26%) and *Discus rotundatus* (13%) were the predominant elements (Fig. 114), and these three species also appear to characterize other cave faunas of Postglacial origin; for example, Aveline's Hole, Burrington Coombe

(Kennard and Woodward, 1922–1925a), Merlin's Cave, Symond's Yat (Kennard and Woodward, 1922–1925b), Kilgreany Cave, Co. Waterford (Jackson, 1926–1929) and Sun Hole Cave, Cheddar (Davis, 1953–1956). There is too, a similarity between these and faunas from open-air limestone sites such as Glebe Low (McMillan *in* Radley, 1966: 66), South Cadbury and Court Hill Cairn (Fig. 114), (Evans, J. G. and Jones, H., unpublished), which contrasts them with sites on the Chalk, for example Beckhampton Road (p. 248). The most noticeable

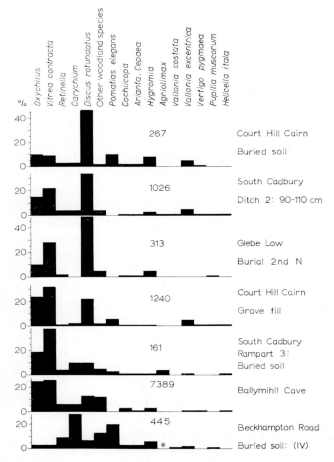

Figure 114. Faunas from cave deposits and limestone rubble, with one from the Chalk (Beckhampton Road) for comparison. Court Hill Cairn and South Cadbury (Evans and Jones, unpublished); Glebe Low (McMillan *in* Radley, 1966); Ballymihil (Williams and Williams, 1966); Beckhampton Road (p. 248). *=present but not counted.

differences are in the abundance, on limestone sites, of *Vitrea* and *Oxychilus cellarius* and the corresponding paucity of *Retinella* among the Zonitidae, and in the abundance of *Discus* and the paucity of *Carychium*. There is a suggestion too, that *Vallonia excentrica* may predominate over *V. costata* on this type of site, for in spite of its "woodland" affinities (p. 157), *V. costata* is totally absent from Ballymihil (*V. excentrica*: 75 examples) and Court Hill Cairn (*V. excentrica*: 65 examples), and rare by comparison with *V. excentrica* at South Cadbury. Whether these differences with sites on the Chalk are a function of the limestone nature of the sites, or of their more westerly situation in Britain, one cannot say. Kennard and Woodward (1917) drew attention to the fact that *Oxychilus cellarius* is often subterranean and particularly common in all cavern deposits; this does not appear to apply to *Retinella* spp.

The peculiar way in which limestone screes provide suitable microhabitats for "woodland" species of snail has already been discussed (p. 287), and it will be appreciated that a snail fauna living in the interstices of a deposit after it has formed amounts to contamination. The loosely packed, angular rock débris of caves (thermoclastic scree) is particularly prone to this type of contamination, and not only do snails live a few centimetres below the surface of the deposits, but at death their shells may be washed down to some depth. This process does not necessarily diminish the value of a fauna for environmental reconstruction so long as it is realized that it may not be coeval with the deposit in which it occurs. The situation is analogous to the downward movement of pollen in mineral soils (Dimbleby, 1961). Thus at Sun Hole Cave, Cheddar (Davis, 1953–1956) solifluxion deposits of angular gravel with comparatively little admixture of fine earth were assigned on stratigraphical, faunal (Mammalia) and archaeological grounds to the Late Weichselian or early Post-glacial. They contain a flint industry of Cheddarian (Cresswellian) type. But the molluscan fauna throughout the deposits down to a depth of at least 2 m is fully temperate in character and almost certainly of later origin, open-country species and the more characteristic cold climate indicators being sparse or wholly absent.

In spite of these difficulties, however, I feel that the molluscan faunas of cave deposits deserve more attention than they have yet received. Future work may show them to be without value in environmental reconstruction in some cases due to the difficulties outlined above, but until this has been demonstrated cave faunas must be considered an untapped source of information. Most valuable, perhaps, would be an investigation of the Mollusca from cave deposits of Late Weichselian age, particularly if made in conjunction with a study of adjacent valley deposits. Thus the sequence of change which took place in the terrestrial environment on the Chalk during the Late Weichselian is now well known, with the mild Allerød Interstadial widely reflected as a fossil soil horizon (p. 223). It is disappointing that no Upper Palaeolithic or Mesolithic industries have been adequately recorded from these deposits, although they undoubtedly occur (Burchell, 1957: 238), for the Allerød soil in particular is an excellent example of a fossilized land surface of Pleistocene age. But the recognition of these deposits on the Chalk should prompt a search for similar sites in the limestone areas of Britain (cf. ApSimon, *et al.*, 1961) especially in the vicinity of those caves from which cultural material has been obtained (Garrod, 1926; McBurney, 1959).

Dry Valleys

Dry valleys in chalk and limestone country provide ready catchment areas for a variety of subaerial hillwash deposits (Fig. 115). Frequently these show a twofold division with material of periglacial origin such as coombe rock and brickearth beneath, and Post-glacial sediments—generally ploughwash—above. Within and beneath these deposits may be preserved former land surfaces—occupation horizons and soils—

Figure 115. Incombe Hole. A dry valley in the Chiltern escarpment. Beyond, the Vale of Aylesbury and the Pitstone Tunnel Cement Works.

which have escaped the destructive processes at work in upland situations (Fig. 71). Naturally, such deposits provide a good opportunity for studying the environmental history of an area in terms of climate, land use and local habitat change as has been done at Pitstone (Evans, 1966b), Brook (Kerney *et al.*, 1964), Barrington (Sparks, 1952) and Pegsdon (Sparks and Lewis, 1957).

The site discussed below comprises the Post-glacial fill of a small dry valley in the Chiltern escarpment near Princes Risborough which illustrates particularly clearly the environmental changes accompanying prehistoric forest clearance and agriculture.

PINK HILL

Location: 2·0 km south-east of Princes Risborough, Bucks. SP 823023 (Fig. 117).

Situation and geology: The section, which was exposed during the construction of a reservoir, was in the upper part of a scarp-slope coombe on the east side of the road leading up to Pink Hill. The geological solid is Middle Chalk, overlain by coombe rock.

Archaeology: Between 55 and 80 cm (Fig. 116) were found four sherds of prehistoric, probably Iron Age, pottery, eleven flint flakes, two of which were retouched, and some animal bone fragments and charcoal. It is possible that this is the remains of domestic rubbish spread on the fields as manure. Above 50 cm two Romano-British sherds were found.

The section through the deposits showed the following stratigraphy (Fig. 116):

Depth below surface (cm)	
0–20	Dark humic loam, stone-free at the surface, but becoming increasingly stony with depth. Mull-humus horizon of modern soil.
20–50	Brown chalky loam becoming darker with depth; sharp boundary at 50 cm. Ploughwash.
50–80	Very stony dark-grey loam. Ploughwash.
80–100	Black chalky loam with chalk lumps throughout. Fossil soil.
100–120	Pale-grey, compact chalky loam, with a large amount of small grit-size particles, filling irregular pockets. Subsoil hollows.
120+	Coombe rock.

10. CAVES, DRY VALLEYS, LYNCHETS AND DITCHES

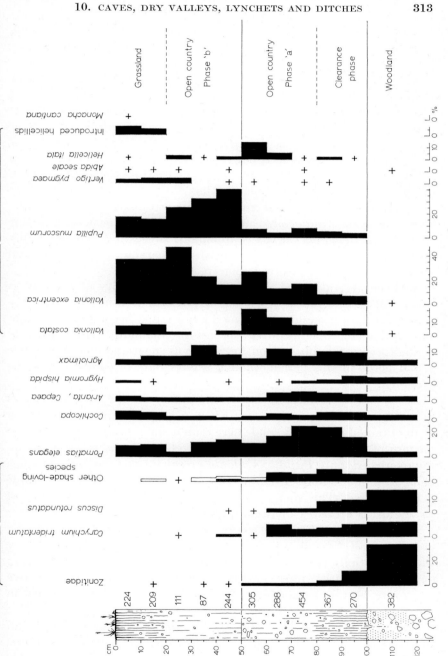

Figure 116. Pink Hill. Dry valley fill.

A series of samples from this section was analysed for land snails, and the results presented as a histogram of relative abundance (Fig. 116).

The fauna from the subsoil hollows (100–120 cm) reflects a closed-woodland environment. Shade-loving species predominate, and the presence of *Vertigo alpestris*, a snail now restricted to a few sites in northern England and Wales, and unknown in the south later than the Neolithic period, suggests that the fauna dates from Atlantic times. The origin of this material is not clear, but is probably similar to that of the fill of other subsoil hollows (e.g. Coombe Hole, p. 241) as suggested by its virtually pure woodland fauna.

In the buried soil (80–100 cm) the fauna is a mixed one with shade-loving species still important, but with open-country species comprising about 20%. *Pomatias elegans* is abundant, suggesting a loose, rubbly soil surface, which is born out by the stony character of the profile and the absence of a worm-sorted zone (p. 133).

Between 50 and 80 cm, an increase in the quantity of chalk fragments suggests a phase of increased physical weathering, probably associated with ploughing. The molluscan fauna becomes more open, and by 50 cm, open-country species predominate. This phase of agriculture is probably of Iron Age origin as suggested by the presence of potsherds of this period, though a later date of course cannot be excluded.

At 50 cm, the sharp horizontal boundary in the stratigraphy is probably an erosion surface caused by a greater depth of ploughing than obtained previously, and there is an equally marked break in the pattern of faunal change. Above this level, the fauna is considerably impoverished, only *Pupilla*, *Vallonia excentrica* and the limacid slugs being at all abundant. A habitat dry and disturbed—probably arable—is suggested. This phase of agriculture, again on the evidence of potsherds, is probably of Romano-British origin.

The examples of *Pomatias elegans* in this upper ploughwash horizon are almost all worn apices, and, as at Badbury (p. 341), are probably the differentially preserved remains of the fauna from the underlying deposits. Similarly, most of the other shade-loving species above 50 cm are worn apices of *Clausilia*, and these have been indicated in the histogram as an open graph.

The introduced species of *Helicella* and *Monacha cantiana* appear in the modern top soil (0–20 cm), strengthening the attribution of the underlying ploughwash deposits to the Roman and prehistoric periods. In this level too there is a tendency towards surface stability as suggested by the decrease of *Pupilla muscorum*, and it is likely that accumulation of ploughwash had ceased by medieval times at the latest, and that the grassland which now covers the site is of long-standing origin.

Thus, at Pink Hill, there is a similar sequence of events to that at Pitstone (Evans, 1966b)—an episode of forest clearance, followed by

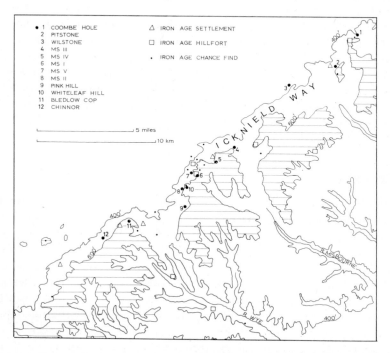

Figure 117. Part of the Chiltern escarpment (cf. Fig. 142) showing the location of sites mentioned in the text, and the distribution of Pre-Belgic Iron Age sites and finds. (After Saunders, 1971.)

two periods when the environment was essentially open. In the first of these, open-country phase "A", *Vallonia costata* and *V. excentrica* are present in equal numbers, while in the second, open-country phase "B", *V. excentrica* predominates. Just what these changes mean in terms of habitat change is not clear. One possibility is that there was a decrease in the water-retaining capacity of the soil from phase "A" to phase "B"; another, that there was a decrease in the available organic content—a change from eutrophic to oligotrophic conditions (p. 104).

Both would be associated with an increase in the intensity of arable cultivation.

The initial clearance phase is of Iron Age date or earlier, and at three other dry valley sites along the Chiltern scarp a similar date is likely. These are Chinnor (Evans, unpublished), Pitstone (Evans, 1966b) (Fig. 53) and Pegsdon (Sparks and Lewis, 1957). The pre-Belgic Iron Age is the earliest period for which there is evidence of permanent settlement in the area (Saunders, 1971) (Fig. 117), and it is probable that the main episode of ploughwashing relates to this time, even though clearance itself may have taken place locally at an earlier stage. Thus while the fauna from beneath the Neolithic barrow on Whiteleaf Hill is a woodland one (Kennard *in* Childe and Smith, 1954: 230), that from beneath the Bronze Age barrow on Bledlow Cop on the opposite side of the Risborough Gap (Fig. 117) reflects, unequivocally, an environment of dry grassland (Kennard *in* Head, 1938).

Lynchets

Lynchet deposits usually comprise little more than ploughwash or soil material which has moved down a cultivated slope and accumulated against an artificial barrier such as a fence, hedge or wall (Fig. 118). The snail faunas of such deposits have been little studied, which is surprising in view of the keen interest of archaeologists in the formation and function of Celtic fields. Moreover, the detailed sequences of land-use which are preserved in dry-valley deposits (p. 313), and their frequent proximity to field systems in which equally valuable deposits may be preserved should prompt archaeologists to make a more comprehensive study of the environs of their sites than has previously been done.

From sites on Overton and Fyfield Down (Fowler and Evans, 1967), the snail faunas of lynchet deposits have given us a certain amount of data about the environmental changes which accompanied the build-up of the lynchets. Thus at Fyfield Down, a well-defined soil horizon lay beneath the lynchet deposits, and the fauna of this was largely of shade-loving type. Within the lynchet itself an open-country fauna was prevalent. At Overton Down the pre-lynchet soil had been severely disrupted by ploughing and little remained of the original snail fauna; open-country forms predominated throughout. At both sites, the fauna of the modern soil differs markedly from that of the lynchet deposits, in the abundance of *Pupilla*, and the virtual absence of *Vallonia costata* and *Helicella itala*.

These two examples were not investigated in detail but indicate that future studies of lynchet deposits by snail analysis would be profitable, particularly if done in conjunction with a study of adjacent valley deposits. Careful sampling of the pre-lynchet soil might reveal former phases of land use; and shells from the lynchet deposits might indicate the environment of standstill phases in lynchet accumulation

Figure 118. Southern Italy. Small hurdles erected to prevent erosion. Structures such as these might lead eventually to the formation of lynchets.

—for instance, whether the land was given over to grazing and kept open, or was totally abandoned and scrub allowed to regenerate.

FYFIELD DOWN I

Location: Fyfield Down, 2·3 km north-north-west of Fyfield, Wilts. SU 141710 (Fig. 143).

Situation and geology: On virtually level ground between 650 ft and 700 ft OD at the head of a dip-slope valley on the Marlborough Downs.

The geological solid is Upper Chalk, and drift is absent, though sarsen stones are numerous in the vicinity.

Archaeology: The lynchet is part of a system of Celtic fields untouched by medieval ploughing; the main lynchet accumulation is assigned to the Early Iron Age. "Sherds range from 'Windmill Hill' ware through Bronze Age and Iron Age fabrics to Roman pottery, and were generally stratified in chronological order in the lynchet. But this fact is probably misleading . . . since three Early Iron Age 'A' sherds were found in the buried soil at the base of the section. Nevertheless, the number of pre-Iron Age sherds indicates earlier settlement or manuring—or both" (Bowen and Fowler, 1962).

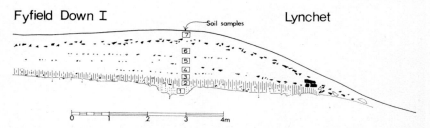

Figure 119. Fyfield Down, lynchet section. (After Fowler and Evans, 1967: Fig. 3C.)

Seven samples taken by P. J. Fowler from the lynchet deposits were analysed for snails, and the results plotted as a histogram of relative abundance (Fig. 120). The description of the samples is as follows (Fig 119):

Sample
7   Dark stone-free loam. Modern turf or mull humus horizon.
6–4   Pale chalky loam ⎫
3   Dark chalky loam ⎬ Lynchet deposits.
2   Dark chalky loam with some flints. Remains of pre-lynchet soil, disturbed by ploughing.
1   Humic chalk rubble. Fill of a roothole (subsoil hollow) below the pre-lynchet soil.

The fauna in 1 and 2 is dominated by shade-loving species, and reflects a relatively undisturbed habitat. Cultivation of the pre-lynchet soil, if it took place at all, was on a gentle scale. In the main body of the lynchet (samples 3–6) the fauna is that of an open environment probably free of shade. The accumulation of the lynchet material clearly took place in an open landscape as is to be expected if cultivation was taking place. The fauna of the modern turf—probably representative of the present-day fauna—is markedly different in its composition from that below, emphasizing the length of time since accumulation of lynchet material ceased.

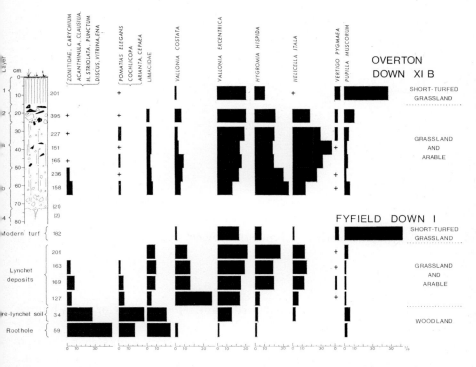

Figure 120. Overton Down and Fyfield Down lynchets. (After Fowler and Evans, 1967: Fig. 4.)

This series of land-snail faunas demonstrates well the effect of man on their diversity. In samples 1 and 2 where the environment is congenial to molluscan life no species (apart from the composite group Limacidae) attains more than 14% abundance. In samples 3 and 6, *Vallonia costata* and *V. excentrica* respectively reach 30% abundance; while in sample 7, reflecting the present-day bleak and unfavourable environment of the Downs, only two species, *V. excentrica* and *Pupilla* are at all abundant, the latter attaining 56%.

OVERTON DOWN XI/B

Location: 2·2 km north of West Overton, Wilts. SU 130703 (Fig. 143).

Situation and geology: The lynchet is in a similar situation to the Fyfield Down site, on the Upper Chalk plateau at the south-west corner of the Marlborough Downs; it lies at *ca.* 700 ft OD.

Archaeology: The lynchet is part of an Iron Age field system which continued in use into the Roman period. A line of postholes was present beneath the lynchet which probably supported the original field boundary along which the lynchet formed (Fowler and Evans, 1967).

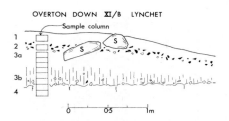

Figure 121. Overton Down, lynchet section. (After Fowler and Evans, 1967: Fig. 3F.)

A section through the lynchet showed the following stratigraphy (Fig. 121):

Depth below surface (cm)	
0–13	Dark stone-free loam. Modern turf or mull-humus horizon. Layer 1.
13–*ca.* 25	Zone of flints, with a few chalk lumps, accumulated at the base of the modern soil by worm action. Some Romano-British sherds were recovered from this layer. Layer 2.
ca. 25–72·5	Lynchet deposits.
	ca. 25–50 Dark chalky loam with many flints and chalk lumps. Layer 3a.
	50–72·5 Dark chalky loam becoming increasingly chalky with depth. Layer 3b.
72·5+	Weathered chalk with slight penetration of humus into fossil rootholes.

A series of samples was analysed for snails (Figs 120 and 121), and the results presented in the form of a histogram of relative abundance.

The overall picture is of an open environment throughout. Open-country species reach values of over 30% while shade-loving species are virtually absent. *Vallonia costata* barely attains 10%. No trace of a pre-lynchet soil was visible in the section, which, with the paucity of shells below 65 cm and the overall lack of shade-loving species probably implies that the site was thoroughly ploughed prior to the formation of the lynchet.

The fauna in the modern turf is closely similar to that from the same horizon at Fyfield Down.

Ditches

The land-snail faunas of soils and sediments filling ancient ditches can tell us about the function of a ditch and about past episodes of land use and environmental change which took place after its excavation.

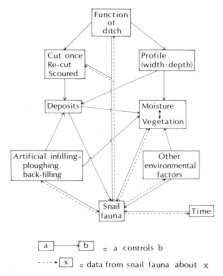

Figure 122. Factors controlling the composition of the snail fauna in a ditch.

But the type and value of information preserved varies, and depends very much on factors such as the use to which the ditch was put, its profile in cross section, and on the level in the ditch and the kind of deposits from which the fauna is extracted. The inter-relationships of the various factors affecting, either directly or indirectly, the snail fauna in a ditch are shown in Fig. 122.

Basically, the fill of a ditch consists of four horizons: the primary fill, the secondary fill, the buried soil and the tertiary fill. The ditch of the South Street Long Barrow can be used as an example (Figs 123 and 127).

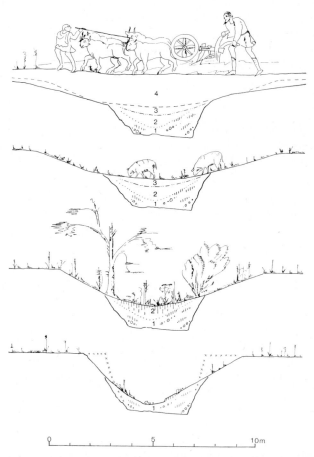

Figure 123. Stages in the infilling of a ditch. 1, primary fill. 2, secondary fill and buried soil. 3, buried soil. 4, ploughwash (tertiary fill).

The **primary fill** consists of angular rock débris, or scree, accumulated largely in the corners of the ditch, and derived from the ditch sides by frost shattering (Crabtree, 1971). Topsoil and turves derived from the contemporary soil surface from which the ditch was cut are also incorporated as a result of undermining by frost-weathering of the ditch sides (Jewell and Dimbleby, 1966). Snails are usually absent from the coarse rock débris, but may occur in the fallen turves. Analysis of these can provide information about the snail fauna and environment *prior* to the construction of the site—an important consideration in the case of sites totally levelled by ploughing, in which all other pre-site horizons have been destroyed.

Figure 124. Chalk-cut ditch, at an early stage of infilling. This particular example has been used, and kept open, as a hollow way.

The **secondary fill** comprises much finer material which frequently shows traces of bedding, and may be more or less humic. Its origin is various. Much derives, at any rate in the lower levels, from the ditch sides by gentle weathering and erosion, when the rock surface is covered by only a patchy plant cover (cf. Fig. 124). At this stage, the snail fauna is a specialized one with species suited to open habitats and bare ground predominant. The similarity of some of these faunas to those of the Late Weichselian is remarkable. Later, when the sides of the ditch have been stabilized by vegetation and the formation of a thin soil cover, the downwashing of material brought up as worm casts probably constitutes a sizable fraction of the secondary fill. It has also been suggested that the upper part of the secondary fill may be of wind-blown origin, in view of its predominantly silty grain size (Cornwall *in* Morgan, 1959: 49), but as already explained (p. 291) this

Figure 125. Ditch at a later stage of infilling than that shown in Fig. 124. The surface is colonized by vegetation, and further natural infilling would take place extremely slowly.

is not in complete accord with the molluscan evidence. The profile of the ditch, which was initially steep-sided, becomes increasingly less so as material accumulates. The patchy vegetation cover of the early stages gives way to one more continuous and luxuriant (Fig. 123; cf. Fig. 125).

The rate of accumulation of the secondary fill slows down and eventually ceases. Chemical weathering acts upon the deposits of the secondary fill, creating a **soil horizon**. This process in fact begins during the later stages of secondary infilling so that colluvial deposition and *in situ* soil formation are taking place contemporaneously for a time. In the absence of human interference and grazing, a rich vegetation of tall grasses and herbs becomes established, and eventually scrub and trees colonize the ditch (Fig. 123). In some cases, as at South Street, only the uppermost layers of the secondary fill are weathered; in others, as at Badbury (p. 339), the deposits may be weathered throughout, so that the buried soil rests directly on the primary fill (Fig. 133).

At this point in its history, the profile is still markedly concave and remains so indefinitely if the environment allows, as is shown by the present-day steep profiles of the ditches of monuments such as Avebury and Maiden Castle. Snail shells are generally present in profusion due to the very favourable conditions of moisture, shelter and shade, and rich assemblages have been recorded from a number of sites. Because of the greater moisture-retaining capacity of the soil, denser vegetation and more sheltered aspect of the habitat, the snail fauna of a ditch at this stage of its infilling is generally richer in numbers and possibly also in species than that in the immediate surrounds. Extreme habitats such as these are foreign to the Chalk and give rise to extreme molluscan faunas. Kennard did not appreciate the full implications of this point, and he often interpreted ditch faunas in terms of general environment and, in some cases, of climate. Thus in a footnote to the "Report on the non-marine Mollusca" from the Thickthorn Down Long Barrow (Kennard *in* Drew and Piggott, 1936: 95), J. G. D. Clark comments as follows: "There is ... a possible alternative explanation for the differences in the two faunules ... Mr. Kennard has grouped the loci into two, one of Neolithic A age, the other of Neolithic B and Beaker age ... When we recall that the younger samples come from the slow silting of an open ditch, where one might expect to find damp-loving species, if general climatic conditions permitted them, whereas the older samples all came from situations less favourable, it seems ... hazardous to draw general conclusions as to climate change between the two periods represented. In fact one would venture to suggest that the differences between the two groups may, to some extent, be interpreted as due to the different loci from which the samples were derived."

The detailed composition of the fauna depends on a number of factors. For example, a steep-sided profile creates a locally denser vegetation and more humid atmosphere than one more gently sloping, and the molluscan fauna is correspondingly richer. The distinction between continuous ditches, such as field boundaries, and isolated lengths, such as those of barrows, is also important. In the latter there is less chance of colonization by shade-loving species than if the ditch forms part of a network of boundaries traversing a variety of habitats and along which dispersal can rapidly take place. The character of the landscape in the area as a whole must also be considered for it is from the various habitats in this that the ditch fauna ultimately derives. And here the time element becomes important. It may take a few centuries for a newly-formed habitat to acquire the total compliment of species it is capable of supporting, if refugia for shade-loving species in the area are sparse and the land across which they have to travel is inimical to molluscan life.

Subsequent infilling and environmental change are generally brought about by man. At South Street (Fig. 128) an episode of ploughing during the Beaker period truncated the soil and flattened the profile by biting into the sides and dragging down chalky rubble into the bottom of the ditch. Subsequently the land was kept open by grazing animals (Fig. 123). Thus the fauna in the buried soil is of a composite nature; in the lower levels it belongs to the later stages of the secondary fill, and reflects the development of a moist and shaded habitat, while above, it is one of dry and open short-turfed grassland. The stratigraphical break is sharp; the faunal changes correspondingly so (Fig. 129); and the need for fine and careful sampling manifest.

Deliberate backfilling and ploughwashing are two processes which bring about the total levelling of a ditch. At South Street pottery from the ploughwash (Fig. 128) suggested a medieval date. In some cases, as at Badbury (p. 339), both processes occur on one site, the ditch being partially levelled before ploughing took place. This material is the **tertiary fill**. Shells in material used deliberately to backfill a ditch are of strictly local origin; but shells in ploughwash come from a wider area and, initially at any rate, reflect a variety of microhabitats. This difference is well brought out at Badbury where the fauna in the backfill (110–140 cm) differs but little from that in the soil below, reflecting an almost totally shaded habitat, while that in the ploughwash reflects an open landscape. Human interference is not always manifest in the snail fauna.

It is important to be absolutely clear about the origin of the deposits and soils at the various levels of an infilled ditch, the origin of their shell assemblages, and the kind of information they may be expected to yield at different levels. From the primary fill we get information from fallen turves about the pre-ditch environment. From the secondary fill we get information about the local environment in the ditch and it is from this level too that we may expect to find out something of the function of the ditch. From the buried soil we may get information about the surrounding environment as a whole or solely

Figure 126. Cherhill, Wilts. Neolithic ditch cut through, successively, a tufa deposit, a buried soil of Boreal age containing Mesolithic flints at its surface, and a periglacial marl. (Scale, 1·0 m)

about the local environment of the ditch, while from the tertiary fill we are most likely to get information about later land use.

Another factor to be taken into account is the type of deposits through which a ditch is cut, for if these are of Pleistocene or Postglacial origin, they too may contain shells. Figure 126 shows a Neolithic ditch cut through, successively, a layer of tufa, a Postglacial soil of Boreal age and an underlying Late Weichselian marl each of which contains a different molluscan fauna—potential sources of contamination of the ditch deposits. It is wise to be aware of these difficulties.

From the point of view of working out ancient land-use patterns, the broad ditches of large sepulchral and ritual monuments generally provide the clearest data. The ditches of Neolithic long barrows and henges are often (though not always) dug for the sole purpose of quarrying rock. Once dug they are left undisturbed and fill up through natural weathering processes, and later through ploughing. Other types of ditch are not so satisfactory. For example, in those of Neolithic causewayed enclosures, much of the fill may have been thrown back deliberately, and consist of a heterogeneous mixture of rock rubble and top soil. The ditches of Iron Age hillforts may have been scoured or re-cut, their fill consisting of a confused series of tip lines. But it is difficult to be more specific on this point, and each site must be considered individually.

In the five sites described below, some of the problems which arise in the investigation of ditch deposits are analysed.

SOUTH STREET

Location: 1·4 km south-west of Avebury, Wilts. SU 091693 (Fig. 143).
Situation and geology: More or less flat, lowlying land on Middle Chalk, overlain by 1·0 m of periglacial drift (p. 257).
Archaeology: The site is a Neolithic long barrow much denuded by ploughing (Smith and Evans, 1968). Antler and bone from the

Fig. 127. South Street Long Barrow. Transverse section through the south quarry ditch. (Scale, 2·0 m.) (N.B. This is not the section from which Fig. 129 is constructed.)

10. CAVES, DRY VALLEYS, LYNCHETS AND DITCHES

bottom of the two quarry ditches yielded radiocarbon dates of about 2700 BC (Evans and Burleigh, 1969). The relationship of the ditches to the mound is shown in Fig. 89a.

A section through the north ditch showed the following stratigraphy (Figs 127 and 128):

Depth below surface (cm)	
0–20	Pale-grey chalky loam. Modern plough soil (h).
20–120	Pale-brown chalky loam becoming slightly darker with depth; lines of small chalk fragments. Medieval ploughwash. (Tertiary fill) (g).
120–140	Dark-brown chalky loam, relatively stone-free, with a well-developed crumb structure. Mull-humus horizon of buried soil (f).
140–150	Zone of chalk rubble. Beaker pottery (e).
150–180	Pale-brown, fine chalky loam, humified at the top. Base of buried soil, weathered into upper levels of secondary fill (d).
180–185	Zone of chalk rubble (c).
185–215	Pale-grey chalky loam, becoming paler and coarser with depth. Lower levels of secondary fill (b).
215–260	Coarse angular chalk rubble with incorporated turves (e.g. at 215–225 cm). Primary fill (a).

A series of samples from this section (Fig. 128) was analysed for

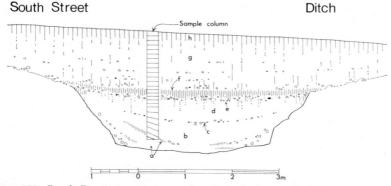

Figure 128. South Street, transverse section through the north ditch. a = primary fill; b = lower part of secondary fill; c = horizon of chalk lumps; d = upper part of secondary fill; e = Beaker clearance horizon; f = buried soil; g = ploughwash (tertiary fill); h = modern plough soil.

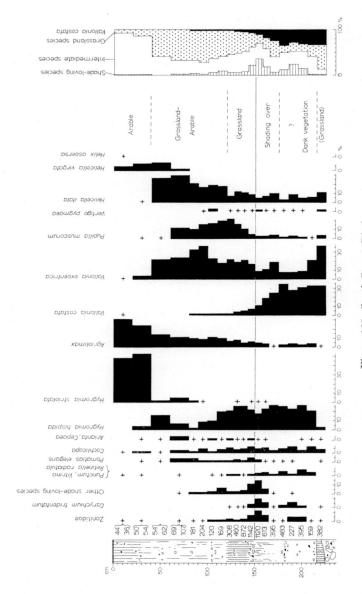

Figure 129. South Street, Ditch.

snails, and the results plotted as a histogram of relative abundance (Fig. 129).

The fauna from a fallen turf (215–225 cm) in the primary fill is closely similar to that from the buried soil beneath the mound (Fig. 90a), being dominated by the two species of *Vallonia* and reflecting a dry grassland environment probably being kept open by grazing animals.

In the secondary fill (up to *ca.* 165 cm) the fauna is a relatively open one, but with more shade-loving species than before. The large number of *Hygromia hispida* suggests that the habitat may have been humid, while the abundance of *Vallonia costata* in relation to *V. excentrica* suggests much broken ground, and this is supported by the non-humic character of the deposits. A zone of chalk rubble between 180 and 185 cm probably indicates some form of disturbance on the sides of the ditch, and this is reflected by a slight increase in the proportion of open-country species at this level.

The similarity of the fauna in the lower levels of the secondary fill to that of Late Weichselian assemblages is remarkable, and probably due to the similarity of the two environments—an unstable, chalky surface colonized here and there by plants but with little soil cover. The presence of the *Punctum–Vitrina–Retinella radiatula* group and the absence of the "Other shade-loving species"—*Discus*, *Ena*, *Acanthinula* and Clausiliidae—is particularly marked. This is a good example of the way in which habitat similarities create similar faunas the age differences between which is very great—here in the order of 6000 years.

Between 175 and 150 cm there is an increase in shade-loving species —with the notable exception of the *Punctum* group—and a decrease in open-country species. Shading over of the habitat is indicated, though to what extent is not clear. Open-country species do not fall below 30% (150–157·5 cm), and it is likely that some open ground remained. In this horizon there are a number of Neolithic sherds (Ebbsfleet/ Mortlake type) which are contemporary with its formation, suggesting that the site had not been totally abandoned by man, even though neither cultivation nor grazing were taking place.

Above 150 cm the fauna begins to revert to its former more open character, suggesting clearance of the vegetation in the ditch. Between *ca.* 143 and 150 cm there is a zone of chalk rubble which extends across the entire width of the ditch as a gentle curve, continuing without break onto the weathering ramps. On the latter are a series of criss-cross ploughmarks—grooves incised into the chalk rock and filled with humic material. These continue across the fill of the ditch at the level of the zone of chalk rubble which can thus be interpreted as being due to ploughing (Fowler and Evans, 1967), and the numerous sherds of Beaker pottery in the material suggest that this process took place around 2000 BC.

Above is the mull-humus horizon of the buried soil, stone-free and with a well-developed crumb structure. The fauna is dominated by open-country species, particularly *Vallonia excentrica*, *Helicella itala* and, in its upper levels, *Pupilla*. Curiously, *Vallonia costata* does not recover its former abundance and by 120 cm is virtually absent. An environment of short-turfed, dry grassland, probably kept open by grazing animals, is envisaged at this stage.

The ploughwash above is largely of medieval origin to judge by the few sherds recovered from the deposit, though at its base (*c.* 120 cm) some Roman sherds were present. The fauna throughout is an open one, dominated by *Vallonia excentrica*, *Helicella itala* and *Pupilla*. Towards the surface these die out, and others such as *Agriolimax*, *Hygromia striolata* and *Helicella virgata* which characterize the present-day fauna become predominant.

HEMP KNOLL

Location: 4·3 km south-west of Avebury, Wilts. SU 068674 (Fig. 143).

Situation and geology: The site is at 625 ft OD on a spur known as Hemp Knoll overlooking a broad dry valley, and only 400 m south-east of the Beckhampton Road Long Barrow (p. 248). Upper Chalk is close to the surface, but is overlain by a thin layer of quartzose calcareous drift (Evans, 1968b).

Archaeology: A Beaker round barrow, built over a Neolithic occupation horizon which consisted of five pits containing a flint industry and Windmill Hill pottery (M.P.B.W., 1966: 7). A radiocarbon date of 1795 ± 135 BC (NPL 139 C 154) was obtained for the primary burial (R. Robertson-Mackay, personal communication).

Today the site is under the plough. The barrow ditch is completely infilled and the mound not more than 0·5 m high.

Beneath the barrow was a pronounced but rather thin buried rendsina soil (max. thickness 15·0 cm). The snail fauna from this (Fig. 130b) is of open-country type with the two species of *Vallonia* predominant—as is characteristic of other Neolithic and Early

10. CAVES, DRY VALLEYS, LYNCHETS AND DITCHES

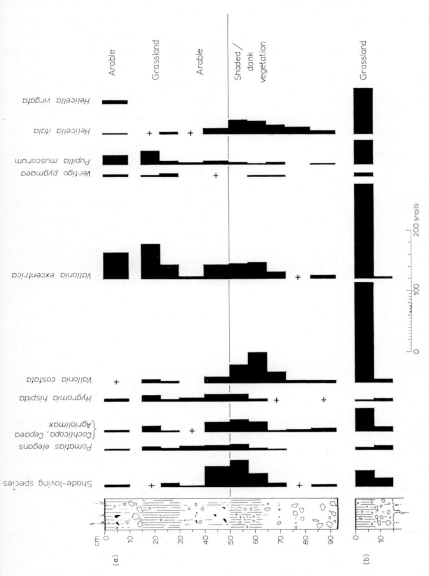

Figure 130. Hemp Knoll. (a) Ditch. (b) Buried soil beneath barrow.

Bronze Age grassland faunas in north Wiltshire (Fig. 144). Clearance of woodland may safely be assumed to have taken place on this site in view of the evidence for tree cover on other sites such as Beckhampton Road (p. 248) and South Street (p. 257) close by. If this took place in the Neolithic period as is suggested by the occupation horizon beneath the barrow, the environment may well have remained open right up to the time when the Beaker barrow was built.

The barrow ditch was shallow (*ca.* 0·9 m), flat-bottomed and with sloping sides, the latter probably caused largely by frost weathering. The fill at the point sampled showed the following stratigraphy (Fig. 130a):

Depth below surface (cm)	
0–10	Modern plough soil.
10–15	Coarse chalk rubble. Material derived from chalk of barrow mound and dragged down by ploughing.
15–30	Dark chalky loam (10 YR 3/1), with a few chalk fragments, and a well-developed crumb structure. Recent mull-humus horizon.
30–50	Pale chalky loam (10 YR 5/1), with flints and chalk lumps throughout. Ancient plough soil.
50–*ca.* 65	Dark-grey (10 YR 3/1) humic loam with numerous chalk fragments and redeposited calcium carbonate (pseudomycelium). Fossil soil horizon.
ca. 65–93	Coarse chalk rubble in a fine chalky matrix; slightly humic above, becoming clean towards the base. Primary fill.

A series of samples from this section was analysed for snails and the results plotted in terms of *absolute* abundance since there were generally too few shells present for percentage values to be meaningful.

The primary fill (up to 65 cm) contained few shells and probably accumulated too rapidly to permit colonization of its surface by snails.

The fauna in the fossil soil horizon (50–65 cm) is a mixed one, of shade-loving and grassland species, suggesting a stable surface with a continuous cover of vegetation, neither grazed nor cultivated.

Above (30–50 cm), there is evidence once more of rapid physical weathering and accumulation, this time probably of a directly artificial nature and caused by ploughing. The number of snails decreases. Later, cultivation gave way to pasture and a worm-sorted soil formed (15–30 cm) in which the snail fauna indicates an extremely open and dry environment—probably short grass grazed by sheep or cattle. *Vallonia excentrica* and *Pupilla muscorum* are the only important elements.

More recently, probably in this century, deep ploughing resulted in the accumulation of a layer of coarse chalk rubble, dragged down from the barrow mound. Above this, the impoverished fauna in the modern plough soil reflects the present-day austere environmental conditions on the Downs.

The faunas from the three buried soils at Hemp Knoll demonstrate neatly the way in which varying congeniality of the environment for molluscan life influences their diversity. Thus in the buried soil beneath the barrow—a downland environment—only five species are at all common, and two of these, *Vallonia costata* and *V. excentrica* each attain 30% abundance (Fig. 132a). In the ditch (50–57·5 cm), the maximum abundance attained by any one species is 18%, by *V. costata* and there is a more even distribution of numbers among the various species present (Fig. 132b). In the recent grassland soil (Fig. 132c), one species alone predominates, *V. excentrica* at about 60%, with *Pupilla* (23%) as the only other abundant element of the fauna.

ROUGHRIDGE HILL

Location: Roughridge Hill, 2·8 km north-east of Bishops Cannings, Wilts. SU 060660 (Fig. 143).

Situation and geology: The site is on the floor of a dry valley. The geological solid is Middle Chalk and there is a thin cover of calcareous periglacial drift (Evans, 1968b).

Archaeology: Two round barrows, probably of Early Bronze Age date, and contemporaneous. Barrow "A" corresponds to Grinsell's Bishops Cannings 61 and barrow "B" to Bishops Cannings 62. Barrow "A" comprised a ditch and a turf mound, beneath which was an indistinct soil horizon. Barrow "B" sealed a well-defined buried soil profile of rendsina type, but had no ditch. The barrows overlay a Neolithic settlement area of the Windmill Hill culture (Proudfoot, 1965).

The fauna in the buried soil beneath barrow "B" is dominated by *Vallonia costata* and *V. excentrica* each at about 35%. It is similar to that at Hemp Knoll, and reflects an open and dry environment (Fig.

132d). As at Hemp Knoll, woodland clearance was probably accomplished during the Neolithic occupation of the site.

The ditch fill of barrow "A" showed the following stratigraphy (Fig. 131):

Depth below surface (cm)	
0–15	Modern plough soil (Fig. 131e).
15–30	Dark, stone-free chalky loam. Recent mull-humus horizon (d).
30–56	Stony loam. Ploughwash (c).
56–80	Fine, humified chalky loam. Buried soil (b).
80–110	Fine chalk rubble becoming coarser in the angles of the ditch (primary/secondary fill) (a).

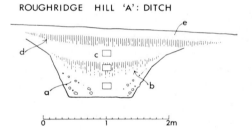

Figure 131. Roughridge Hill, Barrow "A". Transverse section through barrow ditch. a = primary fill; b = buried soil; c = ploughwash; d = recent grassland soil; e = modern plough soil.

Three samples were analysed for snails (location in Fig. 131) and the results of two of these are presented graphically in Fig. 132e and f, where they are compared with the fauna from the buried soil beneath barrow "B".

The fauna from the primary fill (not represented graphically; see Appendix, Table 13), is similar to that in the buried soil of barrow "B", though, as would be expected, with a slightly higher abundance of shade-loving species. In the buried soil horizon (56–80 cm), however, there is a distinct increase in shade-loving forms and an altogether more uniform distribution of numbers among the various species present. Only *Vallonia costata* is noticeably more abundant than the other species. In the ploughwash horizon (30–56 cm) the fauna once more becomes restricted and of open-country type (Fig. 132f).

As at Hemp Knoll, these changes illustrate the reaction of the land-snail population to the changing geniality of the habitat, the most sensitive indicators being the overall diversity of the fauna and the behaviour of the two species of *Vallonia*. Thus prior to the construction of the barrows in the Early Bronze Age, the environment was very dry and open, and the snail fauna (Fig. 132d), dominated almost to the exclusion of all other species by *Vallonia costata* and *V. excentrica*, reflects these austere conditions. The ditch of barrow "A" provided a

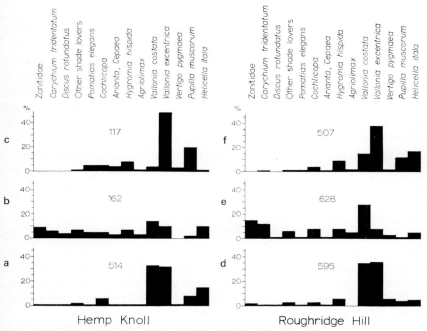

Figure 132. Hemp Knoll and Roughridge Hill. Molluscan faunas from selected levels.

refuge for many of the mesophile elements of the fauna which flourished for a time in an environment more humid than previously (Fig. 132e); *V. costata* was particularly favoured by these conditions, *V. excentrica* less so—a pattern repeated at Hemp Knoll and South Street. Finally, the site was cleared and ploughed, and the fauna once more became restricted and of open-country type, but now with *V. excentrica* predominant (Fig. 132f).

BADBURY EARTHWORK

Location: 7·7 km south-east of Blandford Forum, Dorset and 0·4 km west-south-west of the Iron Age site of Badbury Rings. ST 956030 (Fig. 142).

Situation and geology: The site is on the plateau area between the Rivers Allen and Stour at 225 ft OD. The geological solid is Upper Chalk; drift cover is absent.

Archaeology: The site is a late Iron Age defensive earthwork of linear type comprising a bank and ditch (M.P.B.W., 1966: 4). Archaeological evidence for occupation prior to the construction of the earthwork stretches back over a millennium, and includes Deverel-Rimbury and Beaker material.

Beneath the bank there was a thin but well-defined rendsina soil (Fig. 133b) consisting of black loam with small chalk lumps throughout. The molluscan fauna is of open-country type, *Pomatias elegans*, *Agriolimax*, *Pupilla*, *Vallonia costata* and *Helicella itala* being the important elements. The abundance of *P. elegans* in a fauna which otherwise reflects open conditions is anomalous at first sight, in view of its more usual association with scrub or woodland habitats. However, it does occasionally occur in fields or on open downland at the present day (M. P. Kerney and R. A. D. Cameron, personal communication) and was apparently doing so in this context. It has also been recorded alive in sand-dune habitats (Boycott, 1921a).

It is not possible to say further what form of land use was being practised prior to the construction of the bank. The paucity of both species of *Vallonia* by comparison with their abundance on other sites in Wiltshire and Dorset, and the abundance of *Pupilla*, suggest an arable environment, but it is impossible to be sure about this.

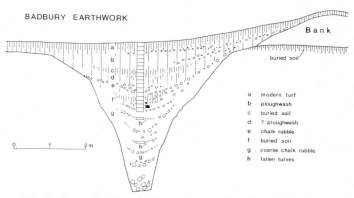

Figure 133. Badbury Earthwork. Transverse section through bank and ditch.

10. CAVES, DRY VALLEYS, LYNCHETS AND DITCHES

A section through the ditch fill showed the following stratigraphy (Figs 133 and 134a):

Depth below surface (cm)	
0–15	Dark stone-free loam. Modern turf.
15–25/30	Small chalk pellets lining earthworm burrows. Pea-grit zone.
25/30–70	Pale-brown chalky loam with numerous chalk fragments. Ploughwash.
70–90	Dark-grey humic loam. Buried soil horizon.
90–110	Pale-brown chalky loam, with numerous chalk fragments. Probably a ploughwash.
110–140	Chalk rubble and finer débris. Deliberate backfill.
140–180	Grey chalky loam. Secondary fill, humified throughout and essentially a soil horizon.
180–400	Coarse chalk rubble and turves. Primary fill.

A series of samples was cut from this section and analysed for snails. The results are presented in graphical form in terms of relative abundance (Fig. 134a).

The fauna in two fallen turves (267-272 and 197·5–202·5 cm) incorporated into the primary fill is almost identical to that from the buried soil beneath the bank, thus confirming, as at South Street, that these turves originate from the pre-bank soil.

The lowest sample of the secondary fill (170–180 cm) yielded a sparse fauna containing open-country and shade-loving species. Above, is a layer of grey chalky loam (140–170 cm) which although not possessing the usual characteristics of a rendsina soil—dark colour and well-developed crumb structure—nevertheless probably accumulated very slowly as indicated by its rich snail fauna. Open-country species are virtually absent (less than 2%), and the fauna is dominated by zonitids, *Carychium* and *Discus*. An environment heavily shaded and free from disturbance is indicated. At this level in the ditch, it is not possible to be certain how representative this fauna is of the environment in the area as a whole, but the apparently rapid spread and colonization by shade-loving species and their virtual dominance suggest that substantial refuges were close at hand.

Between 110 and 140 cm, there is a layer of coarse chalk rubble concentrated on the bank side of the ditch, and probably the result of a deliberate attempt to level the site, or at least reduce its contours,

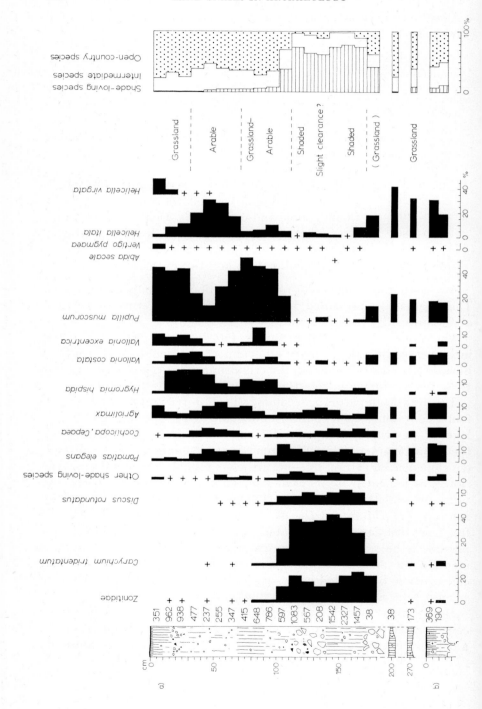

prior to cultivation. Apart from a slight increase of open-country species to about 10%, however, this episode is not reflected in the molluscan fauna.

Above 110 cm the fauna changes abruptly to one of open-country type, and by 100 cm the abundance of shade-loving species has fallen to less than 10%. An episode of woodland clearance is indicated by these changes. Subsequently the site was cultivated, resulting in the formation of a layer of ploughwash, and has remained open to the present day.

The faunal changes which take place above 110 cm are probably representative of the area as a whole, and may be subdivided into a number of stages. Thus between 90 and 110 cm there is a layer of what is probably ploughwash—pale in colour and stony in texture. Above, between 70 and 90 cm is a darker horizon which is relatively stone free, and may be a soil formed when ploughing ceased. The fauna is dominated by *Pupilla*, which at the soil surface (70–80 cm) exceeds 50%, suggesting a very dry and open environment.

Above this level there is a further accumulation of ploughwash (25/30–70 cm) indicating renewed cultivation of the site. *Pupilla* rapidly declines and *Helicella itala* becomes the dominant element. Today, the latter is generally absent from cultivated land, but it has been found in abundance in a few other ploughwash deposits (p. 182) and may formerly have thrived in arable habitats. The introduction since the Iron Age of three species of *Helicella* which today are often found abundantly in cultivated land may also have contributed to the present restriction of *H. itala*. In this layer too, there is a marked increase of *Pomatias elegans*, another species out of keeping in an arable context, but here the problem is not an ecological one. *P. elegans* has a tough shell, and apices are often preserved differentially with respect to the more fragile shells of other species (p. 212). When the examples of *P. elegans* between 40 and 70 cm were compared with those below 140 cm it was found that while the latter comprised a large proportion of adult examples, those from the ploughwash were all small and weathered apices. The same applies to a lesser extent to *Cepaea* and *Cochlicopa*, and it will be noticed that these two behave in sympathy with *Pomatias elegans*.

Finally (0–15 cm) the fauna reverts to the kind which occurred below the ploughwash horizon (70–90 cm), suggesting a return to more stable soil conditions, which do in fact prevail on the site today.

Figure 135. Ascott-under-Wychwood. Transverse section through the infill of a Roman ditch. (Scale, 1·0 m)

ASCOTT-UNDER-WYCHWOOD

Location: 1·2 km south of Ascott-under-Wychwood, Oxon. SP 299175 (Fig. 142).

Situation and geology: On gently sloping ground on the side of a tributary valley of the River Evenlode. The geological solid is Inferior Oolite limestone, overlain by extensive areas of calcareous loamy material (p. 251) into one of which the Roman ditch is cut.

Archaeology: A Roman ditch, uncovered during the excavation of a Neolithic long barrow (D. G. Benson, personal communication).

A section through the ditch showed the following stratigraphy (Figs 135 and 136):

Depth below surface (cm)	
0–7·5	Dark-grey (10 YR 3/2) stone-free loam with a pronounced crumb structure. Worm-sorted mull-humus horizon (Fig. 136e).
7·5–10	Pea-grit zone.
10–ca. 26	Grey (10 YR 4/2) stony loam, weakly bedded. Early-twentieth-century plough soil (d).
ca. 26–30	Zone of coarse limestone rubble, the majority of the stones lying flat. Probably due to ploughing.
30–60	Brown stony loam. Ploughwash (c).
60	Zone of flat limestone slabs. Deliberate infill or ploughing, possibly during the Roman period.
60–70	Dark-brown (10 YR 4/3) stone-free loam with a well-developed blocky structure. Shells abundant. Fossil soil horizon of possible Roman age (b).
70	Stone horizon. Probably brought down by worm action from the soil above.
70–100	Yellowish-brown (10 YR 5/4) fine limestone rubble. Occasional tip lines of stones and humic material. Primary/secondary fill (a).

A series of samples was cut from this section and analysed for snails (Fig. 136). The results have been presented as a histogram of relative abundance (Fig. 137).

The fauna of the primary fill (up to 70 cm) is dominated by open-country species, particularly *Vallonia excentrica* (*ca.* 40%) and *Helicella itala*. The *Punctum–Vitrina–Retinella radiatula* group is well represented, in contrast to its paucity in the horizon above. A dry, open environment is to be envisaged at this stage, with perhaps some areas of bare ground. As the ditch is so shallow and broad, it is probable that this fauna differs little from that in the area as a whole.

In the fossil soil horizon the environment becomes stabilized and possibly shaded. Shade-loving species increase to 28% and open-country species fall to about half their former abundance. However, all one can be really sure about is that cultivation was not taking place, nor was the site being grazed.

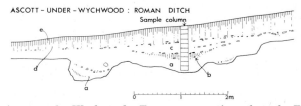

Figure 136. Ascott-under-Wychwood. Transverse section through Roman ditch. a = primary/secondary fill; b = buried soil; c = ploughwash (tertiary fill); d = recent plough soil; e = modern turf.

Above 60 cm the deposits are of artificial origin reflecting two phases of agriculture. In the first, between 60 and 30 cm, shade-loving species are moderately abundant (*ca.* 18%) suggesting that accumulation was gradual, with ploughing perhaps *around* rather than *in* the ditch. At 30 cm there is a well-marked zone of limestone rubble probably caused by recent ploughing of the adjacent long barrow. The fauna becomes extremely open. This phase almost certainly relates to a known episode of ploughing which occurred on the site about forty years ago.

The fauna in the modern turf reflects the revertance of the environment to the present-day conditions of stable grassland.

Ditches on River Gravel

In many of the river valleys of south and east England, the gravels are calcareous, and give rise to conditions suitable for the preservation of shells. Frequently these are in a poor state of preservation due to the low calcium carbonate content of the deposits, and only the thick shells of the larger species such as *Cepaea* remain. Nevertheless, under certain conditions archaeological features such as ditches, pits and

10. CAVES, DRY VALLEYS, LYNCHETS AND DITCHES

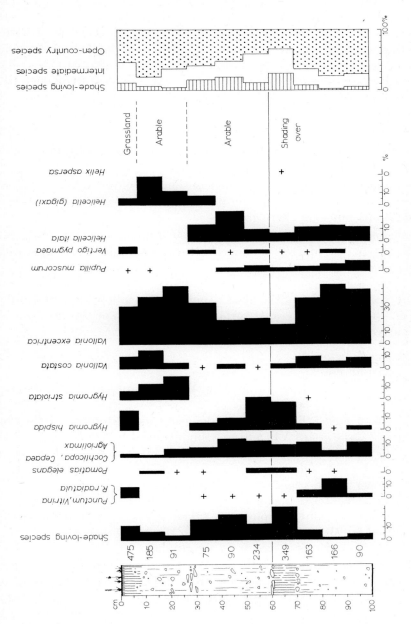

Figure 137. Ascott-under-Wychwood, Roman ditch.

wells cut into river gravel are sometimes rich in shells, particularly when they extend down to the water table where the rate of water percolation, and hence leaching, is reduced. The type of situation obtaining in the two sites discussed below in which waterlogged organic material is overlain by shell-rich loamy clay accumulated in marsh or swamp conditions, may turn out to be a constant feature of such sites (Fig. 138).

MAXEY

Location: Gravel pit at Maxey, Northants., in the Welland Valley. TF 126075 (Fig. 142).

Situation and geology: The geological solid is Oxford Clay, but this is overlain by thick terrace deposits of the River Welland consisting of limestone, flint and chert gravel.

Archaeology: A first/second century AD Roman enclosure ditch. Site OS 124 in "A Matter of Time" (Royal Commission on Historical Monuments, 1960) (W. G. Simpson, personal communication).

A section through the ditch showed the following stratigraphy (Fig. 138) (datum is gravel surface after *ca.* 30 cm of topsoil had been removed):

Depth below datum (cm)	
0–30	Dark-brown (10 YR 3/3) gravelly loam. Plough-wash.
30–45	Partially waterlogged clay, exhibiting yellow/red (5 YR 5/8) mottling. Gleyed horizon.
45–83	Black organic mud, rich in macroscopic plant débris and insect remains, especially towards the base.
83–89	Clean gravel. Primary fill.

A series of eight samples was taken through these deposits and analysed for Mollusca (Fig. 138; Table 16).

No shells were recovered from the primary fill.

In the overlying organic material a few shells were present, but in such low numbers as to suggest their having been accidentally incorporated, rather than their being the remains of snails once living in the ditch. Whether this material was deliberately dumped or whether it accumulated naturally, is uncertain, but its effect was to impede

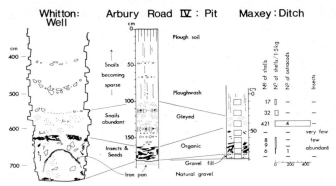

Figure 138. Transverse sections through gleyed and waterlogged deposits, to indicate the antipathetic distribution of snails and organic remains (seeds and insects).

radically the drainage and to create a semi-aquatic environment at its surface. Under these conditions a fauna of land and freshwater snails existed until, at a later stage, the ditch was completely infilled by ploughing and a more terrestrial environment attained.

The semi-waterlogged or gleyed condition of the horizon above the organic layer (30–45 cm) was strikingly apparent in the field as a blue/grey- and orange-mottled clay. The composition of the molluscan fauna suggests that two groups of species are present, each of which would have occupied its own ecological niche in the environment. Thus about 50% of the fauna consists of "slum" species (p. 200), freshwater molluscs showing a preference for, or tolerance of, poor water conditions such as small bodies of water subject to drying, to stagnation and to considerable temperature variations (Sparks, 1961). Included here are *Lymnaea truncatula*, *Aplexa hypnorum*, *Planorbis leucostoma* and *Pisidium casertanum*. A second group comprises those snails which are more terrestrial in their habitat preferences and live on vegetation above the level of the water and at the sides of the ditch, rather than in the ditch itself. Two of these, *Succinea* spp. and *Zonitoides nitidus*, are obligatory marsh species (13%); the remainder are not so restricted. There are too, fragments of two "catholic" freshwater species, *Bithynia* and *Planorbarius corneus*, which prefer permanently aquatic conditions.

The fauna has been presented graphically in terms of these four groups (Fig. 139a). To construct this histogram, the total fauna from the ditch (62–82·5 cm) together with a spot sample from the gleyed horizon were used (Table 16).

The ditch was probably sufficiently waterlogged to allow standing water at least temporarily, and sufficiently undisturbed to permit the growth of reeds and rushes in which *Succinea* and *Zonitoides nitidus* probably lived. The presence of *Succinea*, *Vallonia* and *Vertigo pygmaea* which are generally indicative of open habitats, probably indicates the absence of a dense growth of trees and shrubs along the sides of the ditch but does not preclude their growth altogether.

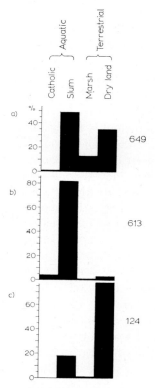

Figure 139. Slum faunas from Roman sites. (a) Maxey. (b) Arbury Road IV. (c) Arbury Road II.

The absence of *Helix aspersa* from this site and others of Roman age in the Welland Valley (Evans, unpublished) is worthy of comment. In the extreme south of England, this species has been found in several first to third century AD contexts (p. 176), while at Arbury Road, Cambridge (p. 349) it occurs in the fourth century, but not before. Ecologically there is no reason why *Helix aspersa* should be absent from the Maxey ditch (the Arbury Road fourth century site is similar)

10. CAVES, DRY VALLEYS, LYNCHETS AND DITCHES

and its absence is therefore probably due to the fact that it had not reached the area by the end of the first century AD.

ARBURY ROAD

Location: 2·6 km north of Cambridge city centre. TL 455612 (Fig. 142).
Situation and geology: The site is on level ground on sands and gravels of the Barnwell terrace of the River Cam (Sparks and West, 1965: 26). The geological solid is Gault Clay.
Archaeology: The site is a Roman farmstead totally levelled by medieval and later ploughing. Like that at Maxey, the site was located as a crop mark from the air (J. Alexander, personal communication).

Two loci were sampled for Mollusca.

Site AR IV. Part of a fourth century ditch, enlarged into a rubbish pit. It showed the following stratigraphy (Figs 15 and 138):

Depth below surface (cm)	
0–17	Modern plough soil.
17–100	Sandy loam, mottled orange towards the base; occasional bands of gravel. Ploughwash.
100–142	Clay loam, yellow- and grey-mottled (gleyed); two lenses of dark organic material. Snails abundant.
142–162	Layer of gravel.
162–182	Organic material.
182	Ditch bottom.
184–185	Iron pan in natural gravel.

A spot sample from the gleyed horizon was analysed for Mollusca (Fig. 139; Table 17). The fauna is dominated by a single species, *Planorbis leucostoma* (ca. 80%), one of the "slum" group of freshwater snails (Sparks, 1961). The presence of *Lymnaea peregra* (10%), a "catholic" freshwater species, probably indicates that the ditch never became completely dry at any time of the year. Otherwise the fauna is composed of a few other slum species and sporadic examples of terrestrial species. The absence of *Lymnaea truncatula* suggests that the surrounding land may have been arable rather than damp pasture, but this is a point on which pollen or insects could provide more definite information.

There is a similarity between this site and the deposits of the first century ditch at Maxey (p. 347) which I have tried to bring out diagrammatically in Fig. 138. In each case, deposits close to the base have remained totally waterlogged and anaerobic since their accumulation. They are rich in macroscopic plant remains, particularly seeds, pieces of wood and mosses, and in insects. Snails are few. Above, in the gleyed horizon aeration is partial and thus organic material is not preserved; shells however are abundant. An almost identical situation was present in a Roman well at Whitton, Glam. (Fig. 138) (M. G. Jarrett, personal communication).

Site AR II. A Roman ditch. A spot sample from gleyed material near the base of the ditch yielded a fauna which differed significantly from that of AR IV in the far greater number of land snails (78%) (Table 17). Nevertheless, "slum" species are still abundant (18%) and *Planorbis leucostoma* is again the dominant member of this group. As at AR IV and Maxey, open-country species are present, precluding the presence of a dense growth of trees and shrubs along the edge of the ditch.

The faunas from Maxey, AR II and AR IV are compared in Fig. 139. The most striking difference between Maxey and Arbury Road is the absence of marsh species at the latter. This possibly indicates that the ditches at Arbury Road were devoid of much vegetation above the water level, whereas at Maxey a growth of reeds is to be supposed.

11
Habitat Change in the Calcareous Regions of Britain

This book concerns essentially the techniques of snail analysis and their application to archaeological horizons of Post-glacial age. In our discussion we have covered a wide range of topics—soils, archaeology, vegetation, snails, and man—often from the historical angle and almost always from the point of view of the individual site. I feel that it will be useful, finally, to draw together some of this material in an attempt to see the pattern of habitat change in the country as a whole.

The Quaternary era, the last two million years or so of the earth's history, is characterized by a series of alternating warm and cold periods—some long, to be measured in tens of thousands of years, others of short duration lasting no more than a few decades or centuries. The last three major cold periods saw the spread of ice sheets over much of Britain, the most extensive as far south as north Devon, the Cotswolds, the Chilterns and the Thames Valley; they are termed the Lowestoft (the earliest), Gipping and Weichselian Glaciations. In reality, the period of ice advance, at least in the Weichselian, occupied a relatively short period of time in the glaciation as a whole (*ca.* 15%), lasting from *ca.* 20 000 to 10 000 BC. At other times, and in the zone beyond the ice sheet during the period of ice advance, an arctic climate generally prevailed, in which the ground was permanently frozen (permafrost) and physical weathering processes such as frost-shattering, solifluxion and loess formation were rife.

The intervening major warm periods, in which a temperate climate similar to that of today prevailed for a sufficient length of time to allow the establishment of deciduous forest in Britain, are termed interglacials. These are the Cromerian (Upper Freshwater Bed), the Hoxnian and the Ipswichian (Eemian). The temperate period which followed the Weichselian, and generally known as the Post-glacial or Holocene, is thought to be yet another interglacial, and is sometimes called the Flandrian.

GLACIAL	INTERGLACIAL
	Post-glacial or Flandrian
Weichselian	
	Ipswichian or Eemian
Gipping	
	Hoxnian
Lowestoft	
	Cromerian (Upper Freshwater Bed)

An up-to-date synthesis of the British Quaternary is given by West (1963).

Molluscan faunas are known from these various glacials and interglacials but are not strictly within the province of this book. Several rich interglacial faunas from the Hoxnian and Ipswichian have been recorded (Fig. 54) (Sparks, 1964b; Kerney, 1959; 1971b) and a number of faunas from the colder periods of the Pleistocene, characterized by a very limited number of species (Fig. 140) (Kennard and Woodward *in* Warren, 1912; Sparks, 1957b; Large and Sparks, 1961; Coope *et al.*, 1961; Sparks *in* Shotton, 1968: 399; Kerney, 1971a). From the Weichselian itself, that is excluding the early and late interstadial horizons of the glaciation, not more than ten marsh and terrestrial species are known, and these mainly from periods of minor climatic amelioration.

The transition period between the Weichselian and the Post-glacial lasted for approximately 4000 years, from 12 000 to 8300 BC, and is known as the Late Weichselian. The sequence of weathering and landscape changes which took place on the Chalk during this time has been established at various sites in southern England—in the North and South Downs (Kerney, 1963; 1965; Kerney *et al.*, 1964), in the Chilterns (Evans, 1966b), in Wiltshire (Evans, 1968b, 1969c), in north Berkshire (Paterson, 1971) and Somerset (ApSimon *et al.*, 1961).

There were two episodes of frost climate (corresponding to zones I and III of the pollen zone scheme) when physical weathering was predominant. Solifluxion processes were active on hillsides causing the wasting of large quantities of chalk débris which were deposited in valley floors and at the foot of the escarpments, often spreading out for many miles on to the plains beyond (Fig. 9). On level ground,

11. HABITAT CHANGE IN THE CALCAREOUS REGIONS 353

Figure 140. Cold-climate faunas from England, with a loess fauna from Czechoslovakia for comparison.

where solifluxion was inactive, frost-heaving caused involutions and other cryoturbation structures to form. These generally contain material of loessic rather than solifluial origin, and their molluscan faunas reflect drier local conditions than those obtaining during the formation of solifluxion deposits (Fig. 141), (Evans, 1968b: Fig. 5). The vegetation was tundra—a sparse cover of grasses, various lichens and mosses, and shrubs such as dwarf birch and juniper—though over much of the Chalk surface the ground was too unstable for a permanent plant cover to be established.

Between zones I and III was a period of warmer climate which lasted for about 1200 years. This is the well-known Allerød Interstadial (zone II), and during this time the Chalk surface became stabilized and soil formed, enabling the establishment of a continuous vegetation cover; locally woodland of pine or birch grew up. In the floors of coombes, and on the plains beyond the escarpments, this soil is fre-

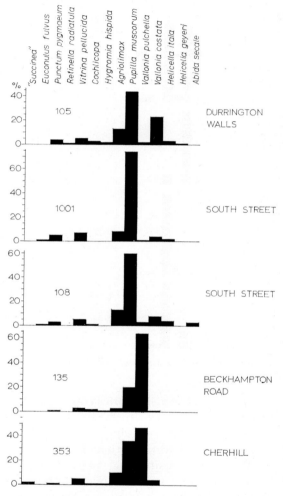

Figure 141. Late Weichselian faunas. (After Evans, 1968b: Fig. 5.)

quently found today, preserved beneath zone III solifluxion débris and later deposits of Post-glacial age (Fig. 71). No archaeological material has ever been found stratified on or in the Allerød soil which is particularly unfortunate since the horizon is an excellent example of a fossilized terrestrial land surface, which is easily located and identified, preserved intact over a wide area of southern England. In the Netherlands where the Allerød Interstadial is represented by the Usselo layer within the Younger Coversands, there is an associated Upper Palaeolithic industry.

As already pointed out, the Allerød Interstadial possibly includes an episode of minor climatic deterioration when solifluxion resumed for a short time (Evans, 1966b; Paterson, 1971). Iversen (1954) suggested that this was the case in Denmark on the basis of pollen analysis; and several pollen diagrams from England show a similar pattern (Smith A. G., 1965; Pennington, 1970).

There was also a short period of relatively mild climate within zone I, the Bølling Interstadial, which lasted for perhaps a few hundred years. This is seen in one or two sites in Kent and Sussex as a thin organic horizon, interstratified in zone I solifluxion deposits (Kerney, 1963; 1965), but is more often reflected solely by a temporary increase in the abundance of Mollusca.

		DATES BC (C–14)
		—— 8300 ——
Solifluxion	Zone III	
		—— 8800 ——
Soil formation (?Minor solifluxion) Soil formation	Zone II—Allerød	
		—— 10 000 ——
Solifluxion	Zone Ic	
Soil formation	Zone Ib—Bølling	
Solifluxion	Zone Ia	
		—— 12 000 ——

The Late Weichselian snail fauna has been described and illustrated in detail by Kerney (1963; Kerney et al., 1964). It comprises just over twenty species of land and marsh Mollusca and a number of freshwater forms, the majority of which are climatically tolerant, having Palaearctic ranges and extending far northwards into Scandinavia. Two others, *Abida secale* and *Helicella itala* are unknown from the Scandinavian mainland and have a more southerly distribution in Europe. *H. geyeri*, now extinct in Britain, is another member of this group. The presence of a few warmth-loving species such as these in an assemblage of otherwise cold-tolerant forms is a phenomenon displayed by other groups of plants and animals (e.g. the insects) during the Late Weichselian, and the problem has been discussed in general terms by West (1968: 124).

As well as establishing some of the major features of the Chalk landscape, the Late Weichselian, and indeed earlier periods of periglacial weathering, brought about the formation of superficial deposits which in many places constitute the parent material of the Postglacial soil. In upland areas of level or gently sloping ground it is to be questioned whether chalk rock *per se* was ever exposed at the surface to weathering at the beginning of Post-glacial time. A cover of calcareous drift, enriched to a greater or lesser degree with quartz and other minerals (Perrin, 1956), and of varying degrees of thickness (Williams, 1964) is more likely. The spread of calcareous solifluxion débris on to the low-lying areas at the foot of the escarpments also altered the parent material of the Post-glacial soil, in effect extending the lime-rich environment on to what would otherwise have been clay or greensand (Fig. 9). In both cases it can be argued that the Post-glacial soil was made more suitable for subsequent cultivation by these processes, a knowledge of which is thus of fundamental importance in understanding the pattern of human settlement in later times on and around the Chalk (Wooldridge and Linton, 1933).

"The Late-glacial Period was the last during which the Chalk landscape was subjected to severe erosion. With the rise of temperature and the spread of forests in the Post-glacial, all processes of rapid physical weathering ceased. Apart from the effects of solution, local accumulations of hillwash, minor spring-sapping, and slight interference by man, the present features of the Chalk escarpment and its associated dry valleys and coombes are essentially those in existence at the end of zone III, approximately 10 000 years ago" (Kerney, 1963: 249).

The climatic improvement of the Post-glacial, which began around 8300 BC, led to many changes in the habitat. Chemical weathering became predominant, resulting in the formation of soil on all but the steepest slopes and unstable areas such as screes and sand-dunes. The dry-flush soils of the Late Weichselian, maintained by frost-heaving and solifluxion, gave way to soils which by the end of the Boreal period had become deep brown earths (Dimbleby, 1965). The open tundra vegetation was soon replaced by forest, first of birch, then of pine and later of a succession of broad-leaved trees—oak, elm, lime and alder—the composition of which, prior to the advent of radiocarbon dating, provided the basis for our Post-glacial chronology (p. 4).

POLLEN ZONE	PERIOD	VEGETATION	SOILS AND SEDIMENTS	YEARS BC (C–14)
VIII	Sub-atlantic	Rise of beech	Hillwashing	
				550
VIIb	Sub-boreal	Forest clearance	Blown sand	
				3000
VIIa	Atlantic	Mixed oak forest Alder	Tufa and blanket bog	
				5300
VI	Boreal	Mixed oak forest	Soil formation	
V		Pine		
				7000
IV	Pre-boreal	Birch		
				8300

The melting ice sheets brought about a rise of the sea from -100 m in the Late Weichselian to approximately its present-day level at 4500 BC, Britain having been cut off from the Continent by *ca.* 5000 BC. By the end of the Boreal, the mean annual temperature exceeded that of the present day by 2–3°C. a situation which persisted through the Atlantic and well into the Sub-boreal. This period of time is known as the Post-glacial climatic (or thermal) optimum (altithermal), and the evidence for it is largely biological, resting on the extension of certain warmth-loving plants and animals beyond their present-day northern limits. For example, the contraction of the ranges of *Pomatias elegans*, *Lauria cylindracea* and *Ena montana* since the Sub-boreal is probably due to temperature decline (Kerney, 1968a).

The early Post-glacial history of the Chalk is obscure due to the paucity of deposits and soils of Pre-boreal and Boreal age. At only one locality, Brook in the North Downs (Kerney *et al.*, 1964), has any detailed work been done, and even this is a valley site, not representative of the Chalk environment as a whole. Here, in the Boreal period, soil formation took place in a marshy environment which, as the snails show, was initially open but later became invaded by woodland.

Towards the very end of this period, hunter-gatherer communities are known to have occupied low-lying sites in England and Wales

(p. 301), which were later overwhelmed by swamping and the deposition of calcareous tufa. The environment in which these groups were living was, by analogy with Brook, probably a forested one, though until detailed data is available from other sites, more cannot be said. At three rather more terrestrial sites, Ascott-under-Wychwood (p. 255) (though not on the Chalk), South Street (p. 257) and Avebury (p. 269), land-snail faunas of probable Boreal age indicate a cover of light woodland. At Ascott-under-Wychwood there is an associated Mesolithic industry.

The most characteristic features of the Boreal fauna appear to be the absence of *Oxychilus cellarius* and *Pomatias elegans*, the paucity of *Discus rotundatus*, and the presence of *Oxychilus alliarius* and of two species, *Vertigo genesii* and *Discus ruderatus*, which have since become extinct.

During the Atlantic period, widespread swamping took place in river valleys and low-lying areas at the foot of the escarpments, often accompanied by the formation of thick deposits of tufa laid down in an environment of tree-shaded swamp. This change was climatically induced, being due to an increase in precipitation, and was reflected in other parts of Britain by a rise in lake levels, the formation of blanket bog in upland regions and by a marked increase of alder (*Alnus*) in the pollen record. Evident standstill phases during the build-up of tufa deposits as indicated by interstratified soil horizons (p. 299) suggest periods of drier climate within the Atlantic period, periods which are reflected also in the stratigraphy of ombrogenous mires. These deposits of tufa now occupy slight terraces above the modern floodplain level, and being well-drained, are suitable for arable land or pasture.

It has been recognized for some time that many areas of Britain which are today characteristically open country were once forested, and would, if left untended by man and his animals, eventually revert to their former state. The East Anglian Breckland and the North York Moors are classic examples—the former having been deforested by Neolithic man (Godwin, 1944), the latter during the Bronze Age (Dimbleby, 1962). In view of this, and on present-day ecological grounds as well, it has generally been assumed that the Chalk and limestone upland regions of Britain too were once forested, having suffered a similar history of forest clearance through the activities of prehistoric farmers. By means of snail analysis this has now proved to have been the case, clearance having taken place on some sites as

11. HABITAT CHANGE IN THE CALCAREOUS REGIONS

early as the fourth millennium BC (Evans, 1971b), or, if we accept the C-14 correction curve (Fig. 1), even in the fifth millennium.

Evidence for former woodland comes from a variety of places. On the Chalk, there are four sites around the headwaters of the River Kennet in north Wiltshire (Fig. 143) where woodland of late fourth or early-third millennium BC date is attested. These are Beckhampton

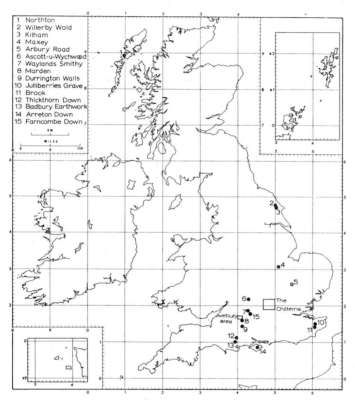

Figure 142. Location map of sites mentioned in Chapter 11.

Road, Windmill Hill, South Street and Avebury. Similar evidence has been obtained from Knap Hill on the northern scarp of the Vale of Pewsey and from Durrington Walls on the edge of Salisbury Plain. In the Oxfordshire Cotswolds there is Ascott-under-Wychwood and in the Chilterns, Whiteleaf Hill. At Northton in the Outer Hebrides, the snail fauna from the base of the machair indicates the former presence of forest in a landscape now quite open; and the fauna from the base of blown sand deposits at Newquay and Perranporth on the north coast of Cornwall also reflects woodland.

Woodland faunas are known from the floors of certain dry valleys on the Chalk, the best documented of which is at Brook in Kent, at the foot of the North Downs. Other sites of this kind occur in the Chilterns at Pink Hill, Pitstone and Coombe Hole, and are dated to the Atlantic period on general faunistic grounds.

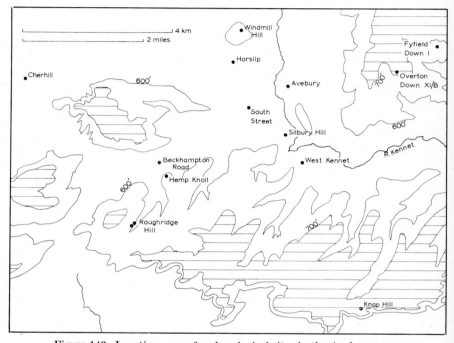

Figure 143. Location map of archaeological sites in the Avebury area.

These sites, where primary woodland faunas occur, are listed below (Figs 142 and 143).

Buried soils below field monuments
   Windmill Hill (p. 244)
   Beckhampton Road (p. 249)
   South Street (p. 257)
   Avebury (p. 269)
   Knap Hill (p. 246)
   Durrington Walls (Wainwright and Longworth, 1971: 329)

11. HABITAT CHANGE IN THE CALCAREOUS REGIONS

Ascott-under-Wychwood (p. 251)
Whiteleaf Hill (Childe and Smith, 1954: 230)

Dry valleys

Brook (Kerney et al., 1964: Fig. 15)
Pitstone (Evans, 1966b: Figs 6 and 7) (Fig. 53)
Pink Hill (p. 314)
Coombe Hole (p. 241)

Blown sand

Northton (p. 295)
Newquay (Kennard and Warren, 1903)

A number of faunas from tufa deposits (p. 299) reflect heavily-shaded woodland, and, although generally in more low-lying situations, where one would reasonably expect forest, they are worthy of note in that they relate to calcareous habitats. They also attest to the essentially terrestrial origin of tufa which is thus all the more convincingly attributed to an increase in climatic wetness rather than a local shift in the configuration of river courses or drainage patterns.

The snail faunas thus bear out in a convincing manner the pattern of Post-glacial habitat development as deduced from the pollen record (Godwin, 1956), at any rate in broad terms. From the Late Weichselian environment of open ground there developed woodland which in its early stages was open and later on became densely shaded. Radiocarbon dates from several sites indicate that this was achieved at the latest by 3000 BC, and in all probability much earlier, as suggested by the richness of the faunas both in species and numbers. Naturally the increase in faunal diversity, which these habitat changes saw, was due in some measure also to the climatic amelioration, and not only with respect to the transition from the Late Weichselian to the Post-glacial, but within the Post-glacial itself. Thus Kerney has pointed out (1966b) that certain species with northerly European ranges, such as *Carychium tridentatum*, *Cepaea nemoralis*, *Clausilia bidentata* and *Vitrea contracta*, became common in south-east England early on in the Post-glacial, while more southerly species such as *Acicula fusca*, *Azeca goodalli*, *Lauria cylindracea* and *Pomatias elegans* did not arrive until late Boreal or early Atlantic times. Increased oceanicity and the onset of milder winters were probably additional factors in enabling the second group to invade, for some, such as *Lauria cylindracea*, have a decidedly western distribution. But climate apart, there is no doubt that these faunas were living in an environment of forest, and that this was all-pervasive, blanketing valley and hill top alike.

At seven sites, an episode of deforestation has been attested, generally in the third millennium BC, and although the evidence is circumstantial, it strongly suggests that the process was an artificial one, brought about either deliberately, or incidentally through the influence of ranging stock, by early farming communities. Only in the case of sites buried beneath blown sand is it likely that deforestation was a natural process, the forest being directly destroyed by the overwhelming action of the sand.

There are too a number of Neolithic sites where an open-country snail fauna is present but where no evidence of an earlier forest phase was located (Fig. 144). Nevertheless, it can be safely assumed, I think,

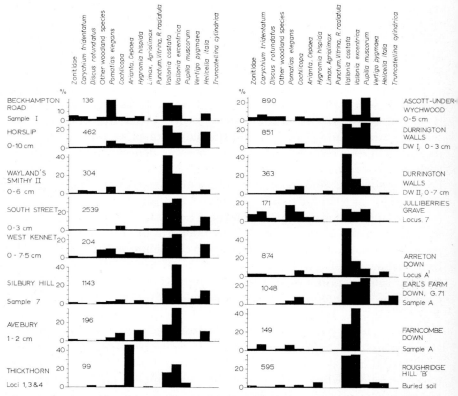

Figure 144. Representative Neolithic and Bronze Age faunas from grassland environments (references in the text, p. 363). * = present but not counted.

that at these, forest clearance has taken place. And at three sites, Kilham, Willerby Wold and Marden, while no snail fauna was present due to the low pH of the soil, an open landscape early on in the third millennium is indicated on pedological grounds.

The dates at or before which forest clearance took place on these sites are listed below, sites at which woodland has been attested being marked with an asterisk.

Beckhampton Road (p. 248)*	3250 ± 160 BC
Horslip (p. 261)	3240 ± 150 BC
Willerby Wold (Manby, 1963: 200)	3010 ± 150 BC
Ascott-under-Wychwood (p. 251)*	2943 ± 70 BC
Kilham (p. 277)	2880 ± 125 BC
Wayland's Smithy (p. 265)	2820 ± 130 BC
South Street (p. 257)*	2810 ± 130 BC
Marden (p. 274)	2654 ± 59 BC
Durrington Walls (Wainwright et al., 1971)*	2635 ± 70 BC
Brook (Kerney et al., 1964; Barker et al., 1971)*	2590 ± 105 BC
Silbury Hill (p. 265)	2725 ± 110 BC
Northton (p. 291)*	3rd millennium BC
West Kennet (p. 263)	3rd millennium BC
Thickthorn Down (Drew and Piggott, 1936: 94)	3rd millennium BC
Julliberrie's Grave (Jessup, 1939: 279)	Neolithic
Avebury (p. 268)*	ca. 2000 BC
Earl's Farm Down, G. 71 (Christie, 1967: 365)	2010 ± 110 BC
Earl's Farm Down, G. 70 (Christie, 1964: 43)	? Neolithic
Arreton Down (Alexander et al., 1960: 299)	Neolithic
Hemp Knoll (p. 332)	1795 ± 135 BC
Roughridge Hill (p. 335)	Neolithic
Farncombe Down (Rahtz, 1962: 22)	Neolithic
Pink Hill (p. 312)	Iron Age
Pitstone (Evans, 1966b)*	? Iron Age
Coombe Hole (p. 240)*	Unknown
Newquay (Kennard and Warren, 1903)*	Unknown
Perranporth (Evans, unpublished)*	Unknown

Following on from clearance, two kinds of Neolithic land use can be recognized, one in which disturbance or tillage was confined to the surface of the soil, the other involving total disturbance to its base. Ascott-under-Wychwood and Beckhampton Road are examples of the former; South Street, Wayland's Smithy, Horslip and Silbury Hill of the latter. A number of sites such as Durrington Walls and Avebury appear to be intermediate between these two extremes, there having been some disturbance of the soil below the turf line, but not sufficient to destroy entirely the shells of the woodland fauna. Ploughmarks were recorded at South Street, Earl's Farm Down (G. 71) and Avebury, from the base of the soil. At South Street there was a complex sequence comprising a non-arable phase of open country between the initial clearance episode and the phase of cross-ploughing, and, following on from the latter, two phases of less vigorous tillage between which there intervened a further non-arable phase.

Further interpretation of these types of land use is at the moment totally speculative. All that can really be said is that deforestation took place, in most cases was probably humanly induced, and in some was followed by tillage. Whether the two types of soil treatment reflect different forms of cultivation—for example different methods of tillage in a system of shifting agriculture—or whether they reflect a distinction between pastoral and arable farming, it is just not possible to say. Many more detailed investigations of buried soils are needed.

One point which is clear, and of considerable importance, is that with the exception of Brook where regeneration took place, Whiteleaf Hill, and the two causewayed enclosures, Knap Hill and Windmill Hill where clearance was not demonstrated, the environment immediately prior to burial of the soil profile was one of stable and dry short-turfed grassland at all the Neolithic sites studied. Cultivation was not taking place. At Ascott-under-Wychwood and Durrington Walls, there was prolific occupation débris over part of the soil surface, and occupation *per se* may have been instrumental in maintaining an open environment at these two sites at a late stage in their history. Otherwise it is most likely that regeneration of scrub and woodland was prevented by grazing, which at Marden, Durrington Walls and Beckhampton Road continued for many centuries. Quite clearly arable land was not at a premium, a state of affairs which continued into the second millennium judging by the thick and mature turf horizons so frequent under barrows of the Bronze Age. Indeed, from the ditch sequence at South Street (p. 331) there is evidence that scrub clearance and ploughing took place in the Beaker period solely for the purpose of creating grazing land, which may have been kept as such right up to the Roman conquest.

During the first half of the second millennium, a number of sites on the Chalk around Avebury and Stonehenge were very open and very dry (Cunnington, 1933b; Evans, 1968a; Fleming, 1971). The density of barrows in these areas suggests a generally open landscape, but it is difficult to extrapolate beyond this to the Chalk as a whole. Refuges for mesophile and shade-loving species of snail persisted as is shown by their presence in the ditches of Bronze Age barrows, e.g. at Roughridge Hill (p. 336) and Avebury, G. 55 (Evans *in* Smith, I. F., 1965a: 44), and indeed, such is the density of barrows on the Chalk that they may have been a major element in maintaining the diversity of the snail fauna.

Nevertheless, as Turner has recently pointed out (1970), by the beginning of the Bronze Age there had developed, with regard to man's influence on the forest vegetation, a regional contrast between the south-eastern chalklands of England and the whole of the rest of the country. This conclusion is based on the work of Kerney (Kerney *et al.*, 1964), and on two pollen diagrams from Wingham and Frogholt in Kent close to the North Downs which show that parts at least of the chalklands had been largely disforested by the Bronze Age (1700 and 1100 BC respectively) and to a much greater extent than appears to have been the case in other parts of the country at this time. Furthermore, these open areas were maintained as such, there being no evidence of widespread forest regeneration on the Chalk. The land-snail evidence strongly supports the validity of these conclusions.

There are no changes in the snail fauna during the Sub-boreal period which cannot be explained either in terms of changes of land use brought about by man, or of local habitat change. A climatic shift to drier conditions at the beginning of, or at any time within the Sub-boreal has not been detected. Clearly, many of the second millennium faunas reflect dry and open conditions, and would be in keeping with a dry climate with warm summers which it is suggested obtained during at least part of the Bronze Age. Warm summers there most probably were, but the problem of whether or not a dry climate prevailed cannot be resolved on the available evidence (Evans, 1971).

Prehistorians have been quick to recognize that much of the evidence for past environments and types of land use is archaeological.

At no time is this more apparent than in the latter half of the first millennium BC when Iron Age farmers brought into cultivation large tracts of chalk and limestone soils; their Celtic fields, now grassed over, are a familiar sight in the downland landscape today, owing their prominence to the fact that their edges are marked by lynchets (p. 316)—banks of earth built up against the original field boundaries (Fowler and Evans, 1967). The formation of lynchets has already been touched on as one aspect of the process of ploughwashing, and it has been shown that on the Marlborough Downs at two sites the environment became increasingly open as the lynchets built up.

In many lowlying situations, particularly in valleys and coombes, thick deposits of ploughwash are frequent, usually about 2–3 m in depth. These may be of various ages, but are generally of Iron Age or later origin (p. 286). As with lynchets, their formation is due to intensive and continuous cultivation of the surrounding slopes and hill tops, part of a general shift from pastoral to arable farming, which took place in the middle of the first millennium BC. There is always the possibility that the onset of hillwashing was hastened or intensified by the climatic deterioration—an increase in rainfall which took place around 550 BC. But there is no doubt that cultivation was the prime cause, for hillwashing cannot take place from a soil surface stabilised by vegetation. Moreover, hillwashes of Neolithic, Roman, medieval and later origin also occur in isolated instances.

On such sites as have been investigated in detail the evidence suggests that the accumulation of ploughwash is not taking place at the present day, and indeed has long since ceased, for there is generally a pronounced faunal, stratigraphical and cultural unconformity at the base of the modern soil. In other instances, where land is being cultivated, there is no doubt that ploughwashing is an active process and that field banks and lynchets are continuously building up against post-medieval enclosure boundaries.

The downland landscape as we know it (Fig. 145) undoubtedly bears the imprint of the great medieval wool industry which rose to dominance in the fourteenth century. "Though the wool industry, in its typical medieval form, ultimately passed away, the importance of sheep farming in British agriculture is one of its legacies, and the characteristic sheep husbandry has survived on the Downs until our own day" (Wooldridge and Goldring, 1953). Local expansions of arable acreage have since taken place as during the Napoleonic wars, and the present-day "plough up" policy is taking into cultivation areas of land

11. HABITAT CHANGE IN THE CALCAREOUS REGIONS

Figure 145. The Chiltern escarpment, looking across the Tring Gap to Ivinghoe Beacon. In the middle distance, the Pitstone works of Tunnel Cement.

which have been pasture since the medieval period. But we can now begin to see these changes in their historical perspective—as the latest in a long series brought about by man, the origins of which stretch back to the first settlement of farming communities in Britain, six thousand years ago.

Appendix: Tables

Only those sites which have not previously been published are included here. For Ascott-under-Wychwood (buried soil), South Street (buried soil) and Northton, see Tables 1, 2 and 3 in Evans, 1966b, and for Durrington Walls, see Table 30 in Wainwright and Longworth, 1971.

Unless otherwise stated, all samples weighed 1·0 kg, air dry.

+ = Non-apical fragments only

	MS I					MS IV	
cm ...	20–30	10–20	5–10	0–5	SUR-FACE	22·5–30	15–22·5
Pomatias elegans (Müller)	71	86	86	48	5	23	16
Carychium tridentatum (Risso)	3	1	2	9	17	6	20
Azeca goodalli (Férussac)	—	—	—	—	—	1	2
Cochlicopa lubrica (Müller)	—	—	—	1	—	—	—
C. lubricella (Porro)	—	11	11	3	2	6	10
Cochlicopa spp.	3	—	—	13	—	—	—
Columella edentula (Draparnaud)	—	—	—	1	—	—	—
Vertigo pygmaea (Draparnaud)	8	18	23	28	5	1	7
Pupilla muscorum (Linné)	45	93	192	153	30	14	94
Abida secale (Draparnaud)	18	26	55	88	38	+	3
Acanthinula aculeata (Müller)	—	4	—	11	10	3	2
Vallonia costata (Müller)	4	11	20	17	2	2	4
V. excentrica Sterki	11	13	13	7	—	18	30
Ena montana (Draparnaud)	—	1	—	2	—	—	—
E. obscura (Müller)	1	—	—	2	11	1	1
Marpessa laminata (Montagu)	6	8	3	6	4	5	6
Clausilia bidentata (Ström)	2	3	1	1	3	2	2
C. rolphi Turton	—	—	—	—	—	1	—
Cecilioides acicula (Müller)	34	25	11	5	—	—	—
Helicigona lapicida (Linné)	—	1	—	1	—	—	—
Arianta, Helix (Cepaea) spp.	10	14	23	11	5	5	16
Hygromia striolata (C. Pfeiffer)	—	—	—	—	—	1	1
H. hispida (Linné)	1	—	1	4	3	10	10
Monacha cantiana (Montagu)	—	—	—	—	—	—	—
Helicella virgata (da Costa)	—	—	—	—	—	—	—
H. itala (Linné)	13	34	44	22	12	17	29
Punctum pygmaeum (Draparnaud)	4	4	18	27	24	2	7
Discus rotundatus (Müller)	5	8	3	13	33	2	7
Euconulus fulvus (Müller)	—	—	—	—	—	—	—
Vitrea crystallina (Müller)	—	—	—	—	1	1	2
V. contracta (Westerlund)	—	1	3	10	8	—	2
Oxychilus cellarius (Müller)	1	—	—	—	11	4	2
O. alliarius (Miller)	—	—	—	—	—	—	—
O. helveticus (Blum)	—	—	—	—	—	—	—
Retinella radiatula (Alder)	—	—	—	2	—	—	—
R. pura (Alder)	3	2	1	10	21	2	4
R. nitidula (Draparnaud)	2	2	4	2	6	—	3
Vitrina pellucida (Müller)	—	1	1	2	9	—	1
Limacidae	46	46	47	69	14	40	36

Table 1. Modern Soil Profiles.

cm	MS IV		MS II			
	7·5–15	0–7·5	25–35	15–22·5	7·5–15	0–7·5
Pomatias elegans (Müller)	9	8	18	32	39	20
Carychium tridentatum (Risso)	80	436	9	18	19	24
Azeca goodalli (Férussac)	—	—	4	—	—	1
Cochlicopa lubrica (Müller)	—	1	—	—	+	2
C. lubricella (Porro)	12	3	10	15	10	5
Cochlicopa spp.	—	9	—	—	—	15
Columella edentula (Draparnaud)	—	—	—	1	—	—
Vertigo pygmaea (Draparnaud)	15	16	9	10	20	5
Pupilla muscorum (Linné)	65	33	43	95	109	58
Abida secale (Draparnaud)	4	1	24	37	27	15
Acanthinula aculeata (Müller)	8	17	5	5	1	1
Vallonia costata (Müller)	1	—	1	1	4	—
V. excentrica Sterki	91	69	14	37	53	27
Ena montana (Draparnaud)	—	—	—	—	—	—
E. obscura (Müller)	4	19	2	3	1	—
Marpessa laminata (Montagu)	1	4	6	2	4	3
Clausilia bidentata (Ström)	2	4	2	1	3	—
C. rolphi Turton	—	—	—	—	—	—
Cecilioides acicula (Müller)	—	—	6	1	4	4
Helicigona lapicida (Linné)	2	—	2	1	—	—
Arianta, Helix (Cepaea) spp.	7	4	5	11	15	15
Hygromia striolata (C. Pfeiffer)	2	1	—	—	—	3
H. hispida (Linné)	11	31	38	26	21	49
Monacha cantiana (Montagu)	—	—	—	—	—	1
Helicella virgata (da Costa)	—	—	—	2	6	9
H. itala (Linné)	3	2	16	15	9	8
Punctum pygmaeum (Draparnaud)	40	115	2	2	4	13
Discus rotundatus (Müller)	3	10	2	2	—	1
Euconulus fulvus (Müller)	4	8	—	—	—	—
Vitrea crystallina (Müller)	10	50	—	7	12	—
V. contracta (Westerlund)	8	38	5	11	3	13
Oxychilus cellarius (Müller)	6	10	—	4	1	6
O. alliarius (Miller)	—	1	—	—	—	—
O. helveticus (Blum)	—	7	—	—	—	—
Retinella radiatula (Alder)	8	6	8	2	3	3
R. pura (Alder)	5	36	15	10	11	13
R. nitidula (Draparnaud)	—	11	13	20	3	14
Vitrina pellucida (Müller)	1	3	—	2	2	1
Limacidae	43	19	47	48	39	43

	MS III				MS V			
cm	30–37·5	15–30	7·5–15	0–7·5	17·5–20	12–17·5	6·5–12	0–6·5
Pomatias elegans (Müller)	10	12	5	5	3	2	4	2
Carychium tridentatum (Risso)	1	—	—	—	—	—	—	—
Azeca goodalli (Férussac)	1	+	—	—	—	—	—	—
Cochlicopa lubrica (Müller)	—	—	—	—	—	—	—	—
C. lubricella (Porro)	—	1	7	2	1	—	—	1
Cochlicopa spp.	1	—	3	4	1	1	1	7
Columella edentula (Draparnaud)	—	—	1	—	—	—	—	—
Vertigo pygmaea (Draparnaud)	—	5	12	25	1	—	4	15
Pupilla muscorum (Linné)	30	32	64	37	32	18	11	48
Abida secale (Draparnaud)	2	1	—	1	10	5	11	33
Acanthinula aculeata (Müller)	—	—	—	—	—	—	2	—
Vallonia costata (Müller)	3	1	—	3	—	—	5	16
V. excentrica Sterki	13	14	72	104	7	4	19	30
Ena montana (Draparnaud)	—	—	—	—	—	—	—	—
E. obscura (Müller)	—	—	—	—	—	—	—	—
Marpessa laminata (Montagu)	7	1	1	2	—	—	—	—
Clausilia bidentata (Ström)	1	1	2	—	—	—	—	—
C. rolphi Turton	—	—	—	—	—	—	—	—
Cecilioides acicula (Müller)	—	—	—	—	4	5	—	5
Helicigona lapicida (Linné)	—	—	—	—	—	—	—	—
Arianta, *Helix* (*Cepaea*) spp.	2	5	5	—	1	1	2	1
Hygromia striolata (C. Pfeiffer)	—	—	—	—	—	4	—	—
H. hispida (Linné)	3	1	5	7	4	1	3	8
Monacha cantiana (Montagu)	—	—	—	—	—	1	4	4
Helicella virgata (da Costa)	—	—	8	4	11	8	20	4
H. itala (Linné)	25	6	4	—	22	13	8	32
Punctum pygmaeum (Draparnaud)	—	1	—	16	—	1	—	8
Discus rotundatus (Müller)	4	7	—	—	—	—	—	—
Euconulus fulvus (Müller)	—	—	—	—	—	—	—	—
Vitrea crystallina (Müller)	—	—	—	3	—	—	—	—
V. contracta (Westerlund)	—	—	—	4	1	—	1	2
Oxychilus cellarius (Müller)	—	4	—	—	—	—	—	—
O. alliarius (Miller)	—	—	—	—	—	—	—	—
O. helveticus (Blum)	—	—	—	—	—	—	—	—
Retinella radiatula (Alder)	1	—	—	6	—	—	—	14
R. pura (Alder)	1	1	—	—	—	—	—	—
R. nitidula (Draparnaud)	3	1	—	4	—	1	—	—
Vitrina pellucida (Müller)	—	—	—	—	—	—	—	2
Limacidae	22	32	30	18	5	9	14	11

Table 1 Continued

APPENDIX

	SUBSOIL HOLLOW	RECENT ROOTHOLE
Pomatias elegans (Müller)	8	23
Carychium tridentatum (Risso)	48	1
Cochlicopa lubrica (Müller)	—	6
C. lubricella (Porro)	—	1
Cochlicopa spp.	8	5
Vertigo pusilla Müller	2	—
V. pygmaea (Draparnaud)	—	2
Pupilla muscorum (Linné)	1	13
Abida secale (Draparnaud)	+	3
Acanthinula aculeata (Müller)	5	1
A. lamellata (Jeffreys)	1	—
Vallonia costata (Müller)	9	2
V. excentrica Sterki	—	20
Marpessa laminata (Montagu)	—	1
Clausilia bidentata (Ström)	1	4
Helicigona lapicida (Linné)	+	+
Arianta arbustorum (Linné)	+	—
Helix hortensis Müller	?+	+
H. nemoralis Linné	+	+
Arianta, Helix (Cepaea) spp.	2	10
Hygromia striolata (C. Pfeiffer)	10	—
H. hispida (Linné)	—	34
Helicella itala (Linné)	+	10
Punctum pygmaeum (Draparnaud)	1	1
Discus rotundatus (Müller)	6	6
Vitrea contracta (Westerlund)	2	—
Oxychilus cellarius (Müller)	3	2
Retinella radiatula (Alder)	—	4
R. pura (Alder)	4	2
R. nitidula (Draparnaud)	6	1
Vitrina pellucida (Müller)	—	1
Limacidae	10	25

Table 2. Coombe Hole

cm	70–85	55–65	40–50	27·5–37·5	15–25	6·5–12·5	0–6·5
Pomatias elegans (Müller)	17	—	2	1	—	—	1
Carychium tridentatum (Risso)	29	3	—	1	—	—	1
Cochlicopa lubricella (Porro)	1	—	—	3	2	1	—
Cochlicopa spp.	28	1	1	3	3	1	—
Vertigo pygmaea (Draparnaud)	1	1	4	6	8	36	19
Pupilla muscorum (Linné)	5	7	4	11	23	55	15
Abida secale (Draparnaud)	+	—	—	—	—	—	—
Acanthinula aculeata (Müller)	2	—	—	—	—	—	—
Vallonia costata (Müller)	62	7	11	42	15	1	—
V. excentrica Sterki	10	10	20	74	94	169	68
Ena montana (Draparnaud)	2	—	—	—	—	—	—
Marpessa laminata (Montagu)	1	—	—	—	—	—	—
Clausilia bidentata (Ström)	6	1	—	—	2	1	1
Cecilioides acicula (Müller)	—	—	2	3	5	1	—
Arianta arbustorum (Linné)	+	—	—	—	—	—	—
Helix hortensis Müller	+	—	—	—	—	—	—
H. nemoralis Linné	+	—	—	—	—	—	—
Arianta, Helix (Cepaea) spp.	18	—	—	1	—	—	2
Hygromia hispida (Linné)	110	26	39	121	217	138	52
Helicella caperata (Montagu)	—	—	—	—	—	17	4
H. virgata (da Costa)	—	—	—	—	—	7	11
H. itala (Linné)	15	17	19	66	183	33	8
Punctum pygmaeum (Draparnaud)	1	2	1	4	—	4	—
Discus rotundatus (Müller)	38	1	—	1	—	1	2
Vitrea contracta (Westerlund)	17	12	3	4	—	—	—
Oxychilus cellarius (Müller)	12	—	—	—	—	—	—
Retinella radiatula (Alder)	1	1	—	8	2	—	—
R. pura (Alder)	4	—	—	—	—	—	—
R. nitidula (Draparnaud)	11	3	3	5	—	—	1
Vitrina pellucida (Müller)	10	—	—	—	—	—	1
Limacidae	45	11	8	7	13	13	6

Table 3. Windmill Hill Causewayed Enclosure. All samples weighed 1·0 kg except for 70–85 cm (the buried soil) which weighed 5·0 kg

Sample	i	ii	iii	iv
Pomatias elegans (Müller)	32	73	43	93
Carychium tridentatum (Risso)	7	38	105	129
Cochlicopa lubrica (Müller)	2	—	1	—
C. lubricella (Porro)	3	1	—	1
Cochlicopa spp.	—	16	10	13
Vertigo pygmaea (Draparnaud)	—	—	1	—
Pupilla muscorum (Linné)	3	3	—	2
Abida secale (Draparnaud)	—	1	—	—
Acanthinula aculeata (Müller)	1	9	6	5
Vallonia costata (Müller)	24	1	8	2
V. pulchella (Müller)	—	1	—	3
V. excentrica Sterki	15	1	13	1
Vallonia spp.	12	3	4	5
Ena montana (Draparnaud)	—	6	3	7
Marpessa laminata (Montagu)	—	5	1	4
Clausilia bidentata (Ström)	4	31	14	35
Arianta, Helix (Cepaea) spp.	4	10	15	15
Hygromia striolata (C. Pfeiffer)	—	—	4	1
H. hispida (Linné)	7	15	31	26
Helicella itala (Linné)	11	—	2	—
Punctum pygmaeum (Draparnaud)	—	2	—	—
Discus rotundatus (Müller)	3	11	44	32
Vitrea crystallina (Müller)	—	3	4	7
V. contracta (Westerlund)	1	—	9	14
Oxychilus cellarius (Müller)	1	6	11	13
Retinella radiatula (Alder)	2	—	3	1
R. pura (Alder)	1	4	9	9
R. nitidula (Draparnaud)	3	11	32	27
Limacidae	*	*	*	*

Table 4. Beckhampton Road Long Barrow. * = Present but not counted

	WAYLAND'S SMITHY		WEST KENNET			HORSLIP		
cm	A	B	15–22·5	7·5–15	0–7·5	15–22·5	10–15	0–10
Pomatias elegans (Müller)	13	25	9	17	20	1	8	20
Carychium tridentatum (Risso)	17	13	1	1	1	—	2	2
Cochlicopa lubrica (Müller)	1	—	1	—	1	—	—	—
C. lubricella (Porro)	2	1	—	—	—	—	1	12
Cochlicopa spp.	—	—	3	5	7	—	—	—
Vertigo pygmaea (Draparnaud)	5	11	—	1	3	—	—	2
Pupilla muscorum (Linné)	4	3	—	—	2	1	—	7
Abida secale (Draparnaud)	—	—	—	—	—	+	—	+
Acanthinula aculeata (Müller)	1	—	2	1	1	1	1	—
Vallonia costata (Müller)	40	128	5	9	44	2	8	68
V. excentrica Sterki	32	67	3	14	53	2	5	54
Ena montana (Draparnaud)	—	—	—	2	—	—	—	1
E. obscura (Müller)	—	—	—	—	—	?1	—	—
Marpessa laminata (Montagu)	—	—	—	2	1	—	—	—
Clausilia bidentata (Ström)	6	3	5	16	16	—	—	5
Cecilioides acicula (Müller)	—	—	—	—	—	4	4	—
Helicigona lapicida (Linné)	—	1	—	—	—	—	—	—
Helix nemoralis Linné	—	1	—	—	—	—	—	—
Arianta, Helix (Cepaea) spp.	9	7	2	8	13	3	3	10
Hygromia hispida (Linné)	7	7	4	11	10	1	5	7
Helicella itala (Linné)	3	15	1	5	18	2	6	41
Punctum pygmaeum (Draparnaud)	1	4	—	—	2	—	—	2
Discus rotundatus (Müller)	5	9	1	1	1	1	2	5
Vitrea contracta (Westerlund)	3	3	—	—	—	—	—	—
Oxychilus cellarius (Müller)	1	1	—	—	2	—	—	—
Retinella radiatula (Alder)	—	—	—	—	—	—	—	2
R. pura (Alder)	1	—	—	1	—	—	—	—
R. nitidula (Draparnaud)	4	—	1	2	3	—	—	—
Vitrina pellucida (Müller)	—	4	—	—	1	—	—	1
Limacidae	2	1	—	6	5	3	3	12

Table 5. Wayland's Smithy, West Kennet and Horslip Long Barrows. Buried soil profiles. A = 6–12 cm (1·6 kg); B = 0–6 cm (1·14 kg)

Sample	1 and 2	5	6	7
Dry weight (kg)	2·0	1·0	1·0	2·0
Pomatias elegans (Müller)	16	6	19	38
Carychium tridentatum (Risso)	4	2	7	9
Cochlicopa lubricella (Porro)	7	7	18	55
Vertigo pygmaea (Draparnaud)	15	11	15	60
Pupilla muscorum (Linné)	3	—	4	7
Acanthinula aculeata (Müller)	1	—	6	2
Vallonia costata (Müller)	27	12	45	191
V. excentrica Sterki	61	42	161	496
Ena montana (Draparnaud)	—	—	1	2
Marpessa laminata (Montagu)	2	—	—	1
Clausilia bidentata (Ström)	3	6	7	12
Helix (Cepaea) nemoralis Linné	—	—	+	+
H. (Cepaea) spp.	5	1	3	9
Hygromia hispida (Linné)	17	9	21	47
Helicella itala (Linné)	33	23	42	170
Punctum pygmaeum (Draparnaud)	2	2	3	8
Discus rotundatus (Müller)	3	1	1	—
Vitrea spp.	—	—	—	1
Cf. *Oxychilus cellarius* (Müller)	—	—	1	—
Retinella radiatula (Alder)	—	—	1	2
R. nitidula (Draparnaud)	1	2	—	6
Vitrina pellucida (Müller)	1	—	1	1
Cf. *Agriolimax* spp.	7	1	14	26

Table 6. Silbury Hill, turf stack. * = species in sample 7 with periostracum intact. (Weights are approximate)

Sample no./cm	Av V 1	Av III 3	Av III 2	Av III 1	12–22	9–12	6–9	4–6	2–4	1–2	0–
Dry weight (kg)	2·0	3·0	2·0	2·5	2·5	1·94	2·31	1·82	1·05	0·81	0·9
Pomatias elegans (Müller)	—	+	21	195	6	22	85	78	38	7	17
Carychium tridentatum (Risso)	—	7	50	438	18	21	15	1	1	1	1
Cochlicopa lubrica (Müller)	—	1	—	1	—	1	—	—	—	—	—
C. lubricella (Porro)	—	3	—	+	cf. 3	—	—	2	—	2	—
Cochlicopa spp.	1	15	—	29	—	5	15	45	43	16	32
Columella edentula (Draparnaud)	—	—	—	1	—	—	—	—	—	—	—
Vertigo pusilla Müller	—	1	—	2	—	—	—	—	—	—	—
V. pygmaea (Draparnaud)	—	—	—	—	+	1	5	10	15	5	9
V. alpestris Alder	—	—	—	2	—	—	—	—	—	—	—
Pupilla muscorum (Linné)	11	8	10	7	2	1	—	17	36	3	17
Lauria cylindracea (da Costa)	—	—	—	1	—	1	—	—	—	—	—
Abida secale (Draparnaud)	—	6	+	1	+	—	—	+	—	—	—
Acanthinula aculeata (Müller)	—	9	4	21	3	2	4	1	2	—	2
Vallonia costata (Müller)	?1	19	1	13	1	5	11	58	61	35	44
V. excentrica Sterki	—	18	2	9	1	4	38	177	175	72	99
Ena montana (Draparnaud)	—	1	—	8	1	1	—	—	—	—	—
E. obscura (Müller)	—	cf. 1	3	6	—	—	—	—	—	—	—
Marpessa laminata (Montagu)	—	—	1	12	—	1	3	10	—	—	1
Clausilia bidentata (Ström)	—	18	7	74	5	11	25	30	14	1	4
Helicigona lapicida (Linné)	—	—	—	+	—	1	—	—	—	—	—
Arianta arbustorum (Linné)	—	—	—	+	—	+	—	—	—	—	—
Helix hortensis Müller	—	1	+	+	—	—	+	—	—	—	—
H. nemoralis Linné	—	—	+	+	—	+	+	+	+	+	—
Cepaea spp., *Arianta*	—	11	13	51	2	7	14	24	9	1	0
Hygromia striolata (C. Pfeiffer)	—	—	—	5	—	—	—	—	—	—	—
H. hispida (Linné)	—	7	6	101	13	18	36	152	104	24	80
Helicella itala (Linné)	—	5	2	2	4	13	35	154	101	22	61
Punctum pygmaeum (Draparnaud)	—	58	6	13	3	1	2	5	9	—	2
Discus ruderatus (Férussac)	—	?+	—	—	—	—	—	—	—	—	—
D. rotundatus (Müller)	—	5	44	244	11	24	43	11	6	—	4
Euconulus fulvus (Müller)	—	1	—	—	1	—	—	—	—	—	—
Vitrea crystallina (Müller)	—	—	3	19	8	+	4	—	—	—	—
V. contracta (Westerlund)	—	—	4	3	2	1	1	—	1	—	—
Vitrea spp.	—	15	—	20	—	4	—	—	—	—	—
Oxychilus cellarius (Müller)	—	—	18	47	3	9	6	2	1	—	—
Retinella radiatula (Alder)	—	7	—	1	—	1	1	2	—	—	—
R. pura (Alder)	—	1	12	37	2	8	8	5	3	2	1
R. nitidula (Draparnaud)	—	37	5	54	3	10	10	4	3	1	1
Vitrina pellucida (Müller)	—	12	—	3	1	—	—	—	—	1	—
Limacidae	3	18	17	92	13	16	47	83	32	3	11

Table 7. Avebury. Av V 1 = involution; Av III 1–3 = subsoil hollow; 0–22 cm. = buried soil beneath henge bank.

APPENDIX

cm	100–120	90–100	80–90	70–80	60–70	50–60	40–50	30–40	20–30	10–20	0–10
omatias elegans (Müller)	17	44	88	114	48	35	35	11	5	21	21
cicula fusca (Montagu)	—	—	—	—	2	—	—	—	—	—	—
rychium tridentatum (Risso)	48	28	25	29	30	3	4	—	1	—	—
ochlicopa lubrica (Müller)	—	—	1	4	—	2	—	—	—	3	16
. lubricella (Porro)	7	17	13	—	—	—	—	3	3	2	—
ochlicopa spp.	10	—	9	14	14	7	5	—	—	8	—
olumella edentula (Draparnaud)	1	—	—	—	—	—	—	—	—	—	—
ertigo pygmaea (Draparnaud)	—	—	2	2	—	1	2	—	4	10	6
. alpestris Alder	1	—	—	—	—	—	—	—	—	—	—
upilla muscorum (Linné)	—	11	17	37	10	22	101	29	28	32	39
bida secale (Draparnaud)	3	—	—	2	—	—	1	—	1	2	2
canthinula aculeata (Müller)	16	2	9	—	1	—	—	—	—	—	—
allonia costata (Müller)	3	13	11	36	41	67	8	—	2	17	15
. excentrica Sterki	1	19	31	77	37	84	39	21	52	75	85
na montana (Draparnaud)	1	2	1	1	—	1	—	—	—	—	—
. obscura (Müller)	2	3	—	9	—	—	—	—	—	—	—
Iarpessa laminata (Montagu)	2	6	20	15	11	—	2	—	—	—	—
Iausilia bidentata (Ström)	1	2	5	6	2	7	6	3	1	5	—
ecilioides acicula (Müller)	—	—	—	—	—	1	3	1	3	6	—
elicigona lapicida (Linné)	—	—	+	—	—	—	—	—	—	—	—
rianta, Helix (*Cepaea*) spp.	17	17	27	37	21	8	8	3	3	6	8
ygromia striolata (C. Pfeiffer)	2	—	—	—	—	—	—	—	—	—	—
. hispida (Linné)	17	11	11	11	2	—	2	—	—	3	4
Ionacha cantiana (Montagu)	—	—	—	—	—	—	—	—	—	—	2
elicella caperata (Montagu)	—	—	—	—	—	—	—	—	—	4	6
. gigaxi (L. Pfeiffer)	—	—	—	—	—	—	—	—	—	5	9
. virgata (da Costa)	—	—	—	—	—	—	—	—	—	—	1
. itala (Linné)	—	3	7	6	13	43	6	1	3	—	1
unctum pygmaeum (Draparnaud)	8	—	2	2	—	—	2	—	—	—	—
iscus rotundatus (Müller)	70	28	34	12	8	2	1	—	—	—	—
uconulus fulvus (Müller)	3	—	—	—	—	—	—	—	—	—	—
itrea crystallina (Müller)	20	3	2	2	2	—	—	—	—	—	—
. contracta (Westerlund)	9	—	—	—	—	2	—	—	—	—	—
xychilus cellarius (Müller)	13	3	5	4	2	—	—	—	—	1	—
etinella radiatula (Alder)	26	1	1	—	—	1	—	—	—	—	—
. pura (Alder)	28	7	5	—	—	—	—	—	—	—	—
. nitidula (Draparnaud)	30	17	—	3	3	2	1	1	—	—	—
itrina pellucida (Müller)	8	—	—	1	3	1	—	—	—	—	—
imacidae	17	27	40	30	38	17	21	15	8	15	9

Table 8. Pink Hill. All samples, 1·0 kg except 100–120 cm, 2·5 kg

Sample	1	2	3	4	5	6	7
Dry weight (kg)	0·32	0·35	0·41	0·49	0·42	0·70	0·61
Pomatias elegans (Müller)	3	4	3	4	1	—	—
Carychium tridentatum (Risso)	6	2	2	6	1	—	—
Cochlicopa lubricella (Porro)	—	—	—	—	2	—	—
Cochlicopa spp.	1	3	1	—	1	—	—
Vertigo pygmaea (Draparnaud)	—	—	1	4	2	2	7
Pupilla muscorum (Linné)	2	2	3	4	3	8	113
Acanthinula aculeata (Müller)	—	—	—	1	—	—	—
Vallonia costata (Müller)	2	—	50	27	27	27	4
V. excentrica Sterki	1	5	31	42	48	65	41
Ena montana (Draparnaud)	?1	1	—	—	—	—	—
Clausilia bidentata (Ström)	4	2	—	—	2	—	—
Arianta, Helix (Cepaea) spp.	6	2	3	4	—	—	—
Hygromia striolata (C. Pfeiffer)	3	—	—	—	—	—	—
H. hispida (Linné)	1	1	14	34	31	54	13
Helicella itala (Linné)	—	1	8	21	25	26	4
Punctum pygmaeum (Draparnaud)	—	—	1	2	2	—	—
Discus rotundatus (Müller)	7	4	1	1	—	—	—
Euconulus fulvus (Müller)	?1	—	—	—	—	—	—
Vitrea contracta (Westerlund)	2	—	—	1	—	—	—
Oxychilus cellarius (Müller)	2	—	—	—	1	—	—
Retinella radiatula (Alder)	—	—	—	1	—	—	—
R. pura (Alder)	—	—	1	2	—	—	—
R. nitidula (Draparnaud)	2	—	1	—	—	—	—
Vitrina pellucida (Müller)	—	—	1	—	—	—	—
Limacidae	15	7	6	15	17	19	—

Table 9. Fyfield Down, Lynchet

APPENDIX

cm ...	80–90	72·5–80	65–72·5	57·5–65	50–57·5	42·5–50	35–42·5	27·5–35	17·5–25	5–12·5
Pomatias elegans (Müller)	—	—	—	3	—	1	1	1	—	—
Cochlicopa lubricella (Porro)	—	—	—	—	1	1	—	5	1	1
Vertigo pygmaea (Draparnaud)	—	—	—	2	2	—	1	7	13	13
Pupilla muscorum (Linné)	—	—	4	9	6	7	5	11	45	95
Vallonia costata (Müller)	—	1	5	10	19	15	6	8	29	5
V. excentrica Sterki	—	—	4	26	54	44	37	66	126	62
Clausilia bidentata (Ström)	—	—	—	—	1	1	—	—	—	—
Helix (Cepaea) spp.	—	—	1	—	1	1	1	—	—	—
Hygromia hispida (Linné)	—	—	3	58	72	33	29	48	90	22
Helicella itala (Linné)	—	—	2	22	62	54	64	68	75	3
Punctum pygmaeum (Draparnaud)	—	1	—	2	1	—	—	—	—	—
Oxychilus cellarius (Müller)	—	—	—	6	6	—	—	1	—	—
Retinella radiatula (Alder)	—	—	—	1	—	1	—	—	2	—
Limacidae	—	—	2	9	12	7	7	12	14	—

Table 10. Overton Down, XI/B.

cm	215–225	205–215	195–205	185–195	175–185	165–175	157·5–165	150–157·5
Pomatias elegans (Müller)	2	—	—	1	1	2	3	30
Carychium tridentatum (Risso)	1	—	7	15	18	6	39	115
Cochlicopa lubrica (Müller)	—	—	—	3	8	—	2	6
C. lubricella (Porro)	12	2	1	3	3	2	9	6
Cochlicopa spp.	5	10	10	11	15	12	30	49
Vertigo pusilla Müller	—	—	—	—	—	—	—	1
V. pygmaea (Draparnaud)	7	—	3	1	6	5	8	19
Pupilla muscorum (Linné)	17	1	9	6	8	17	18	38
Acanthinula aculeata (Müller)	—	—	—	—	—	2	8	36
Vallonia costata (Müller)	128	53	125	64	175	100	121	126
V. excentrica Sterki	142	34	38	20	44	76	70	97
Ena montana (Draparnaud)	—	—	—	—	—	—	1	2
E. obscura (Müller)	—	—	1	—	—	—	1	3
Marpessa laminata (Montagu)	—	—	—	—	—	—	—	4
Clausilia bidentata (Ström)	—	—	—	—	—	—	4	16
C. rolphi Turton	—	—	—	—	—	—	—	2
Cecilioides acicula (Müller)	—	—	—	—	—	—	—	—
Helicigona lapicida (Linné)	—	—	—	—	—	—	—	—
Arianta arbustorum (Linné)	—	—	—	—	—	—	—	+
Helix hortensis Müller	—	—	—	—	—	—	—	—
H. nemoralis Linné	—	—	—	—	—	—	—	—
Arianta, Helix (*Cepaea*) spp.	3	1	1	—	—	—	3	11
Helix aspersa Müller	—	—	—	—	—	—	—	—
Hygromia striolata (C. Pfeiffer)	—	—	—	—	—	—	1	5
H. hispida (Linné)	12	34	110	52	132	111	144	209
Helicella virgata (da Costa)	—	—	—	—	—	—	—	—
H. itala (Linné)	45	9	21	21	17	42	46	53
Punctum pygmaeum (Draparnaud)	—	1	24	4	15	3	18	18
Discus rotundatus (Müller)	—	—	—	—	—	1	24	100
Euconulus fulvus (Müller)	—	—	1	—	—	—	—	—
Vitrea crystallina (Müller)	—	—	—	—	—	2	1	11
V. contracta (Westerlund)	2	—	7	7	2	—	8	12
Oxychilus cellarius (Müller)	—	1	14	4	1	—	2	15
Retinella radiatula (Alder)	2	1	7	2	7	—	2	11
R. pura (Alder)	—	—	—	2	—	—	7	31
R. nitidula (Draparnaud)	—	—	—	—	7	3	20	51
Vitrina pellucida (Müller)	2	5	3	2	9	5	3	2
Limacidae	2	7	13	9	15	6	20	41

Table 11. South Street Long Barrow: Ditch

APPENDIX

cm	142·5–150	135–142·5	127·5–135	120–127·5	110–120	100–110	90–100	80–90
Pomatias elegans (Müller)	53	17	14	6	3	2	4	5
Carychium tridentatum (Risso)	47	8	2	3	—	—	—	—
Cochlicopa lubrica (Müller)	6	3	1	1	—	—	—	—
C. lubricella (Porro)	15	8	2	2	—	—	—	—
Cochlicopa spp.	50	34	14	15	7	2	8	2
Vertigo pusilla Müller	—	—	—	—	—	—	—	—
V. pygmaea (Draparnaud)	9	8	5	4	—	4	1	—
Pupilla muscorum (Linné)	53	103	71	68	34	20	29	13
Acanthinula aculeata (Müller)	29	9	4	1	2	—	1	1
Vallonia costata (Müller)	80	42	19	5	4	2	3	3
V. excentrica Sterki	196	196	89	66	30	27	66	39
Ena montana (Draparnaud)	5	—	—	—	1	1	—	—
E. obscura (Müller)	1	—	—	—	—	—	—	—
Marpessa laminata (Montagu)	4	1	1	—	1	—	—	—
Clausilia bidentata (Ström)	28	10	5	2	3	2	1	2
C. rolphi Turton	?2	—	—	—	—	—	—	—
Cecilioides acicula (Müller)	—	6	1	3	3	7	25	65
Helicigona lapicida (Linné)	1	1	—	—	—	—	—	—
Arianta arbustorum (Linné)	—	+	—	—	—	—	—	—
Helix hortensis Müller	+	+	—	—	—	—	—	—
H. nemoralis Linné	+	—	—	—	—	—	—	—
Arianta, Helix (Cepaea) spp.	20	4	4	5	2	1	1	1
Helix aspersa Müller	—	—	—	—	—	—	—	—
Hygromia striolata (C. Pfeiffer)	2	—	1	—	—	—	1	3
H. hispida (Linné)	246	234	119	64	33	12	17	2
Helicella virgata (da Costa)	—	—	—	—	—	—	—	—
H. itala (Linné)	100	98	53	25	21	22	30	24
Punctum pygmaeum (Draparnaud)	14	7	1	3	2	2	—	—
Discus rotundatus (Müller)	54	16	5	6	4	1	1	—
Euconulus fulvus (Müller)	—	—	—	—	—	—	—	—
Vitrea crystallina (Müller)	3	—	1	—	—	—	—	—
V. contracta (Westerlund)	5	2	—	1	—	—	—	—
Oxychilus cellarius (Müller)	5	2	1	—	—	—	—	—
Retinella radiatula (Alder)	9	2	—	2	—	—	—	—
R. pura (Alder)	16	6	2	1	—	1	—	—
R. nitidula (Draparnaud)	34	10	5	4	3	2	1	1
Vitrina pellucida (Müller)	4	1	2	1	—	—	—	—
Limacidae	51	39	31	20	16	12	15	10

cm	70–80	60–70	50–60	40–50	30–40	20–30	10–20	0–10
Pomatias elegans (Müller)	—	4	—	1	—	1	—	—
Carychium tridentatum (Risso)	—	—	—	—	—	—	—	—
Cochlicopa lubrica (Müller)	—	—	—	—	—	—	—	—
C. lubricella (Porro)	—	—	—	—	—	—	—	—
Cochlicopa spp.	2	1	—	1	2	—	1	—
Vertigo pusilla Müller	—	—	—	—	—	—	—	—
V. pygmaea (Draparnaud)	—	—	—	—	—	—	—	—
Pupilla muscorum (Linné)	11	11	1	—	—	1	—	—
Acanthinula aculeata (Müller)	—	—	—	—	—	—	—	—
Vallonia costata (Müller)	—	—	—	—	—	—	1	—
V. excentrica Sterki	19	5	12	8	4	—	1	—
Ena montana (Draparnaud)	—	—	—	—	—	—	—	—
E. obscura (Müller)	—	—	—	—	—	—	—	—
Marpessa laminata (Montagu)	—	—	—	—	—	—	—	—
Clausilia bidentata (Ström)	—	—	—	—	—	—	—	—
C. rolphi Turton	—	—	—	—	—	—	—	—
Cecilioides acicula (Müller)	48	24	16	14	3	1	—	1
Helicigona lapicida (Linné)	—	—	—	—	—	—	—	—
Arianta arbustorum (Linné)	—	—	—	—	—	—	—	—
Helix hortensis Müller	—	—	—	—	—	—	—	—
H. nemoralis Linné	—	—	—	—	—	—	—	—
Arianta, Helix (Cepaea) spp.	2	2	1	—	1	—	—	—
Helix aspersa Müller	—	—	—	—	—	—	—	1
Hygromia striolata (C. Pfeiffer)	3	2	—	2	22	31	19	21
H. hispida (Linné)	1	6	9	6	4	—	—	—
Helicella virgata (da Costa)	2	—	—	8	5	3	4	4
H. itala (Linné)	11	19	19	6	1	—	—	—
Punctum pygmaeum (Draparnaud)	1	—	—	—	—	1	1	—
Discus rotundatus (Müller)	1	—	—	—	—	—	—	—
Euconulus fulvus (Müller)	—	—	—	—	—	—	—	—
Vitrea crystallina (Müller)	—	—	—	—	—	—	—	—
V. contracta (Westerlund)	—	—	—	—	—	—	—	—
Oxychilus cellarius (Müller)	—	—	—	—	1	—	—	—
Retinella radiatula (Alder)	—	—	—	—	—	—	—	—
R. pura (Alder)	—	—	—	—	—	—	—	—
R. nitidula (Draparnaud)	1	—	—	—	—	—	1	—
Vitrina pellucida (Müller)	—	—	—	—	—	—	—	—
Limacidae	5	5	4	8	11	12	8	17

Table 11 Continued

APPENDIX 385

cm	Buried soil		Ditch									
	55–62.5	47.5–55	82.5–92.5	72.5–82.5	65–72.5	57.5–65	50–57.5	40–50	30–40	22.5–30	15–22.5	0–10
Pomatias elegans (Müller)	5	3	—	—	2	3	8	7	5	4	5	2
Carychium tridentatum (Risso)	1	7	1	—	1	11	10	5	—	—	—	—
Cochlicopa lubrica (Müller)	—	3	—	—	—	—	—	—	—	—	—	—
C. lubricella (Porro)	—	4	2	—	—	6	2	2	—	—	3	—
Cochlicopa spp.	—	24	2	—	—	2	6	3	—	—	2	2
Vertigo pygmaea (Draparnaud)	—	6	—	—	2	2	—	1	—	5	3	4
Pupilla muscorum (Linné)	—	41	2	—	4	2	3	5	4	6	24	15
Acanthinula aculeata (Müller)	—	—	—	—	—	—	6	2	—	—	—	—
Vallonia costata (Müller)	6	164	3	3	18	51	31	10	—	3	5	1
V. excentrica Sterki	2	158	7	1	12	27	24	23	4	22	58	43
Ena montana (Draparnaud)	1	1	—	—	—	—	—	—	—	—	—	1
E. obscura (Müller)	—	—	—	—	1	—	—	—	—	—	—	—
Marpessa laminata (Montagu)	—	1	—	—	—	—	1	3	—	1	1	—
Clausilia bidentata (Ström)	3	2	—	—	1	—	4	4	—	2	—	—
Helicigona lapicida (Linné)	1	—	—	—	—	—	—	—	—	—	—	—
Arianta arbustorum (Linné)	—	—	—	—	—	—	—	1	—	—	—	—
Helix hortensis Müller	—	—	—	—	—	—	1	—	—	—	—	—
H. nemoralis Linné	—	—	—	—	—	1	—	—	—	—	—	—
Arianta, Helix (Cepaea) spp.	3	4	1	1	2	—	5	7	1	2	4	1
Hygromia hispida (Linné)	4	2	1	—	1	4	11	11	5	3	10	5
Helicella virgata (da Costa)	—	—	—	—	—	—	—	—	—	—	—	5
H. itala (Linné)	—	76	6	13	15	21	24	10	1	4	1	2
Punctum pygmaeum (Draparnaud)	—	1	—	1	—	2	1	—	—	—	—	—
Discus rotundatus (Müller)	7	3	1	—	2	2	7	7	1	1	—	2
Vitrea crystallina (Müller)	—	—	—	—	—	1	—	—	—	—	—	—
V. contracta (Westerlund)	—	—	—	—	—	—	3	3	—	—	—	—
Oxychilus cellarius (Müller)	—	2	—	—	2	2	4	6	1	—	—	—
Retinella radiatula (Alder)	—	2	—	—	—	3	—	—	—	—	—	—
R. pura (Alder)	1	—	—	—	—	1	—	—	—	1	—	—
R. nitidula (Draparnaud)	1	—	1	—	—	—	7	3	—	—	—	—
Vitrina pellucida (Müller)	—	6	—	—	1	—	1	—	—	—	—	—
Limacidae	5	4	—	2	—	7	5	4	—	1	1	—

Table 12. Hemp Knoll Round Barrow.

Samples ...	Ai	Aii	Aiii	Aiv	Av	Bi	Bii	Biii
Pomatias elegans (Müller)	14	4	12	4	3	5	7	5
Carychium tridentatum (Risso)	3	5	6	72	1	5	5	2
Cochlicopa lubrica (Müller)	6	3	—	32	8	5	2	3
C. lubricella (Porro)	—	—	7	3	—	—	—	—
Cochlicopa spp.	10	18	11	14	13	10	14	8
Vertigo pygmaea (Draparnaud)	3	15	11	16	9	33	14	12
Pupilla muscorum (Linné)	42	57	25	6	61	21	21	13
Acanthinula aculeata (Müller)	—	1	1	—	—	1	1	8
Vallonia costata (Müller)	85	116	104	176	76	204	215	240
V. excentrica Sterki	143	174	148	49	189	209	221	191
Ena montana (Draparnaud)	1	—	1	1	—	—	—	—
Marpessa laminata (Montagu)	3	2	2	—	—	—	3	2
Clausilia bidentata (Ström)	3	3	3	5	1	2	4	2
Helicigona lapicida (Linné)	—	—	—	—	—	—	1	—
Arianta arbustorum (Linné)	+	—	+	+	+	—	+	+
Helix hortensis Müller	+	—	—	+	—	—	—	+
H. nemoralis Linné	+	+	—	+	—	—	+	+
Arianta, Helix (Cepaea) spp.	4	5	9	4	3	4	4	8
Hygromia hispida (Linné)	20	29	27	50	44	37	42	15
Helicella itala (Linné)	36	42	28	33	83	28	21	23
Punctum pygmaeum (Draparnaud)	2	—	1	3	1	11	4	2
Discus rotundatus (Müller)	1	—	—	5	—	5	6	4
Vitrea crystallina (Müller)	—	—	2	9	—	—	—	—
V. contracta (Westerlund)	—	—	—	26	—	—	—	—
Oxychilus cellarius (Müller)	—	1	4	16	—	3	2	1
Retinella radiatula (Alder)	2	3	1	20	—	2	—	2
R. pura (Alder)	—	1	—	15	—	3	1	—
R. nitidula (Draparnaud)	1	1	2	7	—	3	6	—
Vitrina pellucida (Müller)	2	1	6	31	3	4	4	2
Limacidae	3	5	5	31	12	—	—	4

Table 13. Roughridge Hill Round Barrows A and B. Barrow A: Ai and Aii, turf stack; Aiii, primary fill of ditch (Fig. 131a); Aiv, buried soil in ditch (Fig. 131b); Av, ploughwash in ditch (Fig. 131c). Barrow B: Bi and Bii, buried soil beneath mound; Biii, turf stack

APPENDIX

	Buried soil						
cm ...	7·5–15	0–7·5	267–272	197·5–202·5	170–180	160–170	150–160
Pomatias elegans (Müller)	26	57	19	4	2	105	239
Carychium tridentatum (Risso)	7	4	4	—	4	392	910
Lymnaea truncatula (Müller)	—	—	—	—	—	—	—
Cochlicopa lubrica (Müller)	—	—	—	—	—	17	18
C. lubricella (Porro)	—	1	1	—	—	13	7
Cochlicopa spp.	10	13	5	—	2	37	52
Columella edentula (Draparnaud)	—	—	—	—	—	—	—
Vertigo pygmaea (Draparnaud)	2	2	1	—	—	1	1
Pupilla muscorum (Linné)	31	63	34	9	5	25	13
Abida secale (Draparnaud)	—	—	—	—	—	—	—
Acanthinula aculeata (Müller)	1	—	1	—	—	19	21
Vallonia costata (Müller)	19	31	12	4	3	9	12
V. excentrica Sterki	8	—	3	—	—	—	—
Ena obscura (Müller)	—	—	—	—	—	11	24
Marpessa laminata (Montagu)	1	—	1	—	—	3	1
Clausilia bidentata (Ström)	6	10	7	+	—	23	26
Cecilioides acicula (Müller)	—	—	—	—	—	—	—
Helicigona lapicida (Linné)	2	—	—	—	—	5	4
Helix hortensis Müller	+	+	—	—	—	—	2
H. nemoralis Linné	—	—	—	—	—	7	3
H. (Cepaea) spp.	5	15	3	1	—	29	34
Hygromia hispida (Linné)	4	5	3	—	1	80	60
Helicella virgata (da Costa)	—	—	—	—	—	—	—
H. itala (Linné)	36	115	57	16	7	113	30
Punctum pygmaeum (Draparnaud)	—	—	—	—	—	7	14
Discus rotundatus (Müller)	2	1	2	—	1	152	269
Euconulus fulvus (Müller)	—	—	—	—	—	1	2
Vitrea crystallina (Müller)	—	—	—	—	2	+	+
V. contracta (Westerlund)	1	2	—	—	2	+	+
Vitrea spp.	—	—	—	—	—	117	155
Oxychilus cellarius (Müller)	—	—	1	—	—	29	41
Retinella radiatula (Alder)	—	—	1	—	2	11	25
R. pura (Alder)	—	1	—	—	1	126	185
R. nitidula (Draparnaud)	3	1	—	—	2	57	91
Vitrina pellucida (Müller)	—	—	—	—	—	3	7
Limacidae	26	48	18	4	4	65	81

Table 14. Badbury Earthwork. Buried soil and ditch

	140–150	130–140	120–130	110–120	100–110	90–100	80–90
Pomatias elegans (Müller)	128	19	49	103	89	41	15
Carychium tridentatum (Risso)	642	77	212	422	86	19	14
Lymnaea truncatula (Müller)	—	—	—	—	—	—	—
Cochlicopa lubrica (Müller)	9	—	2	3	2	—	—
C. lubricella (Porro)	6	—	—	2	1	1	—
Cochlicopa spp.	57	5	12	24	8	6	6
Columella edentula (Draparnaud)	—	—	3	—	—	—	—
Vertigo pygmaea (Draparnaud)	—	1	1	1	5	7	3
Pupilla muscorum (Linné)	3	8	5	10	134	371	305
Abida secale (Draparnaud)	+	—	—	—	—	—	—
Acanthinula aculeata (Müller)	31	2	15	27	7	4	—
Vallonia costata (Müller)	6	6	6	7	14	50	25
V. excentrica Sterki	—	—	—	2	4	37	101
Ena obscura (Müller)	7	2	2	5	5	—	1
Marpessa laminata (Montagu)	1	—	—	—	—	1	—
Clausilia bidentata (Ström)	20	1	4	12	14	6	6
Cecilioides acicula (Müller)	—	—	—	—	—	—	—
Helicigona lapicida (Linné)	3	—	1	4	3	—	1
Helix hortensis Müller	3	—	—	—	+	—	—
H. nemoralis Linné	1	—	—	+	1	—	—
H. (Cepaea) spp.	35	6	2	11	4	7	1
Hygromia hispida (Linné)	34	8	17	41	30	99	73
Helicella virgata (da Costa)	—	—	—	—	—	—	—
H. itala (Linné)	25	7	26	15	30	84	39
Punctum pygmaeum (Draparnaud)	18	3	9	24	3	1	1
Discus rotundatus (Müller)	152	16	59	76	43	13	4
Euconulus fulvus (Müller)	1	—	—	1	—	—	—
Vitrea crystallina (Müller)	+	+	+	+	+	—	—
V. contracta (Westerlund)	+	+	+	+	+	3	1
Vitrea spp.	66	11	43	84	19	—	—
Oxychilus cellarius (Müller)	14	1	7	11	12	—	—
Retinella radiatula (Alder)	10	—	2	7	2	—	—
R. pura (Alder)	121	12	33	92	28	8	8
R. nitidula (Draparnaud)	42	4	18	43	17	2	2
Vitrina pellucida (Müller)	4	—	—	—	2	1	—
Limacidae	102	19	39	56	34	25	42

Table 14 Continued

	70–80	60–70	50–60	40–50	30–40	20–30	10–20	0–10
Pomatias elegans (Müller)	24	31	21	25	28	20	28	7
Carychium tridentatum (Risso)	—	2	—	1	—	—	—	—
Lymnaea truncatula (Müller)	—	1	—	—	—	—	—	—
Cochlicopa lubrica (Müller)	—	—	—	—	—	—	1	—
C. lubricella (Porro)	—	2	2	—	1	5	2	—
Cochlicopa spp.	5	7	7	6	9	8	4	4
Columella edentula (Draparnaud)	—	—	—	—	—	—	—	—
Vertigo pygmaea (Draparnaud)	1	1	1	2	3	5	10	15
Pupilla muscorum (Linné)	219	150	72	32	110	400	422	166
Abida secale (Draparnaud)	—	—	—	—	—	—	—	—
Acanthinula aculeata (Müller)	1	—	—	—	—	—	—	—
Vallonia costata (Müller)	8	9	6	15	47	79	74	6
V. excentrica Sterki	13	9	2	6	31	82	56	37
Ena obscura (Müller)	—	—	—	—	—	1	—	—
Marpessa laminata (Montagu)	—	—	1	—	—	1	—	—
Clausilia bidentata (Ström)	9	6	10	1	6	2	6	5
Cecilioides acicula (Müller)	1	—	2	—	8	8	6	—
Helicigona lapicida (Linné)	—	—	—	—	—	—	—	—
Helix hortensis Müller	—	—	—	—	—	—	—	—
H. nemoralis Linné	—	—	—	—	—	1	+	—
H. (Cepaea) spp.	8	10	7	10	12	4	4	1
Hygromia hispida (Linné)	51	23	22	40	97	186	187	9
Helicella virgata (da Costa)	—	—	—	1	1	2	39	47
H. itala (Linné)	22	61	69	73	96	100	75	13
Punctum pygmaeum (Draparnaud)	2	—	—	—	—	—	—	2
Discus rotundatus (Müller)	3	2	1	—	—	—	—	—
Euconulus fulvus (Müller)	—	—	—	—	—	—	—	—
Vitrea crystallina (Müller)	—	—	—	—	—	—	—	—
V. contracta (Westerlund)	—	—	—	—	—	—	—	—
Vitrea spp.	—	—	—	—	—	—	—	—
Oxychilus cellarius (Müller)	1	—	—	—	—	—	—	—
Retinella radiatula (Alder)	—	1	—	—	—	—	—	—
R. pura (Alder)	2	1	—	—	—	—	1	—
R. nitidula (Draparnaud)	1	—	—	1	—	1	—	—
Vitrina pellucida (Müller)	1	—	—	—	—	—	—	—
Limacidae	43	31	32	24	28	33	48	39

cm ...	90–100	80–90	70–80	60–70	50–60	40–50	27·5–37·5	17·5–27·5	7·5–17·5	0–7·5
Pomatias elegans (Müller)	—	2	1	13	9	—	1	+	3	—
Carychium tridentatum (Risso)	1	2	6	32	8	4	2	—	3	—
Cochlicopa lubricella (Porro)	2	6	2	7	3	1	—	—	—	—
Cochlicopa spp.	1	1	2	6	6	4	1	1	—	
Vertigo pygmaea (Draparnaud)	—	5	1	5	8	1	2	—	—	2(
Pupilla muscorum (Linné)	8	10	7	12	13	3	—	—	1	1
Acanthinula aculeata (Müller)	1	—	—	6	5	2	2	—	—	—
Vallonia costata (Müller)	8	10	17	16	3	4	1	5	27	4(
V. pulchella (Müller)	—	—	—	—	—	—	—	—	3	
V. excentrica Sterki	41	81	62	59	50	17	26	44	68	147
Ena obscura (Müller)	—	—	1	—	—	1	—	—	—	
Marpessa laminata (Montagu)	—	—	—	+	—	—	—	—	—	—
Clausilia bidentata (Ström)	1	—	3	10	10	1	—	—	—	1
Cecilioides acicula (Müller)	2	2	2	4	9	2	4	2	3	
Helix hortensis Müller	—	—	—	—	1	—	—	—	—	—
H. nemoralis Linné	—	—	—	1	—	+	—	—	—	—
Cepaea spp.	1	1	2	7	2	2	1	1	1	4
Helix aspersa Müller	—	—	—	1	—	—	—	—	—	—
Hygromia striolata (C. Pfeiffer)	—	—	1	—	—	—	—	17	25	3
H. hispida (Linné)	3	1	8	81	64	9	5	1	—	78
Helicella caperata (Montagu)	—	—	—	—	—	—	—	—	—	1
H. gigaxi (L. Pfeiffer)	—	—	—	—	—	—	—	—	14	
H. virgata (da Costa)	—	—	—	—	—	—	cf. 6	—	2	
Helicella spp.	—	—	—	—	—	—	—	—	24	—
H. itala (Linné)	11	21	20	25	24	22	11	11	—	—
Punctum pygmaeum (Draparnaud)	—	15	7	3	1	1	1	—	—	27
Discus rotundatus (Müller)	—	+	1	20	3	2	2	—	4	2
Vitrea contracta (Westerlund)	—	—	—	3	1	1	2	1	1	
Oxychilus cellarius (Müller)	1	—	2	12	—	1	2	1	2	12
Retinella radiatula (Alder)	—	2	2	1	—	—	—	—	—	12
R. pura (Alder)	1	—	1	4	2	1	1	—	—	3
R. nitidula (Draparnaud)	—	1	—	9	3	6	1	2	3	3(
Vitrina pellucida (Müller)	3	7	1	—	—	—	—	—	—	6
Limacidae	7	1	14	16	18	7	4	5	4	6

Table 15. Ascott-under-Wychwood. Roman ditch

APPENDIX

cm y weight of sample (kg)	83– 89 1·46	75– 82·5 0·75	67·5– 75 0·85	60– 67·5 0·93	50– 57·5 1·38	37·5– 45 1·42	22·5– 30 1·30	7·5– 15 1·43	Spot sample 1·50
rychium tridentatum (Risso)	—	—	—	—	—	9	1	—	1
mnaea truncatula (Müller)	—	—	—	—	—	12	—	—	5
olexa hypnorum (Linné)	—	—	—	—	—	5	—	—	—
anorbarius corneus (Linné)	—	1	+	—	—	—	—	—	—
anorbis leucostoma Millet	—	—	—	+	—	161	1	3	77
ccinea cf. pfeifferi Rossmässler	—	1	—	—	—	21	—	—	3
chlicopa lubrica (Müller)	—	—	—	—	—	7	—	—	—
lubricella (Porro)	—	—	—	—	—	—	—	—	1
chlicopa spp.	—	—	—	—	—	22	2	+	9
ertigo pygmaea (Draparnaud)	—	—	—	—	—	3	—	—	—
allonia costata (Müller)	—	—	—	—	—	20	2	1	7
pulchella (Müller)	—	—	—	—	—	2	—	—	—
excentrica Sterki	—	—	—	—	—	11	—	—	3
cilioides acicula (Müller)	—	—	—	—	—	1	22	67	—
rianta arbustorum (Linné)	—	—	—	+	—	+	—	—	+
elix nemoralis Linné	—	—	+	—	—	+	+	—	+
rianta, Helix (Cepaea) spp.	—	—	2	+	—	31	2	+	7
ygromia hispida (Linné)	—	—	—	—	—	—	1	3	—
. liberta (Westerlund)	—	—	—	—	—	33	3	4	11
iscus rotundatus (Müller)	—	—	—	—	—	—	—	—	+
ychilus cellarius (Müller)	—	—	—	—	—	4	10	cf. 1	2
etinella radiatula (Alder)	—	—	—	—	—	1	—	—	—
. nitidula (Draparnaud)	—	—	cf. 1	—	—	3	—	1	—
onitoides nitidus (Müller)	—	—	—	—	—	34	5	cf. 1	21
itrina pellucida (Müller)	—	—	—	—	—	5	—	—	1
lmacidae	—	—	—	2	—	2	2	1	3
isidium casertanum (Poli)	—	1	—	—	—	33	2	—	21
stracods	—	1	—	—	—	4	—	—	3

Table 16. Maxey, Roman enclosure ditch. The spot sample comes from 30–45 cm

	AR II	AR IV
Carychium minimum Müller	—	1
C. tridentatum (Risso)	23	1
Lymnaea truncatula (Müller)	2	2
L. peregra (Müller)	—	64
Aplexa hypnorum (Linné)	—	5
Planorbis leucostoma Millet	20	496
P. crista (Linné)	—	25
Succinea cf. *pfeifferi* Rossmässler	1	—
Cochlicopa spp.	10	—
Vertigo pygmaea (Draparnaud)	—	1
Vallonia costata (Müller)	12	3
V. excentrica Sterki	7	5
Cecilioides acicula (Müller)	70	1
Arianta arbustorum (Linné)	+	—
Helix nemoralis Linné	+	—
Helix (*Cepaea*) spp.	5	—
H. aspersa Müller	+	+
Hygromia liberta (Westerlund)	25	4
Punctum pygmaeum (Draparnaud)	—	1
Oxychilus cellarius (Müller)	4	—
Retinella nitidula (Draparnaud)	7	—
Limacidae	8	4
Sphaerium lacustre (Müller)	—	3
Pisidium spp.	—	1

Table 17. Arbury Road, Cambridge

Glossary

A/C horizon. The transition zone in a soil between the A and C horizons, usually in a rendsina (p. 209).
Acid. Of a soil (or solution) in which hydrogen ions predominate, and whose base-exchange capacity is generally low.
Aggregate species (agg.). A species which includes more than one closely similar, segregate, species.
A horizon. The surface humus horizon of a soil (p. 208).
Algae. The simplest green plants. Most are aquatic, notably the seaweeds, but some unicellular forms are terrestrial, living on bark and stones.
Alkaline. = basic.
Allerød Interstadial (the). Zone II of the Late Weichselian (p. 355).
Allochthonous. Of material, e.g. soils, sediments or shells, which has been laid down by lateral transport and is not *in situ* (p. 308); cf. autochthonous.
Altithermal. = climatic optimum.
Anthropophobic. Of snails (or other organisms) which shun the presence of man (p. 201).
Aperture. The opening of the gastropod shell through which the animal emerges (Fig. 16).
Apex. An imprecise, but useful, term referring to the upper two or three whorls of the gastropod shell, and always including the protoconch.
Aragonite. A crystalline form of calcium carbonate, less stable than calcite, and of which snail shell is largely composed.
Atlantic. The third period of the Post-glacial, *ca.* 5300–3000 BC, characterized by a warm and humid climate (p. 357).
Autochthonous. Of material, e.g. soils, sediments or shells, which has been laid down by *in situ* processes and not by lateral transport (p. 308); cf. allochthonous.
Barrow. A man-made mound, generally over a burial.
Base. A chemical compound which, in solution, produces a pH above 7·0 and a high concentration of mineral ions such as calcium, potassium and iron.
Base-exchange capacity. The capacity of a soil to exchange bases, and, indirectly, to maintain a good crumb structure and high nutrient status.
Basic. Of a soil (or solution) in which calcium or other metal ions predominate over hydrogen ions, and whose base-exchange capacity is high.
Beaker. Of a Late Neolithic/Early Bronze Age people whose material equipment contains a characteristic form of drinking vessel, the beaker.
B horizon. The horizon of a soil below the A horizon either in which humus, Bh, clay, Bt, or iron, Bfe, has accumulated by downwashing, or in which iron oxides, (B), have been released *in situ* by weathering of the parent material (p. 217).

Blanket peat. Peat spread over a wide area of countryside, and whose formation is brought about by high rainfall rather than by bad drainage.

Body whorl. The final whorl in the adult gastropod shell (Fig. 16).

Bølling Interstadial (the). Zone Ib of the Late Weichselian.

Boreal. The second period of the Post-glacial, *ca.* 7000–5300 BC, during which the climatic optimum was attained (p. 357).

Boulder clay. Material transported by glacial action (= till).

Brickearth. Wind-lain, water-lain or solifluxed fine loam or silt (p. 290).

Bronze Age. The period of prehistory characterized by bronze tools and weapons, in Britain lasting from *ca.* 2000 to 600 BC.

Brown earth. A soil type in which there is a pronounced (B) horizon formed by the *in situ* release of iron oxide from the parent material, and unmasked by humus. The soil solution is generally neutral or weakly acid.

Bt horizon. See B horizon and p. 217.

C-14. See **radiocarbon**.

Calcicole. Of a snail species (or other organism) requiring a lime-rich environment (= calciphile).

Calcifuge. Of a snail species (or other organism) which avoids lime-rich habitats.

Calciphile. = calcicole.

Calcite. A crystalline form of calcium carbonate, more stable than aragonite, and of which the internal shells of slugs are composed.

Calcium carbonate. The chemical compound, $CaCO_3$, of which chalk, limestone and snail shell are composed.

Carbonic acid. The weak acid produced by the interaction of carbon dioxide and water; capable of dissolving limestone and chalk.

Causewayed enclosure. A form of Neolithic embanked enclosure in which the ditch is external to the bank and interrupted by several causeways.

Chambered tomb. A Neolithic sepulchral monument in which the burial chamber is generally constructed of massive stone elements (megaliths).

C horizon. The parent material of a soil.

Clay-with-flints. Non-calcareous drift, predominantly clay with incorporated flints, occurring on the summit plateaux of certain chalk areas, notably the Chilterns and North Downs. It is in part the insoluble residue of the Upper Chalk, and in part the remnants of later geological deposits.

Climatic deterioration. A decrease of temperature and an increase of rainfall—notably that which took place *ca.* 500 BC.

Climatic optimum (the). The period of maximum Post-glacial warmth, *ca.* 5000–1000 BC (= altithermal).

Colluvial. Of soils and sediments which accumulate by lateral transport rather than being formed by weathering *in situ* (p. 280).

Columella. The solid part of the vertical axis of the gastropod shell around the umbilicus, to which the animal is attached.

Columellar. Pertaining to, or situated on, the columella—as of denticles or folds.

Conchology. The study of shells.

Coversand. Sand deposit laid down by aeolian action.

Creswellian. The last of the British Upper Palaeolithic stone industries.

Cromerian Interglacial (the). The ante-penultimate interglacial in Britain; refers specifically to the Upper Freshwater Bed of the Cromer Forest Bed series (p. 352).

GLOSSARY

Cryoturbation. Frost-heaving, generally under periglacial conditions, and caused by alternate expansion and contraction (freezing and thawing) of the ground (p. 221).

Cyclostoma elegans zone. The heavy concentration of *Pomatias elegans* which occurs in the ditches of barrows and other prehistoric earthworks, and referred to by nineteenth and early twentieth century excavators (p. 7).

Danubians. Farming people of central and western Europe in the fifth millennium BC.

Dendrochronology. The dating of wood by tree-ring counting.

Deverel-Rimbury. A Middle Bronze Age culture of southern England, *ca.* 1200 BC.

Dipterous. Pertaining to the Diptera, the true, two-winged, flies.

Drift. Superficial geological material, largely of Pleistocene origin, overlying the solid geological strata.

Dry valley. A valley in chalk or limestone country with no stream in its bottom, or if such is present it is incidental to the main process of valley formation which in most cases was brought about by solifluxion (p. 311).

Ebbsfleet/Mortlake. A Neolithic pottery style.

Eb horizon. A soil horizon which has lost clay particles through downwashing (p. 274).

Ecosystem. The total environment, in particular its dynamic attributes and energy relationships.

Edentulate. Of the gastropod shell aperture; lacking teeth/folds/denticles.

Eemian Interglacial (the). The last interglacial (= the Ipswichian) (p. 352).

Epiphragm. An insoluble mucous secretion, strengthened by calcareous deposits, laid down over the aperture of the gastropod shell in times of adverse weather.

Eutrophic. Of high nutrient status.

Family. A taxonomic group of lower rank than "order" but higher than "genus".

Faunule. A term, now out of use, applied particularly in the nineteenth and early twentieth century to the assemblages of shells extracted from ancient soils and sediments (p. 7).

Flandrian (the). The Post-glacial (p. 352).

Flatworm. (= Platyhelminth). Free-living or parasitic invertebrate, usually flattened dorsoventrally, and of simpler organization than the earthworms and molluscs.

Fluke. A general term for several thousand species of flatworm, which are internal parasites in vertebrates. They have complex life-histories involving several generations of larval forms, the earlier of which are spent in molluscs.

Fold. Lamella-like projection from the inner wall of the gastropod shell, present in some species in the aperture and sometimes, as in *Carychium*, extending well back into the shell (Fig. 21).

Frost hollow. Valley in which cold air accumulates and lingers on still, clear nights, thus creating diurnal extremes of temperature locally.

Gastropod. Molluscs characterized by a spirally twisted shell and commonly known as snails; includes marine, freshwater and terrestrial forms. In some, the slugs, the shell has become internal.

Genus. A taxonomic group of lower rank than "family" consisting of closely related species.

Geophobic. Of creatures spending most or all of their life well above the ground surface; e.g. on trees and walls.

Gipping Glaciation (the). The penultimate glaciation in Britain (p. 352).

Glacial or glaciation. A period of arctic or sub-arctic climate between two interglacials, and generally characterized by at least one episode of ice advance (p. 351).

Gley. A type of soil or soil horizon which is waterlogged or semi-waterlogged.

Hair pit. A minute depression in the surface of the gastropod shell in some species, in which periostracal hairs are embedded.

Heliophile. An organism occurring generally in unshaded habitats; fond of the sun.

Henge. Neolithic ritual enclosure comprising a bank and ditch, the latter generally internal.

Hillfort. A fortified enclosure comprising a bank(s) and ditch(es), generally of Iron Age date.

Hillwash. Soil material washed downhill by rain action and accumulated in a valley or other depression (p. 281).

Holarctic. Of the north temperate and arctic biological region.

Holocene (the). The Post-glacial (p. 352).

Horizon. A zone, roughly parallel with the surface, within a soil profile, having distinct physical and chemical properties—e.g. colour, texture and humus content. Generally designated by letters A, Eb, B and C.

Hoxnian Interglacial (the). The penultimate interglacial (formerly the Great Interglacial) in Britain (p. 352).

Hygrophile. An organism requiring wet conditions for life.

Inner lip. The edge of the aperture, of the gastropod shell, which is appressed to the preceding whorl (Fig. 16).

Interglacial. A period of warm climate during which, in present temperate latitudes, the climate was sufficiently mild to enable deciduous mixed forest to develop (p. 351).

Interstadial. A period of mild climate within a glacial, during which pine/birch forest became established, but in which other, more thermophilous, trees such as oak and elm, were absent (p. 353).

Involution. A structure caused by cryoturbation (p. 221).

Ion. A chemical element or compound in aqueous solution and electrically charged, and thus in a state to react chemically.

Ipswichian Interglacial (the). The last interglacial in Britain (p. 352).

Iron Age. The final stage of prehistory, beginning in Britain *ca.* 650 BC and characterized by the introduction and use of iron implements.

Labial lobes. Extensions of the gastropod mantle around the mouth, sensitive to tactile and chemical stimuli.

Lampyridae. A family of beetles, notably the glow worm, *Lampyris noctiluca*.

Landnam. A form of forest clearance for agriculture, in which the area cleared is generally abandoned after a few years and the vegetation allowed to revert to its former state.

Late Weichselian (the). The final stage of the Weichselian Glaciation, *ca.* 12 000–8300 BC, often simply Late Glacial (p. 355).

Leaching. The removal of water-soluble material from a soil by down-washing.

Lessivation. The downward movement of clay through a soil (p. 277).

Limestone. Rock composed of calcium carbonate; soluble in dilute acids such as carbonic acid.

Loess. Wind-lain silt (p. 290).

Long barrow. Barrow of trapezoidal, rectangular or oval form, generally Neolithic.
Lowestoft Glaciation (the). The ante-penultimate glaciation in Britain (p. 352).
Lungworm. A form of Nematode, parasitic in the lungs of vertebrates.
Lux. A unit of illumination.
Lynchet. A cultivation terrace formed as a result of ploughwashing and build up against a fence, wall or other form of field boundary (p. 316).
Machair. Gaelic term for fixed dune pasture, the stretch of stabilized sand behind the active dune belt of a sand-dune system.
Maglemosian. A Mesolithic culture, the earliest in Britain.
Malacology. The study of molluscs.
Mantle. The soft, fleshy part of the molluscan body immediately underneath the shell and responsible for the secretion of the latter.
Mesic. Of a habitat having no extremes, e.g. of temperature or humidity.
Mesolithic. Hunter-gatherer, non-agricultural communities characterized by the use of microliths; generally of Post-glacial age, *ca.* 8000–4000 BC.
Mesophile. An organism requiring a certain degree of moisture but not to the extent of a hygrophile.
Mesotrophic. Of nutrient status intermediate between eutrophic and oligotrophic.
Microlith. A small flint artefact, characteristic of Mesolithic cultures, and probably used in composite tools such as arrow tips and spear heads.
Middle Weichselian (the). The main, and coldest, episode of the Weichselian, *ca.* 50 000–14 000 BC.
Mollusca. Invertebrates in which the soft parts of the body are generally enclosed in a calcareous shell. The two main groups are the snails (gastropods), and the bivalves, such as the mussel, in which the shell develops as two valves. In a third group, which includes the squid and cuttlefish, the shell is internal or lost altogether.
Morph. One of several forms of a polymorphic animal, transmitted genetically from one generation to the next.
Mortuary house. Building in which bodies are stored prior to final interment (Neolithic in this context).
Mull humus. Humus strongly bound to the clay fraction in water-stable aggregates and devoid of discrete plant or animal débris.
NAP/AP ratio. Non-arboreal pollen/arboreal pollen ratio. A measure of the degree of openness of an area.
Nematodes. Roundworms. Free-living or parasitic invertebrates, unsegmented and of simpler organization than the earthworms.
Nepionic. Of the uppermost whorls of the gastropod shell which display surface features different from, and generally simpler than, those in the adult—usually the first $1\frac{1}{2}$–2 whorls.
Neolithic. Non-metal-using farming communities, in Britain lasting from *ca.* 3500–1800 BC.
Neutral. Of a soil (or solution) which is neither acid nor basic, with a pH of around 7.
Niche. The place of an animal in the ecosystem—its habitat, food and position in the food web.
Nitrification. The conversion, in the soil, of organic nitrogen to inorganic forms such as nitrate which can be utilized by plants.

Oligotrophic. Of low nutrient status.

Operculate. Species of snail with an operculum. Most British species of operculate are freshwater but two, *Pomatias elegans* and *Acicula fusca*, are terrestrial.

Operculum. Calcareous valve situated on the back of the body in certain species of snail, and serving to close the shell when the animal is inside (p. 47; Fig. 19).

Ostracoda. Minute Crustacea—animals related to the shrimp—in which the body is enclosed by two valves. Aquatic.

Outer lip. The free margin of the aperture of the gastropod shell, i.e. where it is not appressed to the preceding whorl (=peristome) (Fig. 16).

Palaearctic. Of the Old World north temperate and arctic biological region.

Palaeolithic. Stone-using hunting communities, in Britain, dying out around the beginning of the Post-glacial.

Palatal. Of the lower, outer, part of the aperture of the gastropod shell (Fig. 16).

Palynology. The science of pollen analysis.

Parietal. Of the inner lip of the gastropod shell (Fig. 16).

Pea grit. = split pea (p. 212).

Pedology. The study of soil-forming processes.

Periglacial. Literally, of the zone around, and influenced by, a glacier. In its wider meaning, the term refers to an area whose climate is arctic or sub-arctic.

Periostracum. The protein coating of the molluscan shell, generally not preserved subfossil (p. 22).

Periphery. The outer margin of the gastropod shell, particularly its profile (Fig. 16).

Peristome. The outer lip of the gastropod shell (Fig. 16).

Petri dish. A shallow glass or plastic dish, 9 cm in diameter, with an overlapping cover (generally used for culturing bacteria).

pH. Literally the negative logarithm of the hydrogen ion concentration of a solution or soil. A measure of acidity, 7·0 being neutral, above 7·0 basic and below 7·0 acidic.

Phototropic. Oriented in response to the stimulus of light.

Pleistocene. The last approx. 2 million years of earth's history characterized by alternate periods of cold and warm climate (p. 351). In some classifications (e.g. West, 1968) it includes the Holocene.

Ploughwash. A variety of hillwash brought about mainly by ploughing (p. 282).

Podsol. A soil showing pronounced leaching in the upper horizons, and pronounced deposition of humus and/or iron lower down. The A horizon is generally poor in nutrients, the humus poorly decomposed, and the soil solution strongly acid.

Pollen analysis. The counting of pollen grains from peat bogs, lake sediments and buried soils for the purposes of reconstructing former vegetation.

Polymorphic. Of a species of animal displaying different forms, or morphs, which are transmitted genetically.

Post-glacial (the). The present period of temperate climate which began *ca.* 8300 BC (= Flandrian or Holocene).

Pre-boreal (the). The first period of the Post-glacial, *ca.* 8300–7000 BC (p. 357).

Protoconch. The young shell as first hatched from the egg.

Pseudomycelium. Re-crystallized calcium carbonate in the humus horizon of a soil, having the appearance of fungal threads.

Pulmonates. The most highly evolved gastropods, lacking an operculum and breathing by means of a lung. They include the majority of the British terrestrial snails and slugs and many of our freshwater forms.

Quaternary. = Pleistocene (p. 351).

Radiocarbon. Radioactive carbon of atomic weight 14. The decay of C–14 takes place at a known rate, so that assay of the C–14:C–12 ratio in organic material can be used as an indication of age (p. 6).

Relative humidity. Ratio of the amount of water vapour in the air to the amount that would saturate it at the same temperature.

Rendsina. A soil developed from lime-rich parent material, and which is highly calcareous throughout. There is no B horizon (p. 208).

Round barrow. Barrow of roughly circular plan. Generally Beaker, Bronze Age or later, though some, particularly in Yorkshire and Strathtay, are Neolithic.

Rupestral. Living on a firm, dry substratum, particularly on rocks, walls and tree trunks.

Sauveterrian. A Mesolithic culture, in Britain coming later than the Maglemosian.

Sciomyzidae. A family of dipterous flies, some of which are parasitic on snails.

Segregate species (seg.). One of two or more closely similar species formerly included as a single, aggregate, species.

Sere. A successional series of plant communities beginning from bare ground (or open water) and culminating in a vegetational climax (p. 88).

Siliceous. Composed of silica.

Slack. A damp hollow in a sand-dune system.

Slug. A gastropod lacking an external shell.

Slug plate. The internal shell of limacid slugs (Fig. 18).

Snail. A gastropod mollusc with an external, spirally coiled, shell.

Soil. Mineral and/or organic matter subjected to physical and/or chemical weathering, and the products combined to a greater or lesser degree with organic matter.

Solid geology. Geological deposits, generally of pre-Pleistocene age.

Solifluxion. The down-slope mass movement of waterlogged soil or bedrock over an impermeable substratum, generally induced by periglacial conditions (p. 289).

Sol lessivé. A brown earth soil in which the B horizon comprises a layer of clay enrichment (p. 216).

Species. A group of closely allied, mutually fertile, individuals showing constant, genetically transmitted, differences from others of the same genus.

Spire. That part of the gastropod shell above (or younger than) the body whorl (Fig. 16).

Split pea. = pea grit (p. 212).

Stomata. The pores in a plant leaf or stem through which respiration and transpiration take place.

Stratification. The state of being layed in strata; of soils, sediments and objects therein.

Stratigraphy. The relationship of layers of sediments one to another.

Stria. Fine streak or line, generally upraised, on the surface of the gastropod shell.

Striate. Of the gastropod shell surface—covered with striae (Fig. 21).

Subaerial. Of deposits formed by processes of transportation which are neither aeolian nor aquatic, but essentially terrestrial, e.g. solifluxion and plough-washing (p. 281).

Sub-atlantic. The fifth, and present, period of the Post-glacial, beginning *ca.* 550 BC, and characterized by a humid and somewhat cooler climate than previously (p. 357).

Sub-boreal. The fourth period of the Post-glacial, *ca.* 3000–550 BC, roughly corresponding to the Neolithic and Bronze Age periods in Britain (p. 357).

Subfossil. Of the dead remains of plants or animals which have been preserved in a soil or sediment but whose chemical and physical structure has not been altered from that in life.

Subsoil hollow. Hollow at the base of a soil, probably the cast of a former tree root (p. 219).

Subspecies. A taxonomic group intermediate in rank between a species and variety.

Suture. The line or incision between adjacent whorls of the gastropod shell (Fig. 16).

Synanthropic. Closely associated with human habitations—of snails or other organisms (p. 201).

Tertiary. The geological era directly preceding the Quaternary (=Cainozoic), extending from *ca.* 70 million to 2 million years ago.

Thermal decline (the). The decrease in temperature which took place during the first millennium BC.

Thermophile. An organism requiring warm conditions for life.

Tooth. Calcareous projection in the aperture of the gastropod shell (Fig. 16). Unlike the teeth of vertebrates, those of snails are not lodged in sockets but are protuberances continuous with the internal lining of the shell. Some extend well back into the shell and are technically known as folds (Fig. 21).

Travertine. An exceptionally hard and compact form of tufa (p. 297).

Trematodes. A group of parasitic flatworms, including the sheep flukes.

Tufa. A calcareous precipitate formed as a result of evaporation of water heavily charged with lime (p. 297).

Turf line. The A, mull-humus, stone-free or worm-sorted horizon of a rendsina soil under grass (p. 208).

Turf-stack. A mound of turfs, generally as a core to a barrow or bank.

Umbilicus. The hollow vertical axis of the gastropod shell around which the whorls are coiled. In some species the umbilicus remains open in the adult, in others it is occluded (Fig. 16).

Variety (var.). A well-marked form of a species, whose characters are transmitted genetically, but of lesser rank than a sub-species; often determined by environmental factors (=morph).

Vice-comital. Of a vice county.

Vice county. Division of a county for the purposes of biological recording.

Weichselian (the). The last glaciation, *ca.* 70 000–8000 BC (p. 352).

Whorl. One complete turn of the gastropod shell (Fig. 16).

Xerophile. An organism capable of living in extremely dry conditions.

Zone. Generally refers to a pollen zone. A period during the Late Weichselian or Post-glacial (p. 357) characterized by a particular combination of tree species, and generally lasting for several centuries.

References

The following abbreviations have been used:

P.M.S. = *Proceedings of the Malacological Society of London*.
J. Conch. = *The Journal of Conchology*.

Otherwise I have used the following:

Brown, P. and Stratton, G. B. (1963). "World List of Scientific Periodicals Published in the Years 1900–1960." London.

Porter, K. I. and Koster, C. J. (1970). "British Union-Catalogue of Periodicals: New Periodical Titles 1960–1968." London.

For archaeological periodicals *not* listed in the above catalogues, "The Abbreviations of Titles of Archaeological Periodicals in C.B.A. Publications" (Council for British Archaeology, 1968) has been followed.

Numbers in square brackets indicate the pages where references occur in the text.

Adam, W. (1960). "Faune de Belgique: Mollusques. Tome I, Mollusques Terrestres et Dulcicoles." Brussels: L'Institute royal des Sciences naturelles de Belgique. [18, 46, 61.]

Adams, L. E. (1896). "The Collector's Manual of British Land and Freshwater Shells." Taylor Bros, Leeds. [45.]

Ager, D. V. (1963). "Principles of Paleoecology." McGraw-Hill, New York. [9.]

Alexander, J., Ozanne, P. C. and Ozanne, A. (1960). Report on the investigation of a round barrow on Arreton Down, Isle of Wight. *Proc. prehist. Soc.* **26**, 263–302. [10, 141, 363.]

Alkins, W. E. (1928). The conchometric relationship of *Clausilia rugosa* (Drap.) and *Clausilia cravenensis* Taylor. *P.M.S.* **18**, 50–69. [111.]

Allison, J., Godwin, H. and Warren, S. H. (1952). Late-glacial deposits at Nazeing in the Lea Valley, North London. *Phil. Trans. R. Soc.* (B) **236**, 169–240. [165.]

Andrewartha, H. G. (1961). "Introduction to the Study of Animal Populations." Methuen, London. [104, 108.]

Ant, H. (1963). Faunistische, ökologische und tiergeographische Untersuchungen zur Verbrietung der Landschnecken in Nordwestdeutschland. *Abh. Landesmus. Naturk. Münster* **25**, 1–125. [90, 95, 143, 146, 147, 156, 161, 162, 180.]

ApSimon, A. M., Donovan, D. T. and Taylor, H. (1961). The stratigraphy and archaeology of the Late-glacial and Post-glacial deposits at Brean Down, Somerset. *Proc. speleol. Soc.* **9**, 67–136. [223, 311, 352.]

Arkell, W. J. (1943). The Pleistocene rocks at Trebetherick Point, North Cornwall: their interpretation and correlation. *Proc. Geol. Ass.* **54**, 141–170. [296.]

Arkell, W. J. (1947). "The Geology of Oxford." Clarendon, Oxford. [26, 305.]

Armstrong, A. L. (1927). The Grime's Graves problem in the light of recent researches. *Proc. prehist. Soc. E. Anglia* **5**, 91–136. [7.]

Armstrong, A. L. (1934). Grime's Graves, Norfolk: report on the excavation of Pit 12. *Proc. prehist. Soc. E. Anglia* **7**, 382–394. [7.]

Ashbee, P. (1966). The Fussell's Lodge Long Barrow excavations 1957. *Archaeologia* **100**, 1–80. [128.]

Ashbee, P. and Smith, I. F. (1960). The Windmill Hill Long Barrow. *Antiquity* **34**, 297–299. [261.]

Ashbee, P. and Smith, I. F. (1966). The date of the Windmill Hill Long Barrow. *Antiquity* **40**, 299. [261.]

Atkinson, R. J. C. (1957). Worms and weathering. *Antiquity* **31**, 219–233. [15, 32, 209, 247.]

Atkinson, R. J. C. (1965). Wayland's Smithy. *Antiquity* **39**, 126–133. [265.]

Atkinson, R. J. C. (1967). Silbury Hill. *Antiquity* **41**, 259–262. [230, 265.]

Atkinson, R. J. C. (1968). Silbury Hill, 1968. *Antiquity* **42**, 299. [265.]

Atkinson, R. J. C. (1969). The date of Silbury Hill. *Antiquity* **43**, 216. [265.]

Atkinson, R. J. C. (1970). Silbury Hill, 1969–70. *Antiquity* **44**, 313–314. [265.]

Avery, B. W. (1964). "The Soils and Land Use of the District around Aylesbury and Hemel Hempstead." H.M.S.O., London. [40, 41, 209, 286, 287.]

Baker, R. E. (1965). *Catinella arenaria* Bouchard Chantereaux at the Braunton Burrows National Nature Reserve, Devon. *P.M.S.* **36**, 259–265. [21, 123, 137.]

Baker, R. E. (1968). The ecology of the wrinkled snail, *Helicella caperata* (Mont.) on the Braunton Burrows sand dune system. *P.M.S.* **38**, 41–54. [21, 97, 111, 119, 122, 123, 179.]

Baker, R. E. (1970). Population changes shown by *Cochlicopa lubrica* (Müller) in a grass sward habitat. *J. Conch.* **27**, 101–104. [21, 103, 120, 123.]

Barker, H., Burleigh, R. and Meeks, N. (1971). British Museum natural radiocarbon measurements VII. *Radiocarbon* **13**, 157–188. [363.]

Barker, H. and Mackey, J. (1961). British Museum natural radiocarbon measurements III. *Radiocarbon* **3**, 39–45. [303.]

Blackburn, E. P. (1941). Distribution of *Clausilia cravenensis* Taylor (*suttoni* Westerlund) in Britain. *J. Conch.* **21**, 289–300. [166.]

Block, M. R. (1964). Rearing snails from the egg. *Conchologists'. Newsletter* No. 10, 55–58. [127, 181.]

Bowen, H. C. and Fowler, P. J. (1962). The archaeology of Fyfield and Overton Downs, Wilts. (Interim Report). *Wilts. archaeol. nat. Hist. Mag.* **58**, 98–115. [318.]

Boycott, A. E. (1920). On the size variation of *Clausilia bidentata* and *Ena obscura* within a "locality". *P.M.S.* **14**, 34–42. [112.]

Boycott, A. E. (1921a). Oecological notes. *P.M.S.* **14**, 128–130. [133, 338.]

Boycott, A. E. (1921b). Notes on the distribution of British land and freshwater Mollusca from the point of view of habitat and climate. *P.M.S.* **14**, 163–167. [92, 94, 126, 195.]

Boycott, A. E. (1921c). Oecological notes. *P.M.S.* **14**, 167–172. [179, 180.]

Boycott, A. E. (1921d). The land Mollusca of the parish of Aldenham. *Trans. Herts. nat. Hist. Soc. Fld Club* **17**, 220–245. [89, 124, 128, 135.]

Boycott, A. E. (1922). *Vitrina major* in Britain. *P.M.S.* **15**, 123–130. [191.]

Boycott, A. E. (1927). Further notes on *Vitrina major* in Britain. *P.M.S.* **17**, 141–148. [78.]

Boycott, A. E. (1929a). The Mollusca of Great Langdale, Westmorland. *NWest. Nat.* **4**, 10–15. [29.]

Boycott, A. E. (1929b). The oecology of British land Mollusca, with special reference to those of ill-defined habitat. *P.M.S.* **18**, 213–224. [94.]

Boycott, A. E. (1934). The habitats of land Mollusca in Britain. *J. Ecol.* **22**, 1–38. [29, 31, 92, 94, 96, 98, 103, 105, 108, 109, 119, 120, 121, 124, 126, 127, 135, 137, 139, 143, 147, 153, 155, 156, 161, 162, 163, 167, 173, 177, 178, 180, 181, 188, 189, 195, 200, 296.]

Boycott, A. E. and Oldham, C. (1930). The food of *Geomalacus maculosus*. *J. Conch.* **19**, 36. [185.]

Brady, G. S. (1868). A monograph of the Recent British Ostracoda. *Trans. Linn. Soc. Lond. (Zoology)* **26**, 353–495. [37.]

Brodribb, A. C. C., Hands, A. R. and Walker, D. R. (1968–1972). "Excavations at Shakenoak Farm, near Wilcote, Oxfordshire." 3 parts. Brodribb, Hands & Walker, Oxford. [201, 305.]

Bullen, R. A. (1902). Notes on Holocene Mollusca from North Cornwall. *P.M.S.* **5**, 185–188. [179, 183.]

Bullen, R. A. (1912). "Harlyn Bay and the Discoveries of its Prehistoric Remains." Padstow (3rd edition). [124, 179.]

Burchell, J. P. T. (1957). Land-shells as a critical factor in the dating of post-Pleistocene deposits. *Proc. prehist. Soc.* **23**, 236–239. [11, 311.]

Burchell, J. P. T. (1961). Land shells and their role in dating deposits of Post-glacial times in south-east England. *Archaeol. News Letter* **7**, 34–38. [11.]

Burchell, J. P. T. and Davis, A. G. (1957). The molluscan fauna of some early Post-glacial deposits in North Lincolnshire and Kent. *J. Conch.* **24**, 164–170. [161, 167.]

Burchell, J. P. T. and Piggott, S. (1939). Decorated prehistoric pottery from the bed of the Ebbsfleet, Northfleet, Kent. *Antiq. J.* **19**, 405–420. [10, 11.]

Burleigh, R., Longworth, I. H. and Wainwright, G. J. (1972). Relative and absolute dating of four late Neolithic enclosures: an exercise in the interpretation of C-14 determinations. *Proc. prehist. Soc.* **38**, (in press).

Bury, H. (1950). Blashenwell tufa. *Proc. Bournemouth nat. Sci. Soc.* **39**, 48–51. [303.]

Bury, H. and Kennard, A. S. (1940). Some Holocene deposits at Box (Wilts.). *Proc. Geol. Ass.* **51**, 225–229. [158, 304.]

Cain, A. J. (1971). Colour and banding morphs in subfossil samples of the snail *Cepaea*. In "Ecological Genetics and Evolution" (R. Creed, ed.). Blackwell, Oxford. 65–92. [12, 173, 174, 175.]

Cain, A. J., Cameron, R. A. D. and Parkin, D. T. (1969). Ecology and variation of some helicid snails in northern Scotland. *P.M.S.* **38**, 269–299. [21, 95, 173, 181.]

Cain, A. J. and Currey, J. D. (1963a). Area effects in *Cepaea*. *Phil. Trans. R. Soc.* (B) **246**, 1–81. [12, 101, 108, 109, 110, 112, 127, 164, 171, 172, 174, 175.]

Cain, A. J. and Currey, J. D. (1963b). Area effects in *Cepaea* on the Larkhill artillery ranges, Salisbury Plain. *J. Linn. Soc. (Zoology)* **45**, 1–15. [175.]

Cain, A. J. and Sheppard, P. M. (1950). Selection in the polymorphic land snail *Cepaea nemoralis*. *Heredity, Lond.* **4**, 275–294. [106.]

Cameron, R. A. D. (1969a). The distribution and variation of three species of land snail near Rickmansworth, Hertfordshire. *J. Linn. Soc. (Zoology)* **48**, 83–111. [21, 110.]

Cameron, R. A. D. (1969b). Predation by song thrushes *Turdus ericetorum* (Turton) on the snails *Cepaea hortensis* (Müll.) and *Arianta arbustorum* (L.) near Rickmansworth. *J. Anim. Ecol.* **38**, 547–553. [21, 105, 106.]

Cameron, R. A. D. (1970a). The survival, weight-loss and behaviour of three species of land snail in conditions of low humidity. *J. Zool.* **160**, 143–157. [21, 170, 171.]

Cameron, R. A. D. (1970b). The effect of temperature on the activity of three species of Helicid snail (Mollusca: Gastropoda). *J. Zool.* **162**, 303–315. [21, 101, 170, 171.]

Cameron, R. A. D. (1972). The distribution of *Helicodonta obvoluta* (Müll.) in Britain. *J. Conch.* **27**, 363–369. [103, 169.]

Cameron, R. A. D. and Palles-Clark, M. A. (1971). *Arianta arbustorum* (L.) on chalk downs in southern England. *P.M.S.* **39**, 311–318. [21, 170, 197.]

Carrick, R. (1942). The grey field slug *Agriolimax agrestis* L., and its environment. *Ann. appl. Biol.* **29**, 43–55. [96, 99.]

Castell, C. P. (1962). Some notes on London's molluscs. *J. Conch.* **25**, 97–117. [167.]

Chaplin, R. E. (1971). "The Study of Animal Bones from Archaeological Sites." Seminar Press, London. [38.]

Chappell, H. G., Ainsworth, J. F., Cameron, R. A. D. and Redfern, M. (1971). The effect of trampling on a chalk grassland ecosystem. *J. appl. Ecol.* **8**, 869–882. [21, 143, 144, 147, 198.]

Chatfield, J. E. (1967). Field meeting to Pitsford, Northampton, 16th September 1967. *Conchologists' Newsletter* No. 23, 26. [163.]

Chatfield, J. E. (1968). The life history of the helicid snail *Monacha cantiana* (Montagu), with reference also to *M. cartusiana* (Müller). *P.M.S.* **38**, 233–245. [21, 95, 96, 106, 119, 120, 122, 123, 179.]

Chatfield, J. E. (1972). Observations on the ecology of *Monacha cantiana* (Montagu) and associated molluscan fauna. *P.M.S.* **40**, 59–69. [179.]

Childe, V. G. and Smith, I. F. (1954). The excavation of a Neolithic barrow on Whiteleaf Hill, Bucks. *Proc. prehist. Soc.* **20**, 212–230. [246, 316, 361.]

Christie, P. M. (1964). A Bronze Age round barrow on Earl's Farm Down, Amesbury. *Wilts. archaeol. nat. Hist. Mag.* **59**, 30–45. [140, 144, 148, 363.]

Christie, P. M. (1967). A barrow-cemetery of the second millennium BC in Wiltshire, England: excavation of a round barrow, Amesbury, G. 71 on Earl's Farm Down, Wilts. *Proc. prehist. Soc.* **33**, 336–366. [140, 363.]

Churchill, D. M. (1962). The stratigraphy of Mesolithic Sites III and IV at Thatcham, Berkshire, England. *Proc. prehist. Soc.* **28**, 362–370. [307].

Clark, J. G. D. (1938). Microlithic industries from tufa deposits at Prestatyn, Flintshire and Blashenwell, Dorset. *Proc. prehist. Soc.* **4**, 330–334. [303, 305.]

Clark, J. G. D. (1939). A further note on the tufa deposit at Prestatyn, Flintshire. *Proc. prehist. Soc.* **5**, 201–202. [305.]

Clark, J. G. D. (1952). "Prehistoric Europe: The Economic Basis." Methuen, London. [38.]

Clark, J. G. D. (1955). A microlithic industry from the Cambridgeshire Fenland and other industries of Sauveterrian affinities from Britain. *Proc. prehist. Soc.* **21**, 3–20. [303.]

Clarke, W. G. (1915). "Report on the Excavation at Grime's Graves, Weeting, Norfolk: 1914." London. [7, 141, 142.]

Clarke, W. G. (1917). Are Grime's Graves Neolithic? *Proc. prehist. Soc. E. Anglia* **2**, 339–349. [7.]

Connah, G. (1965). Excavations at Knap Hill, Alton Priors, 1961. *Wilts. archaeol. nat. Hist. Mag.* **60**, 1–23. [246.]

Connah, G. and McMillan, N. F. (1964). Snails and archaeology. *Antiquity* **38**, 62–64. [10, 178, 263.]

Coope, G. R. (1967). The value of Quaternary insect faunas in the interpretation of ancient ecology and climate. *In* "Quaternary Palaeoecology: Proceedings of the VIIth Congress of the International Association for Quaternary Research, vol. 7." (E. J. Cushing and H. E. Wright, eds.). 359–380. [37.]

Coope, G. R. and Osborne, P. J. (1968). Report on the coleopterous fauna of the Roman well at Barnsley Park, Gloucestershire. *Trans. Bristol Gloucestershire Archaeol. Soc.* **86**, 84–87. [37.]

Coope, G. R., Shotton, F. W. and Strachan, I. (1961). A Late Pleistocene fauna and flora from Upton Warren, Worcestershire. *Phil. Trans. R. Soc.* (B) **244**, 379–421. [352.]

Corbet, G. B. (1964). "The Identification of British Mammals." British Museum (Natural History), London. [37.]

Cornwall, I. W. (1953). Soil science and archaeology with illustrations from some British Bronze Age monuments. *Proc. prehist. Soc.* **19**, 129–147. [15, 33, 291.]

Cornwall, I. W. (1958). "Soils for the Archaeologist." Phoenix, London. [38, 209, 223.]

Crabtree, K. (1971). Overton Down Experimental Earthwork, Wiltshire 1968. *Proc. speleol. Soc.* **12**, 237–244. [322.]

Cranbrook, The Earl of (1970). Snail shells under starling roosts. *J. Conch.* **27**, 139–143. [106.]

Creek, G. A. (1953). The morphology of *Acme fusca* (Montagu) with special reference to the genital system. *P.M.S.* **29**, 228–240. [135.]

Cunnington, M. E. (1929). "Woodhenge." Simpson, Devizes. [7.]

Cunnington, M. E. (1931). The "Sanctuary" on Overton Hill, near Avebury. *Wilts. archaeol. nat. Hist. Mag.* **45**, 300–335. [7, 34.]

Cunnington, M. E. (1933a). Excavations in Yarnbury Castle Camp, 1932. *Wilts. archaeol. nat. Hist. Mag.* **46**, 198–213. [7.]

Cunnington, M. E. (1933b). Evidence of climate derived from snail shells and its bearing on the date of Stonehenge. *Wilts. archaeol. nat. Hist. Mag.* **46**, 350–355. [7, 365.]

Cunnington, R. H. (1935). "Stonehenge and its Date." Methuen, London. [7.]

Currey, J. D. and Cain, A. J. (1968). Studies on *Cepaea*. IV. Climate and selection of banding morphs in *Cepaea* from the climatic optimum to the present day. *Phil. Trans. R. Soc.* (B) **253**, 483–498. [12, 173, 174, 175.]

Dalrymple, J. B. (1955). "Study of Ferruginous Horizons in Archaeological Sections." Unpublished M.Sc. thesis, University of London. [225.]

Dalrymple, J. B. (1958). The application of soil micromorphology to fossil soils from archaeological sites. *J. Soil Sci.* **9**, 199–209. [15, 38.]

Dance, S. P. (1972). *Vertigo lilljeborgi* Westerlund in North Wales. *J. Conch.* **27**, 387–389. [59, 145.]

Darwin, C. (1881). "The Formation of Vegetable Mould through the Action of Worms with Observations on their Habitats." Murray, London. (Republished, 1945, by Faber & Faber as "Darwin on Humus and the Earthworm."). [209.]

Davis, A. G. (1953). On the geological history of some of our snails, illustrated by some Pleistocene and Holocene deposits in Kent and Surrey. *J. Conch.* **23**, 355–364. [135, 137, 138, 141, 146, 164, 167, 175, 179, 180, 284, 285.]

Davis, A. G. (1954). A fauna with *Helicella striata* (Müller) in the Cray Valley, Kent. *J. Conch.* **24**, 1–6. [183.]

Davis, A. G. (1955). *Acanthinula lamellata* in a Holocene tufa at Totland, Isle of Wight. *J. Conch.* **24**, 40. [153, 305.]

Davis, A. G. (1956). *Hygromia* (*Ponentina*) *subvirescens* (Bellamy) in Cornish Holocene deposits. *J. Conch.* **24**, 100. [178.]

Davis, A. G. (1953–1956). Report on the molluscs from Sun Hole Cave, Cheddar. *Proc. speleol. Soc.* **7**, 71 [309, 310.]

Davis, A. G. and Pitchford, G. W. (1958). The Holocene molluscan fauna of Southwell and Wheatley, Nottinghamshire. *J. Conch.* **24**, 227–233. [137, 184, 305.]

Dean, J. D. and Kendall, C. E. Y. (1908). *Vertigo alpestris* (Alder): its distribution in north Lancashire and Westmorland, and its association with *Vertigo pusilla* Müll. *J. Conch.* **12**, 209–211. [145.]

De Leersnyder, M. and Hoestlandt, H. (1958). Extension du gastropode méditerraneen, *Cochlicella acuta* (Müller), dans le sud-est de l'Angleterre. *J. Conch.* **24**, 253–264. [120, 183.]

Dimbleby, G. W. (1955). The ecological study of buried soils. *Advmt Sci., Lond.* **11**, 11–16. [15, 33.]

Dimbleby, G. W. (1961). Soil pollen analysis. *J. Soil Sci.* **12**, 1–11. [15, 36, 310.]

Dimbleby, G. W. (1962). "The Development of British Heathlands and their Soils." Oxford Forestry Memoir No. 23, Oxford. [15, 33, 227, 277, 358.]

Dimbleby, G. W. (1965). Post-glacial changes in soil profiles. *Proc. Roy. Soc.* (B) **161**, 355–362. [356.]

Dimbleby, G. W. (1967). "Plants and Archaeology." Baker, London. [36, 37.]

Dimbleby, G. W. and Speight, M. C. D. (1969). Buried soils. *Advmt Sci., Lond.* **26**, 203–205. [15.]

Drew, C. D. and Piggott, S. (1936). The excavation of long barrow 163a on Thickthorn Down, Dorset. *Proc. prehist. Soc.* **2**, 77–96. [10, 325, 363.]

Duigan, S. L. (1963). Pollen analysis of the Cromer Forest Bed series in East Anglia. *Phil. Trans. R. Soc.* (B) **246**, 149–202. [136, 137, 138, 142, 165, 178, 185, 188, 190, 191, 192, 193.]

Duval, D. M. (1971). A note on the acceptability of various weeds as food for *Agriolimax reticulatus* (Müller). *J. Conch.* **27**, 249–251. [103.]

Ellis, A. E. (1941). The Mollusca of a Norfolk Broad. *J. Conch.* **21**, 224–243. [143, 156, 162, 196.]

Ellis, A. E. (1951). Census of the distribution of British non-marine Mollusca. *J. Conch.* **23**, 171–244. [93, 99, 133, 137, 139, 140, 143, 145, 149, 153, 155, 162, 170, 175, 177, 178, 183, 184, 189.]

Ellis, A. E. (1962). "A Synopsis of the Freshwater Bivalve Molluscs." The Linnean Society of London Synopses of the British Fauna, No. 13. London. [46.]

Ellis, A. E. (1964). Key to land shells of Great Britain. *Conchological Society of Great Britain and Ireland: Papers for Students*, No. 3. [45.]

Ellis, A. E. (1967). *Agriolimax agrestis* (L.): some observations. *J. Conch.* **26**, 189–196. [192.]

Ellis, A. E. (1969). "British Snails." Clarendon, Oxford. [45, 49, 51, 56, 63, 66,

68, 70, 71, 72, 107, 118, 133, 136, 138, 139, 141, 142, 146, 152, 155, 161, 162, 166, 167, 170, 177, 179, 180, 186, 187, 188, 189, 191, 192, 193.]

Elton, C. (1927). "Animal Ecology." Sidgwick & Jackson, London. [9.]

Evans, J. G. (1966a). Land Mollusca from the Neolithic enclosure on Windmill Hill. *Wilts. archaeol. nat. Hist. Mag.* **61**, 91–92. [101.]

Evans, J. G. (1966b). Late-glacial and Post-glacial subaerial deposits at Pitstone, Buckinghamshire. *Proc. Geol. Ass.* **77**, 347–364. [12, 114, 134, 135, 137, 139, 144, 146, 151, 152, 154, 157, 163, 165, 167, 168, 192, 218, 223, 224, 227, 285, 289, 312, 315, 316, 352, 355, 361, 363, 369.]

Evans, J. G. (1966c). A Romano-British interment in the bank of the Winterbourne, Avebury. *Wilts. archaeol. nat. Hist. Mag.* **61**, 97–98. [34.]

Evans, J. G. (1967). "The Stratification of Mollusca in Chalk Soils and their Relevance to Archaeology." Unpublished Ph.D. thesis, University of London. [15, 168.]

Evans, J. G. (1968a). Changes in the composition of land molluscan populations in north Wiltshire during the last 5000 years. *Symp. zool. Soc. Lond.* No. 22, 293–317. [12, 147, 154, 168, 172, 176, 177, 242, 286, 365.]

Evans, J. G. (1968b). Periglacial deposits on the Chalk of Wiltshire. *Wilts. archaeol. nat. Hist. Mag.* **63**, 12–26. [32, 137, 150, 182, 221, 222, 229, 248, 249, 250, 257, 263, 289, 290, 332, 335, 352, 353, 354.]

Evans, J. G. (1969a). Land and Freshwater Mollusca in archaeology: chronological aspects. *World Archaeology* **1**, 170–183. [12, 13, 132.]

Evans, J. G. (1969b). The exploitation of molluscs. *In* "The Domestication and Exploitation of Plants and Animals." (P. J. Ucko and G. W. Dimbleby, eds.). Duckworth, London. 477–484. [38, 297.]

Evans, J. G. (1969c). Further periglacial deposits in north Wiltshire. *Wilts. archaeol. nat. Hist. Mag.* **64**, 112–113. [268, 352.]

Evans, J. G. (1970). Interpretation of land snail faunas. *Univ. London Inst. Archaeol. Bull.* Nos. 8 and 9, 109–116. [7, 12, 170.]

Evans, J. G. (1971a). Notes on the environment of early farming communities in Britain. *In* "Economy and Settlement in Neolithic and Early Bronze Age Britain and Europe." (D. D. A. Simpson, ed.). University Press, Leicester. 11–26. [286, 291, 300, 365.]

Evans, J. G. (1971b). Habitat change on the calcareous soils of Britain: the impact of Neolithic man. *In* "Economy and Settlement in Neolithic and Early Bronze Age Britain and Europe." (D. D. A. Simpson, ed.). University Press, Leicester. 27–73. [12, 33, 37, 38, 39, 173, 188, 189, 251, 254, 257, 259, 292, 293, 297, 359.]

Evans, J. G. and Burleigh, R. (1969). Radiocarbon dates for the South Street Long Barrow, Wiltshire. *Antiquity* **43**, 144–145. [257, 329.]

Evans, J. G. and Smith I. F. (1967). Cherhill. *Archaeological Review* (Dept. Extra-mural Studies, University of Bristol) No. 2, 8–9. [303, 304.]

Fleming, A. (1971). Territorial patterns in Bronze Age Wessex. *Proc. prehist. Soc.* **37**, 138–166. [365.]

Forcart, L. (1965). New researches on *Trichia hispida* (Linnaeus) and related forms. *Proc. Europ. Malac. Congr.* **1**, 79–89. [177.]

Foster, R. (1958a). The effects of trematode metacercaria (Brachylaemidae) on the slugs *Milax sowerbii* Férussac, and *Agriolimax reticulatus* Müller. *Parasitology* **48**, 261–268. [107.]

Foster, R. (1958b). Infestation of the slugs *Milax sowerbii* Férussac and *Agriolimax reticulatus* Müller by trematode metacercaria (Brachylaemidae). *Parasitology* **48**, 303–311. [107.]

Fowler, P. J. (1971). Early prehistoric agriculture in western Europe: some archaeological evidence. *In* "Economy and Settlement in Neolithic and Early Bronze Age Britain and Europe." (D. D. A. Simpson, ed.). University Press, Leicester. 153–182. [128, 286.]

Fowler, P. J. and Evans, J. G. (1967). Plough-marks, lynchets and early fields. *Antiquity* **41**, 289–301. [12, 128, 214, 260, 316, 318, 319, 320, 332, 366.]

Garrod, D. A. E. (1926). "The Upper Palaeolithic Age in Britain." Oxford. [311.]

Germain, L. (1930). "Faune de France **21**. Mollusques Terrestres et Fluviatiles." 2 parts. Lechevalier, Paris. [133, 162.]

Godwin, H. (1944). Age and origin of the 'Breckland' heaths of East Anglia. *Nature, Lond.* **154**, 6. [358.]

Godwin, H. (1956). "The History of the British Flora." University Press, Cambridge. [4, 5, 36, 361.]

Godwin, H. (1962). Vegetational history of the Kentish chalk downs as seen at Wingham and Frogholt. *Veröff. geobot. Inst., Zurich* **37**, 83–99. [5.]

Godwin, H. (1966). Introductory address. *In* "World Climate from 8000 to 0 BC." Royal Meteorological Society, London. 3–14. [4, 175.]

Godwin, H. (1967). Strip lynchets and soil erosion. *Antiquity* **41**, 66–67. [286.]

Godwin, H. and Tansley, A. G. (1941). Prehistoric charcoals as evidence of former vegetation, soil and climate. *J. Ecol.* **29**, 117–126. [36.]

Hall, B. R. and Folland, C. J. (1970). "Soils of Lancashire." Soil Survey of England and Wales, Bulletin No. 5. Harpenden. [296, 297.]

Hayward, J. F. (1954). *Agriolimax caruanae* Pollonera as a Holocene fossil. *J. Conch.* **23**, 403–404. [79, 193.]

Head, J. F. (1938). The excavation of the Cop Round Barrow, Bledlow. *Rec. Buckinghamshire* **13**, 313–357. [316.]

Higgins, L. S. (1933). An investigation into the problem of the sand dune areas on the South Wales coast. *Archaeol. Cambrensis* **88**, 26–67. [293, 296.]

Howe, J. A. and Skeats, E. W. (1903–1904). Excursion to Denham and Gerrard's Cross to the new cutting on the Great Western Railway. *Proc. Geol. Ass.* **18**, 189–190. [304.]

Hunter, P. J. (1966). The distribution and abundance of slugs on an arable plot in Northumberland. *J. Anim. Ecol.* **35**, 543–557. [119].

Hurst, H. R. (1968). Box (ST 824686): Mesolithic occupation and Roman villa. *Wilts. archaeol. nat. Hist. Mag.* **63**, 109. [304.]

Iversen, J. (1954). The Late-glacial flora of Denmark and its relation to climate and soil. *Danm. geol. Unders.* (II) **80**, 87–119. [355.]

Iversen, J. (1964). Retrogressive vegetational succession in the Post-glacial. *J. Ecol.* **52**, Jubilee Symposium Supplement, 59–70. [219.]

Jackson, J. W. (1922). On the tufaceous deposits of Caerwys, Flintshire, and the Mollusca contained therein. *Lancs. Chesh. Nat.* **14**, 147–158. [304.]

Jackson, J. W. (1926–1929). Report on the animal remains found in the Kilgreany Cave, Co. Waterford. *Proc. speleol. Soc.* **3**, 137–152. [309.]

Jackson, J. W. (1956). The Caerwys tufa. *Lpool Manchr geol. J.* **1**, (4), xxiv–xxviii. [304.]
Jessup, R. F. (1939). Further excavations at Julliberrie's Grave, Chilham. *Antiq. J.* **19**, 260–281. [363.]
Jewell, P. A. and Dimbleby, G. W. (1966). The experimental earthwork on Overton Down, Wiltshire, England: the first four years. *Proc. prehist. Soc.* **32**, 313–342. [247, 287, 323.]
Johnson, D. S. and Lowy, J. (1948). Observations on the distribution of *Helicella caperata* and *Trichia striolata*. *J. Conch.* **23**, 9–13. [105.]
Johnson, G. (1959). True and false ice-wedges in southern Sweden. *Geogr. Annlr* **41**, 15–33. [219.]
Jones, J. S. and Clarke, B. C. (1969). The distribution of *Cepaea* in Scotland. *J. Conch.* **27**, 3–8. [173.]

Kennard, A. S. (1897). The post-Pliocene non-marine Mollusca of Essex. *Essex Nat.* **10**, 87–109. [7.]
Kennard, A. S. (1923). The Holocene non-marine Mollusca of England. *P.M.S.* **15**, 241–259. [7, 284.]
Kennard, A. S. (1924). The Pleistocene non-marine Mollusca of England. *P.M.S.* **16**, 84–97. [7.]
Kennard, A. S. (1943). The post-Pliocene non-marine Mollusca of Hertfordshire. *Trans. Herts. nat. Hist. Soc. Fld Club* **22**, 1–18. [158, 163, 184, 304, 305.]
Kennard, A. S. and Musham, J. F. (1937). On the Mollusca from a Holocene tufaceous deposit at Broughton-Brigg, Lincolnshire. *P.M.S.* **22**, 374–379. [145, 184, 189, 304.]
Kennard, A. S. and Warren, S. H. (1903). The blown sands and associated deposits of Towan Head, near Newquay, Cornwall. *Geol. Mag.* (IV) **10**, 19–25. [180, 183, 292, 361, 363.]
Kennard, A. S. and Woodward, B. B. (1901). The post-Pliocene non-marine Mollusca of the south of England. *Proc. Geol. Ass.* **17**, 213–260. [7.]
Kennard, A. S. and Woodward, B. B. (1917). The post-Pliocene non-marine Mollusca of Ireland. *Proc. Geol. Ass.* **28**, 109–190. [7, 150, 292, 296, 308, 310.]
Kennard, A. S. and Woodward, B. B. (1922). The post-Pliocene non-marine Mollusca of the east of England. *Proc. Geol. Ass.* **33**, 104–142. [7.]
Kennard, A. S. and Woodward, B. B. (1923). On the British species of *Truncatellina*. *P.M.S.* **15**, 294–298. [56.]
Kennard, A. S. and Woodward, B. B. (1922–1925a). Report on the non-marine Mollusca from Aveline's Hole, Burrington Combe. *Proc. speleol. Soc.* **2**, 32–33. [309.]
Kennard, A. S. and Woodward, B. B. (1922–1925b). Report on the non-marine Mollusca of Merlin's Cave. *Proc. speleol. Soc.* **2**, 162. [309.]
Kerney, M. P. (1955). On the former occurrence of *Vertigo parcedentata* in Hertfordshire. *J. Conch.* **24**, 55–58. [158, 163, 184, 304.]
Kerney, M. P. (1956). Note on the fauna of an early Holocene tufa at Wateringbury, Kent. *Proc. Geol. Ass.* **66**, 293–296. [37, 158, 305.]
Kerney, M. P. (1957a). Early Post-glacial deposits in King's County, Ireland, and their molluscan fauna. *J. Conch.* **24**, 155–164. [37, 146, 158, 163, 304.]
Kerney, M. P. (1957b). *Lauria sempronii* (Charpentier) in the English Holocene. *J. Conch.* **24**, 183–191. [61, 135, 151, 166, 169, 304.]

Kerney, M. P. (1959). An interglacial tufa near Hitchin, Hertfordshire. *Proc. Geol. Ass.* **70**, 322–337. [46, 48, 51, 66, 135, 136, 138, 139, 141, 152, 166, 167, 187, 188, 190, 191, 301, 304, 352.]

Kerney, M. P. (1962). The distribution of *Abida secale* (Draparnaud) in Britain. *J. Conch.* **25**, 123–126. [153.]

Kerney, M. P. (1963). Late-glacial deposits on the Chalk of south-east England. *Phil. Trans. R. Soc.* (B) **246**, 203–254. [12, 37, 40, 46, 63, 79, 80, 98, 108, 115, 125, 133, 136, 137, 140, 143, 144, 152, 156, 161, 162, 177, 180, 181, 182, 195, 223, 229, 289, 304, 352, 355, 356.]

Kerney, M. P. (1965). Weichselian deposits in the Isle of Thanet, East Kent. *Proc. Geol. Ass.* **76**, 269–274. [223, 290, 352, 355.]

Kerney, M. P. (1966a). Census of the distribution of British non-marine Mollusca: supplement to the 7th Edition. *J. Conch.* **25**, 1–8. [133.]

Kerney, M. P. (1966b). Snails and man in Britain. *J. Conch.* **26**, 3–14. [12, 29, 35, 98, 126, 127, 129, 135, 138, 166, 169, 171, 176, 177, 179, 180, 183, 191, 192, 203, 361.]

Kerney, M. P. (1967). Distribution mapping of land and freshwater Mollusca in the British Isles. *J. Conch.* **26**, 152–160. [21.]

Kerney, M. P. (1968a). Britain's fauna of land Mollusca and its relation to the Post-glacial thermal optimum. *Symp. zool. Soc. Lond.* No. 22, 273–291. [5, 12, 94, 100, 125, 129, 133, 151, 165, 184, 201, 229, 357.]

Kerney, M. P. (1968b). Field meeting to Leicestershire, 27th April 1968. *Conchologists' Newsletter* No. 27, 72–73. [127, 181.]

Kerney, M. P. (1969). Field meeting to Northamptonshire, 22nd September 1968. *Conchologists' Newsletter* No. 28, 84–85. [162.]

Kerney, M. P. (1970). The British distribution of *Monacha cantiana* (Montagu) and *Monacha cartusiana* (Müller). *J. Conch.* **27**, 145–148. [102, 179.]

Kerney, M. P. (1971a). A Middle Weichselian deposit at Halling, Kent. *Proc. Geol. Ass.* **82**, 1–11. [73, 137, 140, 178, 186, 290, 352.]

Kerney, M. P. (1971b). Interglacial deposits in Barnfield Pit, Swanscombe, and their molluscan fauna. *J. Geol. Soc.* **127**, 69–93. [135, 136, 137, 138, 141, 142, 144, 151, 153, 160, 161, 164, 165, 166, 167, 170, 171, 178, 180, 184, 185, 186, 187, 189, 190, 191, 307, 352.]

Kerney, M. P., Brown, E. H. and Chandler, T. J. (1964). The Late-glacial and Post-glacial history of the Chalk escarpment near Brook, Kent. *Phil. Trans. R. Soc.* (B) **248**, 135–204. [12, 38, 46, 49, 59, 72, 98, 117, 128, 133, 134, 137, 142, 144, 145, 147, 150, 157, 158, 161, 165, 166, 176, 179, 181, 184, 187, 188, 189, 223, 225, 285, 286, 304, 312, 352, 355, 357, 361, 363, 365.]

Kerney, M. P. and Carreck, J. N. (1954). Notes on some Holocene chalk rain-washes at Cudham and Keston, near Downe, Kent. *Proc. Geol. Ass.* **65**, 340–344. [285.]

Kerney, M. P. and Fogan, M. (1969). *Vitrea diaphana* (Studer) in Britain. *J. Conch.* **27**, 17–24. [77, 187.]

Kevan, D. K. and Waterston, A. R. (1933). *Vertigo lilljeborgi* (West.) in Great Britain (with additional Irish localities). *J. Conch.* **19**, 296–313. [59, 145.]

Kew, H. W. (1893). "The Dispersal of Shells." Kegan Paul, London. [124.]

Klíma, B., Kukla, J., Ložek, V. and de Vries, H. (1962). Stratigraphie des Pleistozäns und alter des Paläeolithischen Rastplatzes in der Ziegelei von Dolní Věstonice (unter-Wisternitz). *Anthropozoikum* **11**, 93–145. [15.]

Knutson, L. V., Stephenson, J. W. and Berg, C. O. (1965). Biology of a slug-killing fly, *Tetanocera elata* (Diptera: Sciomyzidae). *P.M.S.* **36**, 213–220. [107.]

Kuiper, J. G. J. (1964). On *Vitrea contracta* (Westerlund). *J. Conch.* **25**, 276–278. [77, 187.]

Lane-Fox, A. H. (1869). Further remarks on the hill forts of Sussex: being an account of excavations in the forts at Cissbury and Highdown. *Archaeologia* **42**, 53–76. [6.]
Lane-Fox, A. H. (1876). Excavations at Cissbury Camp, Sussex. *Jl R. anthrop. Inst.* **5**, 357–390. [6.]
Large, N. F. and Sparks, B. W. (1961). The non-marine Mollusca of the Cainscross Terrace near Stroud, Gloucestershire. *Geol. Mag.* **98**, 423–426. [137, 140, 352.]
Lawrence, M. J. and Brown, R. W. (1967). "Mammals of Britain: Their Tracks, Trails and Signs." Blandford, London. [37.]
Long, D. C. (1970). *Abida secale* (Draparnaud) in the north Cotswolds. *J. Conch.* **27**, 117–120. [152.]
Ložek, V. (1964). "Quartärmollusken der Tschechoslowakei." *Rozpr. ústřed. Úst. geol.* **31**, 1–374. [13, 46, 64, 66, 166.]
Ložek, V. (1965a). The relationship between the development of soils and faunas in the warm Quaternary phases. *Sb. Geol. Ved, Rada A* No. 3, 7–33. [15.]
Ložek, V. (1965b). Problems of analysis of the Quaternary non-marine molluscan fauna in Europe. *Geol. Soc. Amer.* Special Paper 84, 201–218. [40.]
Ložek, V. (1967). Climatic zones of Czechoslovakia during the Quaternary. In "Quaternary Palaeoecology: Proceedings of the VIIth Congress of the International Association for Quaternary Research, vol. 7." (E. J. Cushing and H. E. Wright, eds.). 381–392. [15.]

Macan, T. T. (1969). Key to the British fresh- and brackish-water gastropods. *Freshwater Biological Association Scientific Publication* No. 13, (3rd edition). [45].
Machin, J. (1967). Structural adaptation for reducing water-loss in three species of terrestrial snail. *J. Zool.* **152**, 55–65. [92, 95.]
Manby, T. G. (1963). The excavation of the Willerby Wold Long Barrow, East Riding of Yorkshire. *Proc. prehist. Soc.* **29**, 173–205. [37, 226, 363.]
Manby, T. G. (1971). The Kilham Long Barrow excavations 1965–1969. *Antiquity* **45**, 50–53. [278.]
Marr, J. E. and Shipley, A. E. (1904). "Handbook to the Natural History of Cambridgeshire." University Press, Cambridge. [146.]
Mason, C. F. (1970). Snail populations, beech litter production, and the role of snails in litter decomposition. *Oecologia, Berl.* **5**, 215–239. [21, 111.]
McBurney, C. B. M. (1959). Report on the first season's fieldwork on British Upper Palaeolithic cave deposits. *Proc. prehist. Soc.* **25**, 260–269. [311.]
McMillan, N. F. (1947). The molluscan faunas of some tufas in Cheshire and Flintshire. *Proc. Lpool geol. Soc.* **19**, 240–248. [300, 304, 305.]
Megaw, J. V. S. *et al.* (1961). The Bronze Age settlement at Gwithian, Cornwall. *Proc. W. Cornwall Fld Club* **2**, 200–215. [293.]
M. P. B. W. (1966). "Ministry of Public Building and Works: Excavations Annual Report, 1965." H.M.S.O., London. [338.]
Morgan, F. de M. (1959). The excavation of a long barrow at Nutbane, Hants. *Proc. prehist. Soc.* **25**, 15–51. [38, 291, 324.]
Morton, J. E. (1954). Notes on the ecology and annual cycle of *Carychium tridentatum* at Box Hill. *P.M.S.* **31**, 30–46. [21, 136.]

Newall, R. S. (1931). Barrow 85 Amesbury (Goddard's list). *Wilts. archacol. nat. Hist. Mag.* **45**, 432–458. [140.]

O'Kelly, Clair (1969). Bryn Celli Ddu, Anglesey: a reinterpretation. *Archaeol. Cambrensis* **118**, 17–48. [15.]

O'Kelly, M. J. (1951). Some soil problems in archaeological excavation. *J. Cork Hist. Archaeol. Soc.* **56**, 29–44. [15.]

Oldham, C. (1929). *Cepaea hortensis* (Mueller) and *Arianta arbustorum* (L.) on blown sand. *P.M.S.* **18**, 144–146. [108, 173.]

Osborne, P. J. (1969). An insect fauna of Late Bronze Age date from Wilsford, Wiltshire. *J. Anim. Ecol.* **38**, 555–566. [37.]

Osborne, P. J. (1971). An insect fauna from the Roman site at Alcester, Warwickshire. *Britannia* **2**, 156–165. [37.]

Paterson, K. (1971). Weichselian deposits and fossil periglacial structures in north Berkshire. *Proc. Geol. Ass.* **82**, 455–468. [223, 224, 289, 352, 355.]

Paul, C. R. C. (1971). *Columella* in Britain. *Conchologists' Newsletter* No. 38, 215. [56, 139, 140.]

Peake, A. E. (1919). Excavations at Grime's Graves during 1917. *Proc. prehist. Soc. E. Anglia* **3**, 73–93. [7.]

Pearsall, W. H. (1968). "Mountains and Moorlands." Fontana, London. [104.]

Pennant, T. (1812). "British Zoology," vol. IV, 302. [180.]

Pennington, W. (1970). Vegetation history in the north-west of England: a regional synthesis. *In* "Studies in the Vegetational History of the British Isles." (Walker, D. and West, R. G. eds.). University Press, Cambridge. 41–79. [225, 355.]

Perrin, R. M. S. (1956). Nature of "Chalk Heath" soils. *Nature, Lond.* **178**, 31–32. [24, 38, 209, 216, 290, 356.]

Pickrell, D. G. (1965). Field meeting to examine the Mollusca of the River Cray. *Conchologists' Newsletter* No. 15, 101. [163.]

Piggott, S. (1954). "The Neolithic Cultures of the British Isles." University Press, Cambridge. [10.]

Piggott, S. (1962). "The West Kennet Long Barrow: Excavations 1955–56." H.M.S.O., London. [264.]

Proudfoot, E. (1965). Bishop's Cannings: Roughridge Hill (SU/060660): Bronze Age. *Wilts. archaeol. nat. Hist. Mag.* **60**, 132–133. [335.]

Quick, H. E. (1933). The anatomy of British Succineae. *P.M.S.* **20**, 295–318. [49, 137.]

Quick, H. E. (1943). Land snails and slugs of West Glamorgan. *J. Conch.* **22**, 4–12. [146, 180, 181.]

Quick, H. E. (1949). "Synopses of the British Fauna No. 8: Slugs (Mollusca) (Testacellidae, Arionidae, Limacidae)." Linnean Society, London. [45, 66, 79.]

Quick, H. E. (1953). Helicellids introduced into Australia. *P.M.S.* **30**, 74–79. [123.]

Quick, H. E. (1954). *Cochlicopa* in the British Isles. *P.M.S.* **30**, 204–213. [51, 54, 138, 139, 143.]

Quick, H. E. (1960). British slugs (Pulmonata: Testacellidae, Arionidae, Limacidae). *Bull. Br. Mus. nat. Hist. Zool.* **6**, No. 3, 103–226. [45, 79.]

Radley, J. (1966). Glebe Low, Great Longstone. *Derbyshire Archaeol. J.* **86**, 54–69. [309.]

Rahtz, P. A. (1962). Farncombe Down Barrow, Berkshire. *Berkshire Archaeol. J.* **60**, 1–24. [363.]
Rees, W. J. (1965). The aerial dispersal of Mollusca. *P.M.S.* **36**, 269–282. [124.]
Renfrew, C. (1970). The tree-ring calibration of radiocarbon: an archaeological evaluation. *Proc. prehist. Soc.* **36**, 280–311. [6.]
Ritchie, W. (1966). The Post-glacial rise in sea-level and coastal changes in the Uists. *Trans. Inst. Br. Geogr.* No. 39, 79–86. [296.]
Roberts, E. (1958). "The County of Anglesey: Soils and Agriculture." H.M.S.O., London. [40.]
Roscoe, E. J. (1962). Aggregations of the terrestrial snail *Cionella lubrica*. *Nautilus* **75**, 111–115. [111.]
Royal Commission on Historical Monuments (1960). "A Matter of Time." H.M.S.O., London. [346.]
Runham, N. W. and Hunter, P. J. (1970). "Terrestrial Slugs." Hutchinson, London. [99, 103, 107, 119, 120, 122.]
Russell, E. J. (1961). "The World of the Soil." Fontana, London. [284.]

Saunders, C. J. (1971.) The Pre-Belgic Iron Age in the central and western Chilterns. *Archaeol. J.* **128**, 1–30. [315, 316.]
Scharff, R. F. (1907). "European Animals: Their Geological History and Geographical Distribution." Constable, London. [124, 151, 184, 185.]
Schlesch, H. (1951). The north European *Helicella*. *J. Conch.* **23**, 137–141. [182.]
Schmid, E. (1969). Cave sediments and prehistory. *In* "Science in Archaeology" (D. Brothwell, and E. Higgs, eds.). Thames & Hudson. 151–161. [308.]
Shackleton, N. J. (1969). Marine Mollusca in archaeology. *In* "Science in Archaeology" (D. Brothwell and E. Higgs, eds.). Thames & Hudson. 407–414. [297.]
Sheppard, P. M. (1951). Fluctuations in the selective value of certain phenotypes in the polymorphic land snail *Cepaea nemoralis* (L.). *Heredity, Lond.* **5**, 125–134. [106.]
Shotton, F. W. (1968). The Pleistocene succession around Brandon, Warwickshire. *Phil. Trans. R. Soc.* (B) **254**, 387–400. [138, 140, 352.]
Shrubsole, G. (1933). Non-marine Mollusca of the Eastbourne district. *J. Conch.* **19**, 361–368. [181.]
Simmons, I. G. (1969). Evidence for vegetation changes associated with Mesolithic man in Britain. *In* "The Domestication and Exploitation of Plants and Animals." (P. J. Ucko and G. W. Dimbleby, eds.). Duckworth, London. 110–119. [3.]
Simpson, D. D. A. (1966). A Neolithic settlement in the Outer Hebrides. *Antiquity* **40**, 137–139. [293, 296.]
Sissons, J. B. (1967). "The Evolution of Scotlands' Scenery." Oliver & Boyd, Edinburgh. [28.]
Small, R. J., Clark, M. J. and Lewin, J. (1970). The periglacial rock-stream at Clatford Bottom, Marlborough Downs, Wiltshire. *Proc. Geol. Ass.* **81**, 87–98. [290.]
Smith, A. G. (1958a). The context of some Late Bronze Age and Early Iron Age remains from Lincolnshire. *Proc. prehist. Soc.* **24**, 78–84. [304.]
Smith, A. G. (1958b). Post-glacial deposits in south Yorkshire and north Lincolnshire. *New Phytol.* **57**, 19–49. [5, 304.]
Smith, A. G. (1965). Problems of inertia and threshold related to Post-glacial habitat changes. *Proc. R. Soc.* (B) **161**, 331–342. [225, 355.]

Smith, A. G. (1970). The influence of Mesolithic and Neolithic man on British vegetation: A discussion. *In* "Studies in the Vegetational History of the British Isles." (D. Walker and R. G. West, eds.). University Press, Cambridge. 81–96. [3, 256.]

Smith, I. F. (1965a). Excavation of a bell barrow, Avebury G. 55. *Wilts. archaeol. nat. Hist. Mag.* **60**, 24–46. [10, 365.]

Smith, I. F. (1965b). "Windmill Hill and Avebury: Excavations by Alexander Keiller, 1925–1939." University Press, Oxford. [35, 128, 211, 243, 246, 268.]

Smith, I. F. and Evans, J. G. (1968). Excavation of two long barrows in Wiltshire. *Antiquity* **42**, 138–142. [248, 328.]

South, A. (1965). Biology and ecology of *Agriolimax reticulatus* (Müll.) and other slugs: spatial distribution. *J. Anim. Ecol.* **34**, 403–417. [192.]

Southern, H. N. (1964). "The Handbook of British Mammals." Blackwell, Oxford. [37.]

Sparks, B. W. (1952). Notes on some Pleistocene sections at Barrington, Cambridgeshire. *Geol. Mag.* **89**, 163–174. [135, 137, 145, 161, 223, 289, 312.]

Sparks, B. W. (1953a). Fossil and recent English species of *Vallonia*. *P.M.S.* **30**, 110–121. [63, 64, 161, 164, 165.]

Sparks, B. W. (1953b). The former occurrence of both *Helicella striata* (Müller) and *H. geyeri* (Soós) in England. *J. Conch.* **23**, 372–378. [72, 182, 183.]

Sparks, B. W. (1955). Notes on four Quaternary deposits in the Cambridgeshire region. *J. Conch.* **24**, 47–53. [164, 193.]

Sparks, B. W. (1957a). The non-marine Mollusca of the interglacial deposits at Bobbitshole, Ipswich. *Phil. Trans. R. Soc.* (B) **241**, 33–44. [12, 46, 64, 79, 165, 178, 192, 193, 306.]

Sparks, B. W. (1957b). The taele gravel near Thriplow, Cambridgeshire. *Geol. Mag.* **94**, 194–200. [49, 51, 138, 289, 352.]

Sparks, B. W. (1961). The ecological interpretation of Quaternary non-marine Mollusca. *Proc. Linn. Soc. Lond.* **172**, 71–80. [7, 8, 40, 43, 80, 200, 305, 347, 349.]

Sparks, B. W. (1962). Post-glacial Mollusca from Hawes Water, Lancashire, illustrating some difficulties of interpretation. *J. Conch.* **25**, 78–82. [306.]

Sparks, B. W. (1964a). Non-marine Mollusca and Quaternary Ecology. *J. Anim. Ecol.* **33**, 87–98. [12, 40, 41, 80, 94, 112, 169, 178.]

Sparks, B. W. (1964b). The distribution of non-marine Mollusca in the Last Interglacial in south-east England. *P.M.S.* **36**, 7–25. [35, 125, 135, 136, 137, 138, 139, 141, 142, 145, 146, 151, 153, 160, 161, 164, 165, 166, 167, 169, 170, 171, 173, 175, 178, 180, 182, 184, 185, 186, 187, 188, 189, 190, 191, 192, 193, 352,]

Sparks, B. W. (1969). Non-marine Mollusca and archaeology. *In* "Science in Archaeology" (D. Brothwell and E. Higgs, eds.). Thames & Hudson. 395–406. [23, 87, 132.]

Sparks, B. W. and Lambert, C. A. (1961). The Post-glacial deposits at Apethorpe, Northamptonshire. *P.M.S.* **34**, 302–315. [35, 41, 178, 192, 193, 306.]

Sparks, B. W. and Lewis, W. V. (1957). Escarpment dry valleys near Pegsdon, Hertfordshire. *Proc. Geol. Ass.* **68**, 26–38. [285, 312, 316.]

Sparks, B. W. and West, R. G. (1959). The palaeoecology of the interglacial deposits at Histon Road, Cambridge. *Eiszeitalter Gegenw.* **10**, 123–143. [12, 306.]

Sparks, B. W. and West, R. G. (1964). The interglacial deposits at Stutton, Suffolk. *Proc. Geol. Ass.* **74**, 419–432. [12.]

Sparks, B. W. and West, R. G. (1965). The relief and drift deposits. *In* "The Cambridge Region" (J. A. Steers, ed.). 18–40. [349.]

Sparks, B. W. and West, R. G. (1968). Interglacial deposits at Wortwell, Norfolk. *Geol. Mag.* **105**, 471–481. [12.]

Sparks, B. W. and West, R. G. (1970). Late Pleistocene deposits at Wretton, Norfolk. I. Ipswichian Interglacial deposits. *Phil. Trans. R. Soc.* (B) **258**, 1–30. [12, 306.]

Speight, M. C. D. (1969). Insect clues to prehistoric Britain. *Animals*, April 1969, 532–535. [37.]

Stelfox, A. W. (1911). A list of the land and freshwater molluscs of Ireland. *Proc. R. Ir. Acad.* (B) **29**, 65–164. [49.]

Stephenson, J. W. (1966). Notes on the rearing and behaviour in soil of *Milax budapestensis* (Hazay). *J. Conch.* **26**, 141–145. [96, 97, 99.]

Stephenson, J. W. (1968). A review of the biology and ecology of slugs of agricultural importance. *P.M.S.* **38**, 169–178. [107, 119.]

Stephenson, J. W. and Knutson, L. V. (1966). A résumé of recent studies of invertebrates associated with slugs. *J. econ. Ent.* **59**, 356–360. [107.]

Stone, J. F. S. (1931). Easton Down, Winterslow, south Wiltshire, flint mine excavation, 1930. *Wilts. archaeol. nat. Hist. Mag.* **45**, 350–365. [10.]

Stone, J. F. S. (1933). Excavations at Easton Down, Winterslow, 1931–1932. *Wilts. archaeol. nat. Hist. Mag.* **46**, 225–242. [7, 169.]

Stone, J. F. S. and Hill, N. G. (1938). A Middle Bronze Age site at Stockbridge, Hampshire. *Proc. prehist. Soc.* **4**, 249–257. [7, 141.]

Stone, J. F. S. and Hill, N. G. (1940). A round barrow on Stockbridge Down, Hampshire. *Antiq. J.* **20**, 39–51. [7, 170.]

Strahan, A. and Cantrill, T. C. (1902). "Memoirs of the Geological Survey, Sheet 263: The Geology of the South Wales Coal-field. Part III: The Country around Cardiff." H.M.S.O., London. [305.]

Stratton, L. W. (1954). On *Arianta arbustorum* (L). *J. Conch.* **23**, 405–412. [170.]

Stratton, L. W. (1955). *Clausilia dubia* Draparnaud in the Malham area. *J. Conch.* **24**, 41–46. [166.]

Stratton, L. W. (1963). An ecological study. *J. Conch.* **25**, 174–179. [127.]

Stratton, L. W. (1964). The non-marine Mollusca of the parish of Dale. *Fld Stud.* **2**, 41–52. [146.]

Swanton, E. W. (1912). "The Mollusca of Somerset." Somerset Archaeological and Natural History Society, Taunton. [143, 163, 181.]

Taylor, D. W. (1965). The study of Pleistocene non-marine molluscs in North America. *In* "The Quaternary of the United States" (H. E. Wright and D. G. Frey, eds.). University Press, Princeton. 597–611. [103.]

Taylor, J. W. (1894–1921). "Monograph of the Land and Freshwater Mollusca of the British Isles." 3 vols. + 3 parts (unfinished). Taylor Brothers, Leeds. [45.]

Thistleton, B. M. (1966). A display of land and freshwater shells of the Isle of Wight. *Conchologists' Newsletter* No. 19, 134–135. [156.]

Thorley, Anne (1971). Vegetational history in the Vale of the Brooks. *In* "Guide to Sussex Excursions" (R. B. G. Williams, ed.). Sussex University. [5.]

Tratman, E. K., Donovan, D. T. and Campbell, J. B. (1971). The Hyaena Den (Wookey Hole), Mendip Hills, Somerset. *Proc. speleol. Soc.* **12**, 245–279. [308.]

Turner, J. (1970). Post-Neolithic disturbance of British vegetation. *In* "Studies in the Vegetational History of the British Isles." (D. Walker and R. G. West, eds.). University Press, Cambridge. 97–116. [4, 365.]

Turton, W. (1857). "Manual of the Land and Freshwater Shells of the British Islands." (New edition, with additions by J. E. Gray). Longman, London. [45.]

Verdcourt, B. (1945). The Mollusca of Bedfordshire. *J. Conch.* **22**, 124–129. [181.]
Verdcourt, B. (1946). Two years' collecting in the Reading district. *J. Conch.* **22**, 227–228. [163.]
Verdcourt, B. (1951). Land snails of a residential area in N.W. Hertfordshire. *J. Conch.* **23**, 155–157. [156, 163.]

Wade, A. G. (1923). Ancient flintmines at Stoke Down, Sussex. *Proc. prehist. Soc. E. Anglia* **4**, 82–91. [169.]
Wainwright, G. J. (1972). The excavation of a late Neolithic enclosure at Marden, Wiltshire. *Antiq. J.* **52**, 177–239. [141, 274.]
Wainwright, G. J. and Longworth, I. H. (1971). "Durrington Walls: Excavations 1966–1968." (Reports of the Research Committee of the Society of Antiquaries of London No. 29). Society of Antiquaries, London. [34, 35, 39, 83, 140, 148, 149, 165, 167, 183, 211, 222, 272, 277, 360, 363, 369.]
Waldén, H. W. (1955). The land Gastropoda of the vicinity of Stockholm. *Ark. Zool.* (II) **7**, 391–448. [143, 156.]
Waldén, H. W. (1966). Einige Bermerkungen zum Ergänzungsband zu Ehrman's "Mollusca" in "Die Tierwelt Mitteleuropas." *Arch. Molluskenk.* **95**, 49–68. [139.]
Warburg, M. R. (1965). On the water economy of some Australian land-snails. *P.M.S.* **36**, 297–305. [92.]
Warren, S. H. (1912). On a late glacial stage in the Lea Valley, subsequent to the epoch of river-drift man. *Q. Jl geol. Soc. Lond.* **68**, 213–251. [352.]
Warren, S. H. (1945). Some geological and prehistoric records on the north-west border of Essex. *Essex Nat.* **27**, 273–280. [305.]
Watson, H. (1920). The affinities of *Pyramidula, Patulastra, Acanthinula* and *Vallonia. P.M.S.* **14**, 6–30. [63, 118, 163.]
Watson, H. (1943). Notes on a list of the British non-marine Mollusca. *J. Conch.* **22**, 13–22, 25–47, 53–72. [47.]
Watson, H. and Verdcourt, B. (1953). The two British species of *Carychium. J. Conch.* **23**, 306–324. [48, 136.]
Watt, A. S. (1934). The vegetation of the Chiltern Hills, with special reference to the beechwoods and their seral relationships. *J. Ecol.* **22**, 230–270, 445–507. [88.]
West, R. G. (1963). Problems of the British Quaternary. *Proc. Geol. Ass.* **74**, 174–186. [352.]
West, R. G. (1968). "Pleistocene Geology and Biology with especial Reference to the British Isles." Longmans, London. [38, 290, 356, 398.]
Wild, S. V. and Lawson, A. K. (1937). Enemies of the land and fresh-water Mollusca of the British Isles. *J. Conch.* **20**, 351–361. [105.]
Williams, D. W. (1942). Studies on the biology of the larva of the nematode lungworm *Muellerius capillaris* in molluscs. *J. Anim. Ecol.* **11**, 1–8. [107.]
Williams, J. H. (1971). Roman building-materials in south-east England. *Britannia* **2**, 166–195. [302, 303.]
Williams, P. W. and Williams, R. B. G. (1966). The deposits of Ballymihil Cave, Co. Clare, Ireland, with particular reference to non-marine Mollusca. *Proc. speleol. Soc.* **11**, 71–82. [139, 308, 309.]
Williams, R. B. G. (1964). Fossil patterned ground in eastern England. *Biul. peryglac.* **14**, 337–349. [290, 356.]

Williams, R. B. G. (1968). Some estimates of periglacial erosion in southern and eastern England. *Biul. peryglac.* **17**, 311–335. [290.]

Woodward, B. B. (1904). The British species of *Vallonia*. *J. Conch.* **9**, 82. [161.]

Woodward, B. B. (1908). Notes on the drift and underlying deposits at Newquay, Cornwall. *Geol. Mag.* (V) **5**, 10–18, 80–87. [118, 179, 180, 292.]

Wooldridge, S. W. and Goldring, F. (1953). "The Weald." Collins, London. [366.]

Wooldridge, S. W. and Linton, D. L. (1933). The loam-terrains of south-east England and their relation to its early history. *Antiquity* **7**, 297–310. [290, 356.]

Wright, E. V. and Wright, C. W. (1947). Prehistoric boats from North Ferriby, East Yorkshire. *Proc. prehist. Soc.* **13**, 114. [305.]

Wrigley, A. (1948). A. S. Kennard, 1870–1948. *J. Conch.* **23**, 20.

Zeuner, F. E. (1955). Loess and Palaeolithic chronology. *Proc. prehist. Soc.* **21**, 51–64. [290.]

Zeuner, F. E. (1957). "Dating the Past: An Introduction to Geochronology." Methuen, London. [290.]

Zeuner, F. E. (1959). "The Pleistocene Period: Its Climate, Chronology and Faunal Successions." Hutchinson, London. [132, 137, 139, 141, 145, 173, 175, 180.]

Zimmermann, F. (1925). *Z. indukt. Abstamm.- u. Vererb Lehre* **37**, 291–342. [136.]

Index

Primary (individual site) histograms have not been indexed for species or habitats. The Glossary and Appendix have not been indexed.
*=shell drawings in Chapter 3.
A number of important references, in particular the identification and ecology of the snail species, are indicated in bold type.

A

Abida secale, 52*, 53*, 54, 57*, 60*, **61**, 89, 116, 124, **152**, 158, 198, 233, 288, 355
Absolute abundance, 79, 89, 334
Acanthinula
   aculeata, 8, 23, 54, 57*, 60*, **63**, 96, 113, **153**, 195, 213, 331
   lamellata, 54, 57*, 60*, **63**, 92, 94, **153**, 195, 242, 301
A/C horizon, 208, 210, 212 *et seq.*, 222, 244, 248
Acicula, 46, 48
   diluviana, 135
   fusca, 47, **48**, 55*, 92, 99, 116, **135**, 195, 285, 361
   polita, 135
Acid-insoluble material, 209
Acmidae, 48, 135
Activity, of snails, 131
Aeolian deposits, 207, 281, 290, 291, 324
Africa, 13
Aggregations, of snails, 111, 119
Agriculture, 98, 105, 126, 130, 146, 159, 193, 198, 226, 279, 314, 344
Agriolimax, 50*, **79**, **192**, 196, 332, 338, 353
   agrestis, 163, **192**, 196, 199
   caruanae, **193**
   laevis, 107, 156, **193**, 199, 200
   reticulatus, 96, 99, 107, 119, 120, 163, **192**, 197

A horizon, 208, 211, 216, 222
Alder, 357
Algae, 89, 103, 299
Alien species, 179, 187, **200**, 202
Allerød Interstadial, 114, 124, 156, 162, 311, **353**
   soil, 115, 222 *et seq.*
Allochthonous components, 308
Allolobophora, 209
Alluvial pasture, 104
Alluvium, 28, 281
Alnus, 358
Alps, 14, 140
Altithermal, 357
Amino-acids, 104
Amphibia, 37
Amphibious species, 199
Anaerobic conditions, 22, 37, 119, 121, 266, 350
Animals (of man), 127, 130, 131, 358
Anthropophobic snails, 128 *et seq.*, 151, 165, **201**, 203
Ants, 210
Apethorpe, Northants., 178, 192
Apical fragments/apices, worn and pitted, 213, 232, 263, 264, 314, 341
Aplexa hypnorum, 200, 347
Arable farming, 364
Arable habitats/land, 91, 104, 108, 123, 129, 133, 146, 154, 162, 176, 179, 182, 192, 227, 314, 358

INDEX

Aragonite, 23
Arbury Road, Cambridge, 37, 42, 176, 178, **347**–**349**
Archaeological deposits, 31
Arctic climate, 351
Area effect, 175
Arianta arbustorum, 21, 67*, **68**, 101, 106, **170**, 174, 196, 199, 202, 212
Arion, 73, **185**
  ater, 74, 163, 186
  circumscriptus, 163, 186
  fasciatus, 163, 185
  hortensis, 119, 163, 186
  intermedius, 163, 186
  lusitanicus, 186
  rufus, 186
  silvaticus, 185
  subfuscus, 186
Arionidae, 73, 118, 185
Arionid granules/internal shells, 23, 73, 186
Ariophantidae, 74, 186
Arreton Down, Isle of Wight, 8, 141, 148, 359, 362
Arrhenatherum elatius, 87, 113, 236
Ascott-under-Wychwood, Oxon., 124, 135, 147, 149, 154, 157, 162, 182, 219, 223, 226, 230, 246, **251**, 260, **343**, 358, 361, 362
Ashbury, Berkshire, 265
Aspect, 226
Atlantic period, 4, 5, 116, 117, 128, 256, 296, **299**, 357, 361 (not indexed for Chapter 5)
  soils, 226
Auger, 43
Autochthonous assemblages, 299
  components, 308
Avebury, Wilts., 126, 128, 211, 223, **268**, 325, 360, 362
Avebury area, 154, 159, 164, 174, 360, 365
Avebury Trusloe, 174
Aveline's Hole, 308
Azeca, 46
  goodalli, 51, 52*, 53*, 99, 125, **138**, 163, 195, 361
  menkeana, 51, 138

B

Bacombe Hill, Wendover, 239
Bacteria, 223
Badbury Earthwork, Dorset, 21, 213, 325, **337**, 359
Badbury Rings, 338
Balea perversa, **64**, 65*, 105, 112, **167**, 195
Ballymihil Cave, 308, 309
Banding, of shells, 95, 173, 174
Bank material, 247
Banks, 141, 243, 245, 247, 268
Bare ground, 115, 146, 158, 182, 196, 223, 323, 331
Barnes Bridge, London, 156
Barriers to dispersal, 123
Barrington, Cambridge, 145, 312
Barrows, 123, 227, 364, 365 (*see also* Long and Round barrows)
Barry Island, 305
Beachy Head, 125, 178, 181
Beacon Hill, Ellesborough, Bucks., 155, 163
Beaker period, 148, 263, 292 *et seq*., 325, 326, 332, 338, 364
Beaker clearance horizon, 258, 285
Beckhampton Road, Wilts., 214, 217, 221, 230, **248**, 309, 332, 354, 359, 362
Bedfordshire, 181
Beech, 357
Beechwood/woodland, 88, 103, 112, 156, 230, 233, 236, 284
Behaviour patterns, 92, 111, 171
Belgrandia marginata, 46
Bembridge Limestone, 25
Benbecula, 29, 296
Berkshire, 153, 221, 223, 224, 265, 352
  Downs, 149, 265
  Ridgeway, 265
Bexhill, 191
B horizon, 255
Biennial species, 201
Biological control, 108
Birch, 354, 357
  forest, 30
Birds, 91, 105 *et seq*., 122
Bithynia, 347
Blackpatch, Sussex, 151
Blandford Forum, Dorset, 337

Blanket bog, 357, 358
Blashenwell, Dorset, 299 *et seq.*, 303
Bledlow Cop, 315, 316
Blown sand, *see* Sand
Bølling Interstadial, 156, 178, 181, 223, 355
Bones, 37, 44, 299, 303, 312
Borer, 43
Boreal period, 4, 5, 229, 256, 272, 357, 361 (not indexed for Chapter 5)
  fauna, 358
  soils, 225, 300, 301
Boscombe Down, Wilts., 140
Boulder clay, 27, 28
Box, Wiltshire, 158, 301, 304
Bracken, 36
Branches, 96, 103, 115, 194, 301
Brandon, 353
Braunton Barrows, Devon, 120, 137
Brean Down, Somerset, 224, 225
Breckland, 24, 358
Breeding, 119
  cycle, 111
  season, 111, 120
Brickearth, 213, 222, 248, 290, 311
Brigg, Lincolnshire, 300, 304
Broken ground, 133, 135, 149, 187, 261
Bronze Age, 10–12, 98, 109, 125, 148, 164, 172, 174, 200, 291, 358, 365
  barrows, 8, 335, 364
  climate, 149
  faunas, 113, 362
  levels, 32
  soils, 229
Brook, Kent, **116**, 128, 134, 142, 157, 181, 225, 285, 304, 357, 359, 363
Broughton-Brigg, 184, 189
Brown earths, 126, 128, 216, 225, 249, 256, 305, 356
Brundon, 164
Bt horizon, 217, 221, 274
Buckinghamshire, 137, 145, 154–156, 163, 223, 232, 304, 312
Building débris, 31, 34
Building material, 303
Buildings, 29, 31
Bulhary nad Dyji, 353
Burial mounds, 128, 227
Burial, rate of, 230
Burials, 34

Buried soils, 4, 15, 23, 32 *et seq.*, 114, 123, 211, 222, 224—228, 243 *et seq.*, 285, 306, 332, 361
Burning, 129
Burrington Combe, Somerset, 288, 308
Burrowing animals, 210
  species, 97, 133, 168, **201**, 203
Butcombe, Somerset, 163, 217

C

C-14, *see* Radiocarbon
Caddis-fly larvae, 299
Caerwent, 29, 128
Caerwys, 174, 303
Cain, A. J., 43
Cainscross, 353
Cairns, 24, 188
Calcicole species
  of snail, 109, 126, 180, 296
  of plant, 216
Calcifuge species
  of snail, 190
  of plant, 216
Calciphile species, 133
Calcite, 23
Calcium carbonate, 209
Calcium carbonate/silica ratio, 293
Cambridge, 349
  gravels, 182
Cambridgeshire, 137, 146, 176, 180, 223
Cameron, R. A. D., 21
Camouflage, 106
Canals, 136
Carbon dioxide, 299
Carbonic acid, 209
Carboniferous limestone, 23, 25, 27, 216, 217
Carnivores, 103
Carrion feeders, 103
Carychium, 48, 113, 301, 310
  minimum, **48**, 55*, 136, 199, 295
  tridentatum, 8, 20, **48**, 55*, 87, 89, 116, 120, 124, 136, 163, 195, 199, 202, 261, 272, 295, 361
Catholic species, 196 *et seq.*, 295
  freshwater species, 347, 349
Catinella arenaria, 50*, **51**, 123, **136**, 199
Cattle, 104, 131, 155, 156, 273, 291
Causewayed enclosures, 243, 247, 328

INDEX

Cave faunas, 308, 311
Caves, 24, 136, 139, 188, 299, **308**
Cecilioides acicula, 52*, 55*, **66**, 80, 100, 168, 201
Celtic fields, 39, 316, 366
Central Europe, 13, 15, 184, 290
Cepaea, 8, 21, 23, 43, 45, 47, 69*, 118, 159, 170, 202, 212, 341, 344
   hortensis, 67*, **70**, 101, 106, 108–110, **171–174**, 196
   nemoralis, 12, 67*, **70**, 94, 101, 106, 108–110, 112, 124, 127, **173–175**, 196, 199, 361
Chalfont St. Giles, Buckinghamshire, 218
Chalk heath, 24, 38, 216
   meltwater débris, 289
   pits, 124
Chambered tombs, 188
Charcoal, 36, 44, 219, 230, 246, 252, 277, 299, 303, 312
Cheddar, 309
Cheddarian, 310
Cheltenham, 26
Cherhill, Wiltshire, 184, 216, 225, 286, 300, **304**, 327, 354, 360
Chiltern Hills/escarpment, 24, 27, 28, 88, 110, 136, 143, 152, 154 *et seq.*, 190, 196, 232, 241, 284, 311, **315**, 351, 352, 359, 367
Chinnor, Oxfordshire, 286, 315
Cholsey, Berkshire, 224
Chondrinidae, 61, 152
C horizon, 208
Clark, J. G. D., 10, 325
Clausilia, 23, 45
   bidentata, 8, 55*, **64**, 65*, 111, 112, **166**, 195, 213, 264, 314, 361
   cravenensis, 111
   dubia suttoni, 65*, **66**, **166**
   parvula, 166
   pumila, 66, 166
   rolphi, **64**, 65*, 100, **167**, 195
   ventricosa, 66, 167
Clausiliidae, 58, 64, 96, 166, 212, 229, 331
Clay, 209, 274, 278, 346, 356
   -illuviation horizon, 217
Clay-with-flints, 24, 265

Clearance
   fauna, 182, 285
   horizon/phase, 134, 158, 258, 316
Clearing, 116
Cliffs, 153, 287
Climate, 3, 5, 9, 11, 14, 18, 20, 22, **92**, **117**, **125**, 131, 135, 145, 149, 164, 170, 174, 185, 202, 223, 300, 325, 365
Climatic
   amelioration, 124, 125, 352, 361
   change, 79, 124, 184, 296, 365
   deterioration, 98, 125, 129, 149, 156, 366
   factors, 125
   optimum, 357
   selection, 13, 174
   variation, 92
Coast, 141
Coastal
   species, 71, 171, 183
   habitats, 152, 178 (*see also* Maritime habitats)
Cobbling, 209
Cochlicella acuta, 43, 52*, 53*, 64, **72**, 120, **183**, 200, 295
Cochlicopa, 23, 45, 51, 53*, 163, 190, 202, 213, 341
   lubrica, 8, **51**, 52*, 53*, 104, 106, 111, 120, 123, **138**, 163, 196, 199
   lubricella, **51**, 52*, 53*, **139**, 196, 198
Cochlicopidae, 51, 138
"Cold" faunas, 164, 183, **353**
Cold places, 170
Colluvial
   deposits, 27, 31, 34, 41, 116, 207, 213, 280
   soils, 207, 224, 226, 280
Colonization, 123, 124, 339
Colour, of shells, 174
Columella
   alticola, 102
   aspera, **56**, 57*, 60*, **139**
   columella, **56**, 57*, 60*, 101, **140**, 177, 199, 291, 353
   edentula, 54, **56**, 57*, 60*, **139**, 195, 199, 272
Comb's Ditch, Dorset, 144
Compensatory factors, 109

Competition, 96, **108**, 110, 142, 159, 160, 164, 171, 173, 182
Compression, of soils, 222, 230 (*see also* Wastage)
Conchological Society, 13, 21, 45
Contamination, 43, 310, 327
Continentality, 185
Coombe Hole, Bucks., 33, 37, 212, 219, 229, 233, **240**, 255, 315, 360
Coombe rock, 209, 216, 224, 236, 241, 242, 248, 257, 258, 269, **289**, 311, 312
Coombes, 116, 146, 241, 312, 356, 366
Copulation, 18 (*see also* Mating)
Corallian, 24, 25
Cornbrash, 24
Cornwall, 171, 175, 178, 179, 183, 291, 296, 360
Costae, 96
Cotswold hills, 26, 28, 124, 152, 182, 351, 359
County
    Clare, 308
    Cork, 185
    Galway, 292
    Kerry, 185
    Louth, 191
    Offaly, 163, 304
    Waterford, 309
Court Hill Cairn, 309
Courtship, 118
Coversands, Younger, 355
Cowpats, 223
Cracks, 210
Creswellian, 310
Cromerian Interglacial, 351 (not indexed for Chapter 5)
Cross-ploughing, 285, 364
Crumb structure, 284
Cryoturbation, 221, 229
    structures, 251, 253, 290, 353 (*see also* Periglacial structures)
Cultivated
    fields, 111
    land, 127, 188, 191, 227
Cultivation, 116, 136, 148, 149, 159, 195, **214**, 226, 228, 260, 279, **283**, 316, 318, 341, 356, 364, 366
Cwm Mawr, Glamorgan, 305
Cwm Nash, Glamorgan, 305

Cyclostoma elegans, 6, 7, 31, 134 (*see also Pomatias elegans*)
Czechoslovakia, 13, 14, 46, 353

D

Dactylis, 113
Damp habitats, *see* Moist habitats
Danubian farmers, 290
Dating, 17, 39
Davainea proglottina, 107
Death assemblage, 20
Decalcification, of soil, 211, 213, **216**, 219, 224, 226, 255, 277
Decaying bodies, 188 (*see also* Rotting flesh)
Defensive earthworks, 128
Deforestation, 295, 362 (*see also* Forest clearance)
Degradation, 130, 286
Denmark, 355
Department of the Environment, 243, 264, 268
Deposition, phases of, 293
Derbyshire, 146
Destruction, of shells, *see* Preservation
De-turfing, 246, 247
Deverel-Rimbury, 338
Devil's Kneadingtrough, 147
Devon, 70, 120, 137, 141, 176, 351
Devonian Limestone, 20, 25, 27
Dicrocoelium dendriticum, 107
Differential weathering, 32
Dimbleby, G. W., 247
Dipterous pupae, 299
Dipslope, 248, 277, 286
Discus
    rotundatus, 8, 11, 20, 62*, **73**, 94, **95**, 113, 184, **185**, 195, 202, 255, 260, 269, 288, 301, 308, 310, 358
    ruderatus, 62*, **72**, 98, 102, **184**, 195, 272, 300, 304, 358
    shimeki, 102
Diseases, 107, 122
Disjunct distributions, 141, 153
Dispersal, 123–125, 148
Distribution mapping, 4, 5, 13, 22, 125
Disturbance,
    of parent material, 216
    of soil (surface), 134, 214, 221, 226, 265, 284

Disturbance—cont.
  of the habitat, 87, 90, 92, 97, 105, **126**, 136, 146, 163, 166, 195
Ditches, 24, 31–34, 36, 73, 94, 116, 123, 127, 128, 230, 282, 289, **321** *et seq.*
  deposits in, 134, 164, 196, 213, 227
  faunas of, 158, 263
  on river gravel, 344 *et seq.*
  primary fill of, 243, 258, 261, 281, 287, 322
  secondary fill of, 196, 213, 258, 322
  soils in, 134, 223, 228, 258, 322
  tertiary fill of, 322
Diurnal rhythm, 201
Diversity,
  of snail faunas, 90, 319, 335, 337, 361, 365
  of the landscape/habitat, 127, 128
Dog's mercury, 236
Dorset, 141, 144, 299, 337
Downland, 94, 104, 108, 110, 111, 126, 127, 170–174, 196, 197
Drainage, 226, 284, 347
Drift, 25, 28, 32, 128, 209, 216
  periglacial, 126, 215, 217, **221**, 248, 257, 263, 268, 274, 278, 335 (*see also* periglacial deposits)
  siliceous/non-calcareous, 216, 287
Drought, 99, 126, 170
Dryness, 96, 101, 119, 126, 150, 170, 210
Dry places/environments/habitats/conditions, **94**, 119, 120, 130, 140, 143, 146, 150, 152, 155, 161, 167, 177, 180, 183, 186, 190, 365
Dry valleys, 41, 224, 233, 248, 285, 290, **311**, 356, 360
Durness Limestone, 25, 27
Durrington Walls, Wiltshire, 83, 140, 148, 183, 211, 230, 272, 354, 359, 362
Dwarf birch, 353

E

Earl's Farm Down, Wiltshire, 140, 144, 148, 362
Earthworks, 32, 33, 227
Earthworm activity/sorting, 230, 247, 252
  active, 212, 242
  passive, 209, 227, 236, 242

Earthworms, 9, 20, 103, 208, 209, 210, 214
  aestivation chambers of, 212, 233
  burrows of, 241
  casts of, 209, 212
East Anglia, 5, 24, 94, 178, 185, 290, 358
Eastbourne, 181
Easton Down, 7
Ebbsfleet, 11
Ebbsfleet/Mortlake sherds, 331
Eb horizon, 274, 278
Ecological
  amplitude/range, 97, 143
  groups, 79, 80, 97, 117, **194**, 295
  tolerance, 177, 196, 198
  changes of, 130
Economic data, 38
Ecosystems, 90, 104, 108
Eemian Interglacial, 351 (not indexed for chapter 5)
Eggs, 111, 118, 121
Eire, *see* Ireland
Ellobiidae, 48, 136
Ena, 64, 96, 331
  montana, 52*, 53*, **64**, 100, 124, 129, **165**, 195, 201, 357
  obscura, 52*, 53*, **64**, 112, 155, **165**, 195
Endemic species, 178
Endodontidae, 72, 74, 183
Enidae, 64, 165
Ensay, Outer Hebrides, 298
Environmental change, 123, 125, 177, 211
Epiphragm, 92
Erosion, 126, 233, 284, 286, 317
Escarpments, 153, 154, 277, 282, 286, 356
Essex, 164, 179, 182, 305
Euconulus fulvus, 54, 57*, 60*, **74**, 94, 115, 163, **186**, 195, 199
  var. *alderi*, 186
Eutrophic habitats/conditions, 187, 315
Evaporation, 98, 126, 299
Evesham, 26
Evolution, 118, 132, 177
Exposed habitats, 139, 140, 147, 157
Exposure, 147
Extinctions, 125, 130

Extinct species, 72, 77, 102, 133, 169, 180
Extraction, 44

F

Fan gravel, 26, 28
Farming, 126, 127, 154
Farming communities, 3, 362
Farm yards, 128
Farncombe Down, 148–150, 359, 362
Faunal changes/fluctuations, 122, 124, 125, 211
Felling, 129
Fen, 143, 160
  carr, 156
Fences, 316
Fens, the, 303
Fern spores, 36
Ferussaciidae, 66, 168
Fescue, 88, 113
Festuca rubra, 88
Field
  boundaries, 326, 366
  systems, 316, 320
Fields, 139
Fixed-dune pasture, 28, 111, 163, 173, 291
Flandrian, 352
Flatworms, 107
Flint
  flakes, 258, 273, 312
  industry, 332
  -knapping débris, 252
  line, *see* stone line
  mines, 7, 24, 26, 103, 128, 141, 151, 169
Flints, 299, 303, 304
Flintshire, 305
Flood loam, 41, 161
Flooding horizons, 299
Flood-plain, 358
Folds, 48, 49
Food, 89, **103**, 108, 109, 129, 131, 159, 171, 196
  animal débris, 38, 297
  of man, 176
Forest, *see* Woodland
  clearance, *see* Woodland clearance
  glades, 154
Fossil soils, 223

Fowl tapeworm, 107
France, 133, 162
Freshwater
  deposits, 12, 35, 41, 281, 305
  species, 45, 80, 97, 117, 132, 157, 196, 199, 299, 347, 349, 353
Frogholt, Kent, 365
Frost, 99, 175
  climate, 352
  hollows, 109, 127, 159, 164, 171, 172, 174
  weathering, 221, 224, 229, 284, 287, 322, 351, 353
Fruticicola fruticum, 68, 169, 200
Fungi, 103
Fussell's Lodge Long Barrow, 128
Fyfield Down, Wiltshire, 143, 147, 154, 168, 316, **317**, 360

G

Gardens, 128, 156, 175, 176, 191, 193, 201
Gault Clay, 27, 116, 349
Generation interval, 99
Geomalacus maculosus, 73, 185
Geophobic species, 105, 167
Germany, 90, 95, 143, 146, 156, 161, 180
Gerrard's Cross, Buckinghamshire, 304
Gipping Glaciation, 150, 164, 351
Glaciation, 351
Glamorgan, 143, 176, 181, 303, 304, 305
Glazing, 284
Glebe Low, 309
Gleyed deposits, 347, 349
Gley soils, 297, 346
Gloster, Co. Offaly, 304
Gloucester, 26
Gloucestershire, 151, 154, 180
Glow-worm larvae, 106, 107, 122
Godstow, Oxon, 155
Gogmagog Hills, 146
Gotham, Nottinghamshire, 165
Gower Peninsula, 143, 146
Graciliaria, 66, 167
Gradient, 116, 117, 287, 293
  of shell numbers, 210
  of shell sizes, 248
Grain, 39
Granules, *see* Arionid granules

INDEX

Granules/nodules, of *Acanthinula*, 23, 61, 63, 213
Granules, in eggs, 119
Graphical presentation, 79
Grasses, 87, 127, 136, 157, 159, 223, 239, 267, 353
Grassland, 87–91, 98, 104, 106, 110, 112, 126, 127, 131, 136 *et seq.*, 194–196, 216, 236–243, 246, 252, 261, 274, **362**, 364 (*see also* Meadows, Pasture and Downland)
   faunas/species, 104, 109, 126, 130, 131
   soils, 147, 229, 243, 252, 286
Gravel, 248, 346, 349
Graves, 33, 34
Grazing, 126, 129, 130, 136, 180, 195, 267, 274, 326, 364
Great Kimble, Buckinghamshire, 233
Greensand, 27, 274, 356
Gregarious habit, 111
Grimes Graves, Norfolk, 7, 24, 26, 128, 141, 142
Growth rate, 99
Gwithian, Cornwall, 292

H

Habitat,
   development, 131
   preferences, 144
      change of, 130, 182, 190, 202
Habitation, 130
Hair pits, 68, 70–72
Hairs, 45, 70, 95
Halling, Kent, 186, 353
Hampshire, 129, 140, 169, 176, 179
Haresfield Beacon, Gloucestershire, 154, 156
Harlyn Bay, Cornwall, 179, 183
Harris, Outer Hebrides, 142, 173, 183, 294
Hawthorn sere, 88, 105, 136, 147, 152, 154, 162, 190, 196, 238
Hazel, 246
Heat, 95
Heavy-mineral analysis, 38, 290
Hedgehogs, 91, 105
Hedges, 94, 127, 156, 165, 170 *et seq.*, 183, 185, 189, 197, 201, 316

Height, 110, 116, 117, 128
Helicella, 68, 72, 127, 341
   caperata, 72, 97, 105, 111, 119, 122, **179**, 200, 247
   crayfordensis, 180
   elegans 183, 200
   geyeri, 72, 182, 355
   gigaxi, 72, 129, **179**, 182, 200
   itala, 8, 21, 43, 62*, 67*, 69*, **72**, 95, 107, 124–129, **180**, 198, 202, 229, 233, 247, 267, 295, 341, 355
   neglecta, 180 200
   striata, 72, 182
   virgata, 72, 107, **179**, 200, 230, 239, 247, 332
Helicellids, 123
Helicidae, 66, 74, 91, 118, 119, 169
Helicigona lapicida, 67*, **68**, 69*, 96, 100, **170**, 195
Helicodonta obvoluta, 67*, **68**, 103, 129, **169**, 195, 201
Heliophile species, 116, 143, 180
Helix (*see also Cepaea* for *H. hortensis* and *H. nemoralis*)
   aspersa, 35, 67*, 68, 69*, **70**, 111, 118, 121, 129, **175**, 200, 201, 348
   pomatia, 19, 68, 69*, **70**, **176**, 200
Hemp Knoll, Wilts., 140, 147, **332**, 337, 360, 363
Henges, 83, 128, 268, 274, 328
Herbivores, 103, 104
Herefordshire, 31, 163
Hertfordshire, 184, 304, 306
Hibernation, 201
Hillforts, 31, 315, 328
Hillwash, 14, 23, 31, 41, 116, 135, 146, 151, 164, 167, 179, 182, 190, 207, 218, 223, 227, **281**, 305, **311**
Hillwashing, 214, 221, 224, 357
Histogram, 12, 22, 79, 81, 201
Hitchin, Hertfordshire, 135, 167, 300, 304
Holborough, Kent, 156, 304
Hollow way, 323
Holocene, 352
Horslip Long Barrow, Wiltshire, 148, 178, 181, 244, **261**, 360, 362
Hoxnian Interglacial, 160, 300, 307, 351 (not indexed for Chapter 5)

INDEX

Humidity, 92, **94–97**, 131, 133, 136, 159, 171, 289
Humus, 209, 284, 293
Hungerford, Berkshire, 221
Hunter-gatherer communities, 256, 357
Huntingdonshire, 166
Huntonbridge, Hertfordshire, 158, 226, 299, 304
Hurdles, 317
Hydrogen peroxide, 44
Hygromia
  cinctella, 70, 176, 200
  hispida, 8, 11, 67*, 69*, **71**, 95, 98, 106, 163, **177**, 197, 199, 202, 331, 353
    var. *nana*, 177, 197, 198
  liberta, 71, 178, 197
  limbata, 70, 176, 200
  striolata, 35, 67*, 68, **70**, 89, 105, 128, **176**, 195, 201, 242, 332
  subrufescens, 24, 70, 93, 176
  subvirescens, 71, 178
Hygrophile species, 98, 144, 157, 295

I

Ice Age, 3
Ice sheets, 351
Ice-wedge casts, 221
Icknield Way, 315
Identification, 45–79, 196
Incombe Hole, Buckinghamshire, 311
Ingress Vale, Kent, 179
Insects, 5, 37, 42, 44, 105, 107, 346, 347, 350, 356
Insolation, 95, 287
Interglacials, 351
Intermediate species, 80, 194, **196**, 202, 295
Intertidal organic horizon, 296
Introduced species/Introductions, 72, **129**, 130, 133, 151, 168, 171, 176, 179, 180, 182, 183, 191, **200**, 295, 341
Inverness, 31
Invertebrates, 37, 105
Involutions, 150, 183, 215, **221**, 224, 228, 257–260, 268, 270, 275, 290, 353

Ipswichian Interglacial, 351
Ireland, 27, 124, 125, 137, 139, 145, 149, 175, 183, 291, 296
Iron,
  in soils, 222, 224
  pans, 223, 253, 270, 278, 347, 349
Iron Age, 129, 148, 164, 173, 179, 182, 286, 366
  deposits, 176
  field systems, **320**
  occupation horizons, 293
  pits, 37
  settlement, **315**, 318
  sherds, 312
  sites, 178, 307, 338
Isle
  of Thanet, 290
  of Wight, 8, 141, 156, 305
Italy, 133, 184, 283, 317
Ivinghoe Beacon, 233, 240, 367
Ivinghoe, Buckinghamshire, 154
Ivy, 141, 155

J

Julliberries Grave, 11, 148, 359, 362
Juniper, 353
  scrub, 241
  sere, 88, 104, 136, 147, 152, 154, 196, 234, 240
Jurassic
  Limestone, 23, 24, 169
  ridge, 182
Jutland, 219
Juvenile mortality, 96, 120, 121
Juveniles, 46, 111, 118, 121, 201

K

Kennard, A. S., 6, 10, 16, 325
Kent, 5, 137, 142, 151, 156, 157, 160, 165, 177, 182, 186, 223, 290, 304, 305, 365
Kentish Rag, 25
Kerney, M. P., 265
Kilgreany Cave, Co. Waterford, 309
Kilham, 24, 38, 128, 226, **277**, 359, 363
Knap Hill, Wiltshire, 143, 154, 155, 163, 181, 246, 359, 360

INDEX 427

L

Laciniaria biplicata, 65*, **66**, 156, **167**
Lake
  District, 145
  levels, 358
  sediments, 281
Lakes, 136, 305
Laminifera, 66, 167
Lampyris noctiluca, 106, 123
Lancashire coast, 296, 297
Land
  boundaries, 128
  use, 117, 132, 198, 226
Landscape, 227, 230, 366
  change, 352
Last Glaciation, 131, 305
Last Interglacial, 125
Late Weichselian, 98, 125, 130, 196, 219, **223**, 229, 301, 311, **352**
  deposits, 12, 116, 222, 289
  fauna, 11, 46, 79, 80, 114, 249, 269, 331, **354** (not indexed for Chapter 5)
Lateral variation, 111, 115, 117
  of the environment, 130
Lauria, 112
  anglica, 54, 57*, 60*, **61**, 92, 94, **151**, 195, 199, 201, 301, 305
  cylindracea, 54, 57*, 60*, **61**, 92, 125, **151**, 155, 195, 272, 288, 295, 357, 361
    var. *anconostoma*, 57*, 61
  sempronii, 57*, 60*, **61**, **151**
Leached zone, 253, 270
Leaching, 210, 222, 269, 287, 346
Leaf
  fall, 226, 277
  litter, 87, 88, 104, 112, 115, 136, 185, 186, 194, 195, 230, 233, 236
  width, 105, 110, 126
Leaves, 103, 136, 141, 153, 155, 166, 183, 194
Leckwith, Glamorgan, 146, 305
Lessivation, 38, 224, 226, 277, 287
Letchworth, Hertfordshire, 300, 304
Lias/Liassic Limestone, 24, 25, 29, 225, 303, 305
Lichens, 103, 185, 353
Life
  cycles, 91, 118

history, 99
Light, 88, 95, 115, 150, 198
Limacid slugs, 23, 314
Limacidae, 50, 78, 191, 196, 197, 202, 212
Limax, **79**, **192**, 197
  cinereoniger, 79, 192
  flavus, 79, 192, 201
  marginatus, 192
  maximus, 18, 79, 118, 192
  tenellus, 192, 201
Lime, 20, 87, 90, 92, 109, 128, 130, 190, 196
Lime-rich soils, 126
Limestone, 226, 251
  heath, 216
  rubble, 287, 309
  sites, 310
Liming, 128, 130
Lincolnshire, 5, 145, 184, 304
Lithological change, 125
Little Chesterford, Essex, 164
Little Kimble, Buckinghamshire, 240
Little Missenden, Buckinghamshire, 156
Little Oakley, Essex, 182
Living space, 108
Llancarfan, Glamorgan, 304
Loch Tay Limestone, 25, 27
Lodge Hill, Buckinghamshire, 236
Loess, 14, 15, 121, 150, 207, 213, 216, 223, 281, **290**, 351
  fauna, 353
Logs, 111, 112, 151, 153, 162, 166, 185, 188, 195, 295
Long barrows, 34, 248, 251, 257, 258, 261, 264, 265, 275, 328
Longevity, of shells, 230
Longleat House, Wiltshire, 143
Love darts, 118
Low Countries, 15
Lowestoft Glaciation, 351
Lullingstone, Kent, 176
Lungworms, 107
Lux scale, 143, 161
Lycopodium spores, 247
Lymnaea
  glabra, 199
  palustris, 199
  peregra, 349
  truncatula, **49**, 50*, 199, 295, 347, 349

Lymnaeidae, 49
Lynchets, 34, 182, 282, **316**, 366

M

Machair, 28, 158, 291
Macroscopic plant remains, 3–5, 35, 36, 346
Maglemosian, 303
Magnesian Limestone, 23–25
Maiden Castle, 325
Mammals, 132, 310
  small, 106, 122
Man,
  activities of, 256
  assault on the landscape, 130
  habitations of, 129
  influence on the snail fauna, 123, **126**, 130, 246, 319
  interference with the habitat, 129, 165, 166, 177, 356
  spread of snails by, 129, 192
Man-made
  changes, 125
  habitats, 128, 153, 177, 201
Manure, 312
Manuring, 39, 318
Marden, Wiltshire, 38, 141, 221, 226, 230, 231, **274**, 359, 363
Marine
  deposits, 296
  environments, 297
  shells, 291
Maritime
  habitats, 136, 140, 143 (*see also* Coastal habitats)
  distributions, 151, 180, 184
Marl, 115, 209, 221, 222, 224, 248, 249, 283, 301
Marlborough Downs, 101, 109, 110, 171, 172, 317, 320, 366
Marpessa laminata, **64**, 65*, **166**, 195
Marsh
  deposits, 161
  habitats, 130, 136 *et seq.*, 199, 295, 346
  soil, 157
  species, 34, 97, 117, 130, 137, 142, 145, 158, 196, **199**, 202, 347, 350
Matera, Italy, 133, 283
Mating, 118 (*see also* Copulation)

Maxey Northamptonshire, 22, 23, 37, 178, 199, 200, **346**, 359
Meadows, 136, 143, 163, 177, 189
Medieval, 34, 128, 177, 179, 230, 286, 349, 366
Medway Valley, 115
Mercurialis perennis, 236
Mercury beechwood, 88
Merlin's Cave, Symond's Yat, 309
Mesic habitats, 115
Mesolithic, 174, 256, 278, 299, 300, 305, 307, 311, 358
Mesophile species, 113
Mesotrophic habitats, 187, 189
Mice, 37
Micro-environments, 228
Micro-habitats, 115, 125, 135, 293, 310
Microliths, 256
Microscopic plant remains, 36
Middle East, 13
Middleton Stoney, Oxfordshire, 73, 74
Middle Weichselian, 137, 140, 150, 161, 178, 186, 274, 290
Migration routes, 296
Milax, 50*, 79, 156, **191**, 197
  budapestensis, 96, 99, 119, 191
  gagates, 191
  insularis, 191
  sowerbyi, 107, 191
Millpark, Co. Offaly, 146, 158, 163, 304
Mine shafts, 287
Modern soils, 232 *et seq.*, 245, 247, 252 (*see also* MS I—V)
Moist/damp habitats, 133, 143, 156, 161, 170, 172, 177, 178, 186, 189, 194, 295
Moisture, 96, **97**, 101, 135, 152, 171, 188, 198, 199, 325
  loss, 116, 121
  tolerance/requirements, 162, 171
Moisture-loving species, 143
Moles, 105, 210
Monacha, 71
  cantiana, 68, 69*, **72**, 95, 96, 100, 102, 106, 119, 120, 122, 123, **179**, 200, 314
  cartusiana, 67*, **72**, **179**, 198
  granulata, 24, 67*, **71**, 156, **178**, 199, 292
Monolith sampler, 43

Moor Park, Hertfordshire, 306
Morar, Inverness, 292, 297
Morphs, 174, 175
Mortality, 99, 104, 111, 119, 122
Mortuary house, 34
Mosses, 36, 141, 299, 350, 353
Mounds, 128, 249, 252, 258, 264
MS
  I, 143, 146, 211, 213, **233**, 235, 246, 315
  II, 147, 150, 154, 211, 229, **236–238**, 315
  III, 144, 147, 150, 154, 211, 229, 237, **239**, 315
  IV, 211, 213, 230, 235, **236**, 246, 315
  V, 147, 155, 214, 237, **240**, 315
Muds, 223
Muellerius capillaris, 107
Mull-humus horizon, 208, 217
Myxomatosis, 181

N

Napoleonic Wars, 366
NAP/AP ratio, 114
Nash Point, Glamorgan, 225, 226, 300, 305
National Trust, 243
Nematodes, 107
Neolithic period, 10, 94, 109, 125, 131, 148, 150, 151, 154, 164, 172, 175, 177, 180, 182, 198, 286, 291, 293 *et seq.*, 325, **358**
  burials, 34, 103
  faunas, 113, 273, 362
  occupation/settlement, 255, 332, 335
  sites, 246, 362
  soils, 135, 229, 232, 242 *et seq.*
  turf lines, 84, 115
Netherlands, 355
Newquay, Cornwall, 118, 179, 183, 292, 360, 361, 363
Niche, 89, 91
Nitrification, 88, 105
Nitrogen
  compounds, 104
  metabolism, 88, 89, 104, 132
Norfolk, 7, 141, 143
North
  America, 13, 14, 102
  Downs, 24, 352, 357, 360, 365
  Ferriby, Yorks., 305
  Uist, Outer Hebrides, 28, 183, 291, 295
  Wales, 145
  York Moors, 358
Northamptonshire, 178, 192, 346
Northton, Outer Hebrides, 142, 144, 158, 173, 182, 183, 199, **293**, 359, 363
Norton Common, Herts., 158, 163, 184
Nottinghamshire, 137, 165, 184, 305
Nutrients, 130
Nutrient status, of soils, 126, 130–132, 199

O

Occupation, 244, 255, 256, 364
  horizons, 31, 34, 246, 311, 332
Oceanicity, 149, 184, 361
Oligotrophic habitats/conditions, 189, 196, 199, 315
Ombrogenous mires, 358
Oolite, 24, 25, 152, 251, 343, 351
Open-country
  faunas/species, 113, 116, 134, 153, **198**, 202, 213, 230, 246
  habitats, 136, 143, 152, 154, 161, 172, 183, 196, 198, 289, 295, 315, 365
Operculates, 47
Operculum, 47, 92
Orchards, 163
Ordovician Limestone, 25
Organic
  acids, 209
  content, 293, 315
  matter, 214, 222, 223, 265, 286, 346, 349
  mud, 346
  remains, 347
Ostracods, 37, 44, 299, 347
Other woodland (shade-loving) species, 194, **195**, 202, 331
Outer Hebrides, 28–30, 124, 163, 181, 183, 291, 298, 359
Overgrazing, 283
Overton Down
  Experimental Earthwork, 247
  Wiltshire, 147, 155, 182, 316, **320**, 360
Owslebury, Hampshire, 37, 176

Oxford
 area, 110
 Clay, 182, 346
Oxfordshire, 26, 138, 143, 146, 154, 162, 163, 182, 251, 305, 359
Oxidation, 223, 284
Oxted, Surrey, 135, 162, 164
Oxwich, Glamorgan, 146
Oxychilus, 23, **74**, **77**, 112
 alliarius, 45, 75*, 76*, **77**, **188**, 194, 295, 358
 cellarius, 8, 34, 74, 75*, 76*, **77**, 103, 121, **188**, 194, 255, 260, 269, 272, 288, 308, 310, 358
  var. *hibernica*, 187
 draparnaldi, 75*, 76*, **77**, 121, 156, **187**, 200
 helveticus, 75*, 76*, **77**, **189**, 194
Oxygen, 223

P

Palaeolithic, Upper, 311, 355
Parasites, 105, 122, 127, 130, 131
Parasitism, 107
Parent material, 207, 213, 216, 222, 225, 226, 279, 356
Particle-size analysis, 38, 290
Pastoral farming, 364
Pasture, 127, 149, 154–156, 180, 349, 358
Peacock's Farm, 303
Pea grit, 208, 212, 214, 242, 244, 247
Peat, 98, 140, 145, 297, 300, 304
 charcoal, 296
Pedological changes, 126
Pegsdon, Hertfordshire, 312
Pembroke, 178
Periglacial
 deposits, 164, 213, 290, 301, 311, 356 (*see also* Drift)
 environments, 161
 faunas, 221
 gravel, 266
 structures, 219, **221**, 228, 255, 272 (*see also* Involutions)
 weathering, 229, 289
Periostracum, 19, 22, 45, 121, 210, 232, 266
Permafrost, 351
Perranporth, Cornwall, 360, 363

pH, 88, 207, 363
Physiological changes, 132, 150
Phytoliths, 36
Piedish diagram, 10, 11 (*see also* Sector diagram)
Pine, 354, 357
Pink Hill, Buckinghamshire, 134, 146, 147, 154, 158, 211, 213, 285, **312**, 360, 363
Pisidium
 casertanum, 199, 347
 obtusale, 199
 personatum, 199
Pits, 31–37, 42, 230, 252, 255, 332, 344, 347, 349
Pitstone, Buckinghamshire, 114, **134**, 150, 154, 157, 168, 185, 218, 222–224, 227, 305, 315, 361, 363
Pitstone Tunnel Cement Works, 311, 367
Planorbarius corneus, 347
Planorbis
 crista, 392
 leucostoma, 200, 347, 349
Plantations, 163
Plant impressions, 37, 299
Plateaux, 170, 172, 290, 320
Pleistocene, 14, 15, 130, 132, 137–140, 150, 178, 192, 299, 306 (*see also* Quaternary)
 deposits, 12, 32, 35, 43, 186
 snails, 48, 51, 56, 64, 66, 78
Plough soils, **214**, 224, 240, 249, 266
Ploughed fields, 180
Ploughing, 32, 126, 128, **214**, 222, 250, 255, 260, 265, 279, **282**, 314, 326, 344
Ploughmarks, 214, 216, 222, 260, 273, 332, 364
Ploughwash, 31, 34, 47, 168, 197, 213, 214, 216, 218, 224, 225, 227, 258, **282**, 311, 322, 326, 366
Podsolization, 277
Podsols, 297
Pollen, 24, 42, 224, 278
 analysis, 3–5, 9, 13, 15, 33, 35, 36, 246, 287, 300, 304, 310, 355
 record, 361
 zone, 357
Polymorphic snails, 12, 106

Pomatias elegans, 5–8, 11, 12, 20, 23, 47, 52*, 53*, 69*, 92, 100, 124, 125, **133**, 196, 201, 202, 212, 229, 232, 260, 263, 264, 272, 314, 338, 341, 358, 361
Pomatiidae, 47, 133
Ponds, 156
Poor habitats, *see* Oligotrophic habitats
Population
 density, 122
 genetics, 43
Portland Limestone, 24, 25
Post-glacial, 352
 chronology, 357
Post-holes, 33, 34, 320
Pot sherds, 252, 312
Pre-boreal period, 4, 357
Pre-cambrian limestones, 25
Precipitation, 92, 98, 117, 131, 284, 286, 358
 of calcium carbonate, 299
 /evaporation ratio, 92, 98
Predation, 106, 121, 122
Predators, 105
Preservation/destruction, of shells, 8, 18, 19, **22–24**, 46, 47, 89, 92, 109, 121, 166, 210, 212, 263, 264, 314, 341, 344
Prestatyn, Flintshire, 301, 305
Princes Risborough, Buckinghamshire, 152, 154, 232, 284, 312
Protein synthesis, 104
Protoconch, 119
Protozoa, 107
Pulpit Hill, Buckinghamshire, 143, 146, 233, 240
Punctum pygmaeum, 8, 54, 62*, 66, **72**, 87, 94, **183**, 195, 199, 240, 331, 344
Pupilla muscorum, 8, 11, 53*, 54, 57*, 59, 60*, 89, 95, 98, 115, 130, **146**, 158–160, 182, 198, 202, 249, 257, 261, 269, 273, 290, 315, 319, 353
Purbeck Limestone, 25
Pyramidula rupestris, 54, 57*, 60*, 92, **139**, 288
Pyrenees, 175

Q

Quarries, 152, 155, 287
Quarrying, 31, 128, 130
Quaternary period, 132, 351

R

Rabbits, 127, 146, 155, 181
Radiocarbon
 assay, 4, 6, 33, 39, 185, **230**, 231, 244, 248, 251, 274, 277, 278
 correction curve, vii, 6, 359
 dates, 184, 243, 255, 257, 261, 265, 296, 300, 303, 329, 332, 361
Rain, 126
Rainfall, 3, 9, 92–94, 98, 117, 135, 140, 175, 299, 366
Rate of sedimentation/accumulation, 41, 89, 90, 280, 286, 324
Reading, Berkshire, 163
Recurrence surface, 98
Reducing conditions, 223
Reed swamp, 160, 162, 196
Reeds, 348, 350
Reference collection, 46
Refugia/refuges, 116, 141, 172, 326, 339, 365
Refuse heaps, 128, 201
Regeneration, 364
Relative humidity, 90, 95, 99
Relict distribution, 141, 153, 187
Rendsinas, 126, 128, 207, **208**, 222, 223, 226, 229, 230, 248, 279, 287
Reproductive
 activity, 120
 rate, 89, 90
Retinella, 74, 77, 112, 188, 310
 nitidula, 8, 74, 75*, 76*, **78**, 87, 89, 113, 121, **190**, 194
 petronella, 77, 189
 pura, 8, 62*, 74, 75*, 76*, **78**, 87, 163, **189**, 194
 radiatula, 74, 75*, 76*, **77**, 89, 94, 98, 163, **189**, 191, 195, 198, 240
Reworking, of ancient deposits, 281, 293, 299
Rhine Valley, 290
Ribs, 95
Ridge and furrow, 39, 216

432

INDEX

Risborough Gap, 241, 316
Ritual monuments, 128
River
  Allen, 338
  Avon, 26, 274
  banks, 178, 197
  Cam, 349
  Cray, 163
  Evenlode, 251
  gravel, 28, 36, 200, 281, 344
  Kennet, 140, 263, 359, 360
  Severn, 26
  Stour, 338
  Thames, 167
  valleys, 170, 344
  Welland, 346
  Winterbourne, 257, 265
Rivers, 136, 305
Road dust, 128
Roads, 31
Roadside faunas, 128
Rock débris, 310, 322
Rocks/rocky places, 96, 139, 143, 145, 151, 152, 162
Roman period, 124
  buildings, 302
  hillwashes, 286, 366
  introduced species, 176, 179
  sherds, 312
  sites, 29, 34, 42, 128, 201, 303, 307, 343, 346, 349
Root channels, 208
Rootholes, 219, 221, 239, 318, 320
Roots, 37, 103, 210, 219
Rosapenna, Co. Galway, 292
Rotting flesh, 103 (*see also* Decaying bodies)
Roughridge Hill, Wiltshire, 140, 148, **335**, 360, 362
Round barrows, 265, 332, 335
Rubbish dumps, 89, 156 (*see also* Refuse heaps)
Rudston, 137
Run-off, 284
Rupestral
  habitats, 162–164, 181, 194, 288
  species, 89, 111, 112, 116, 151, 158, 165, 166, 170, 195, 288
Rushes, 34, 348

S

St. Andrews, 292
St. Catherine's Hill, Winchester, 150, 154
Salisbury Plain, 128, 359
Sampling, 8, 41, 112, 115
Sanctuary, the, Avebury, 7, 34
Sand dunes/hills, 97, 104, 108, 133, 140, 146, 173, 176, 180–183, 190, 196, 291, 356
Sand, wind-lain, shell, 14, 20, 25, 28, 118, 143, 178–181, 183, 228, 281, **291**, 357, 360
Sanicle beechwood, 88
Sarsen stones, 290
Sauveterrian, 303, 305
Scandinavia, 14, 195, 355
Scarp slopes, 180
Scavengers, 103, 104
Sciomyzid flies, 107
Scotland, 27, 28, 149, 155, 173, 175, 177, 180
Scree, 64, 115, 281, **287**, 310, 322, 356
Scrub, 88, 113, 133, 147, 150, 163, 166, 201
  faunas, 91
Sea level, 125, 296, 297, 357
Sector diagram, 10, 11, 80
Sedimentation, 207 (*see also* Rate of)
Sediments, 207, 221
Seeds, 36, 42, 44, 347, 350
Settlement, 126, 356
Shade, 94, 96, 325
  habitats, 196, 198, 295
  -loving species (*see* Woodland species)
Shakenoak, Oxfordshire, 305
Shapwick Station, Somerset, 163
Sheep, 104, 107, 127, 131, 146, 155, 159, 181, 273, 291
  farming, 366
  fluke, 107
Shell
  fish, 38
  middens, 292, 297, 298
Shells,
  dead, 168, 267
  live, 168, 232
  reworked, 281
  weathered, 232

Shelter, 90, **92**, 103, 108, 129–131, 325
Sheltered habitats, 194
Shrews, 37, 91, 105
Shrubs, 353
Sicily, 185
Silbury Hill, 22, 37, 121, 144, 148, 178, 260, **265**, 360, 362
Siliceous matter, 226
Silt, 209, 290
Silurian Limestone, 25, 27
Sinistrally-coiled species, 58, 64
Skiddaw, 139
Slacks, 136, 143
Slopes, 146, 170, 196, 214, 227, 230, 281 *et seq.*, 316, 356
Slopewash, 28, 90, 281 *et seq.*
Slow-worms, 105
Slug
  plates, 46, 78, 197
  populations, 99, 107, 122
Slugs, 8, 45, 74, 97, 99, 103, 107, 118, 201
Slum species, 45, 199, 200, 202, 347, 348
Soil
  analysis, 33, 287
  creep, 152
  crumbs, 209, 284
  formation, 223, 228, 251, 280, 293, 355–357
  horizons, 15, 31, 40, 41, 114–116, 143, 228, 289, 299 *et seq.*, 358
  map, 31, 40
  moisture, 22, 87
  sample, 20
  structure, 214
  Survey Memoirs, 40
  texture, 96
Soils, 47, 130, 207, 221, 230, 311
  depth of, 159
  fossil features of, 222
  macromorphology of, 38
  microscopic and chemical properties of, 38
  non-calcareous, 190 (*see also* Brown earth and *Sols lessivés*)
  recording of, 40
  representation of, 80
  sampling of, 41
  vertical changes within, 115

water-retaining capacity of, 87–89, 92, 115, 126, 233, 239, 315, 325
Solifluxion, 164, 221, 223, 224, 229, 351 *et seq.*
  débris, 281, **289**, 305, 310
Sols lessivés, 126, **216**, 218, 221, 226, 249, 274, 275 *et seq.*, 278, 287
Solution, 32, 89, 221, 356
  hollows, 219 (see also *Sols lessivés*)
Somerset, 163, 176, 181, 217, 223, 224, 288, 352
Somerset Levels, 5, 307
Sonning, Berkshire, 163
South Cadbury, Somerset, 176, 289, 309
South Downs, 170, 352
South Street Long Barrow, Wilts., 134, 148, 150, 158, 168, 174, 188, 215, 228, 257 *et seq.*, 322, **328** *et seq.*, 354, 360, 362
South Wales, 29, 107, 135, 146, 171, 225, 293, 296
Southwell, Nottinghamshire, 137, 184, 300, 305
South-west England, 71
Sphaerium lacustre, 200
Spicula amoris, 118
Spines, 45, 63, 96
Split pea, *see* Pea grit
Spot samples, 43, 249
Springs, 299
Stakeholes, 217, 258
Stalactites, 299
Stalagmites, 299
Standstill phases, 299, 358
Starlings, 106
Statistical significance, 81
Stems, 103
Stereoscopic microscope, 44, 46
Stockbridge, Hants., 140
Stockholm, 156
Stone line, 208, 214, 227, 236, 255, 277
Stone-free zone, 208, 211, 214, 230, 243
Stonehenge, Wiltshire, 7, 140, 247, 365
Stones, 103, 162, 188, 194, 195, 295
Stratification, of shells, 207 *et seq.*, 212, 224, 225, 240
Streams, 136
Sub-aerial processes, 207, 256

INDEX

Sub-arctic
  conditions, 229
  faunas, 251
Sub-atlantic period, 4, 149, 223, 286, 357
Sub-boreal period, 4, 223, 296, 300, 357, 365 (not indexed for Chapter 5)
Subfossil assemblage, 20
Subsoil, 208, 213, 217, 225, 242, 255, 257
Subsoil hollows, 116 et seq., **219**, 241, 251 et seq., 257 et seq., 270, 312, 318
Substratum, 95, 96, 98, 103, 105, 116
Subterranean species, 168, 201, 310
  (see also Burrowing species)
Succinea, 43, **49**, 136, 200, 347, 353
  elegans, 138
  oblonga, 50*, **51**, **137**, 186, 199
  pfeifferi, 50*, **51**, **137**, 199
    var. schumacheri, 137
  putris, 50*, **51**, **137**, 138, 199
  sarsi, 51, 137, **138**, 199
Succineidae, 49, 136
Suffolk, 179
Summers, 365
Sun, 126
Sun Hole Cave, Cheddar, 309
Surrey, 135, 151, 162, 223
Survival of shells, 230 (see also Preservation)
Sussex, 5, 6, 103, 129, 151, 169, 179, 191, 223
Swamp, 157, 186, 196, 299, 300, 346, 358
  deposits, 223
  carr, 143, 156, 162
Swamping, 358
Swanscombe, Kent, 135, **160**, 161, 164, 307
Sweden, 140, 143, 156, 221
Symond's Yat, 309
Synanthropic species/faunas, 35, 89, 128, 130, 156, 175–177, **201**, 203, 246

T

Tabular Hills, 25
Tackley, Oxfordshire, 154, 155, 162
Takeley, Essex, 305
Taransay, Outer Hebrides, 30
Temperate climate, 351
Temperature, 5, 9, 22, **99**, 110, 146, 163, 164, 171, 196, 200, 289, 357
  decrease/increase, 124, 173, 356
  summer, 101, 125, 129, 131, 165, 170, 175
  winter, 101, 125, 134, 151
Testacella, 66, 103, **168**, 200
  haliotidea, 169
  maugei, 169
  scutulum, 169
Testacellidae, 66
Tetanocera elata, 107
Tewkesbury, 26
Texture, of deposits, 41
Thames Valley, 351
Thatcham, Berkshire, 307
Theba pisana, 70, 171, 200
Thermal
  decline, 125, 140, 159, 175
  maximum/optimum, 124, 229, 357
Thermoclastic scree, 287, 308, 310
Thermophile species, 124, 125, 135, 182
Thickthorn Down, Dorset, 10, 148, 325, 359, 362
Thriplow, Cambridgeshire, 353
Thrushes, 106
Tilia, 36
Tillage, 216, 250, 363
Time, 116, **118**, 124, 231, 326
  range, 115
  scale, 228
Topography, 115–117
Totland, Isle of Wight, 305
Towan Head, Newquay, 292
Trampling, 246
Transcaucasia, 184
Travertine, 281, 297 et seq.
Tree
  -root casts, 217, 242, 251, 277
  roots, 210, 216, 219, 225, 233
  trunks, 96, 103, 165, 166, 195, 256
Trees, 105, 116, 118, 171, 284
Trematodes, 107
Tring Gap, 367
Truncatellina, 56, 58, 59, 160
  cylindrica, 54, 55*, **56**, 95, 125, **140**, 198
  britannica, 55*, **56**, **141**

INDEX

Tufa, 13, 14, 37, 98, 117, 225, 281, **297** *et seq.*, 357, 358, 361
Tundra, 250, 257, 271, 274, 353
Turf, 40, 152, 209, 281, 322
    lines, 84, 149, 208 *et seq.*, 230, 247, 250, 252, 255, 258
    stack, 258, 265

U

Udal, North Uist, 183, 295
Ultraviolet light, 95
Upper Pleistocene, 300
Usselo Layer, 355

V

Vale
    of Aylesbury, 311
    of Glamorgan, 305
    of Pewsey, 141, 274, 282, 359
Valleys, 127, 164, 170–172, 197, 227, 282, 303, 366 (*see also* Dry and River valleys)
Vallonia, **63**, 66, 88, 95, 118, **160**, 307, 337, 348
    costata, 8, 21, 62*, **63**, 89, 95, 96, 98, 104, 109, 110, 152, **153**, 161 *et seq.*, 198, 202, 242, 244, 246, 252, 255, 272, 310, 315
    excentrica, 8, 62*, **63**, 89, 101, 104, 109, 110, 127, 154 *et seq.*, **161**, 198, 202, 266, 310, 315
    pulchella, 62*, **63**, 95, 115, 157, **161**, 199, 249
        var. *enniensis*, 46, 150, 161, 165
    tenuilabris, 164, 353
    tenuilimbata, 165
Valloniidae, 63, 74, 153
Vegetation, 3, **89** *et seq.*, 103, 113 *et seq.*, 130–132, 147, 155, 173, 216, 223, 239, 265, 280, 283, 324, 357
Vertebrates, 37, 105
Vertiginidae, 54, 61, 139
Vertigo, 45, 46, 58, 63
    alpestris, 54, 55*, **59**, 92, 96, 142, **145**, 195, 272, 301, 305, 314
    angustior, 54, 55*, **58**, 64, 142, **146**, 199, 305
    antivertigo, 55*, **58**, **142**, 199
    concinna, 145
    genesii, 55*, **59**, 101, **145**, 184, 199, 300, 304, 305, 358
    geyeri, 55*, **58**, **145**, 184, 199
    lilljeborgi, 55*, **59**, **145**, 199
    moulinsiana, 55*, **59**, **145**, 199, 301, 304
    parcedentata, 145
    pusilla, 54, 55*, **58**, 65*, 96, **141**, 195, 242, 261, 272, 301, 305
    pygmaea, 8, 20, 47*, 54, 55*, **58**, 60*, 95, 130, **143**, 158, 163, 198, 199, 202
    substriata, 54, 55*, 56, **58**, 92, **142**, 195, 199, 295
Vice-comital distribution maps, 99
Vipers, 105
Visual selection, 175
Vitrea, 74, 77, 190, 196, 310
    contracta, 8, 62*, 74, 75*, 76*, **77**, **187**, 194, 255, 288, 308, 361
    crystallina, 74, 75*, 76*, **77**, 163, **186**, 194, 199
    diaphana, 74, 75*, **77**, **187**
Vitrina, 78, 213, 331, 344
    major, 69*, **78**, **191**
    pellucida, 8, 50*, 69*, **78**, 94, 103, 163, 181, **190**, 195, 198, 199, 295
    pyrenaica, 191, 200
    semilimax, 78, 191
Vitrinidae, 78, 190
Voles, 37, 105

W

Waden Hill, Wiltshire, 147, 154, 168
Wales, 71, 180, 183, 191, 358 (*see also* South Wales)
Wall débris, 187, 188
Walls, 29, 31, 94, 96, 105, 128, 195, 295, 316
Waste ground, 156, 175, 176
Wastage, of turf, 222, 230, 252
Water
    loss, 92, 95, 171
    meadows, 170
    regime, 117
    table, 299, 346
Wateringbury, Kent, 158, 305
Water-lain sediments, 207
Waterlogged deposits, 43, 347, 350
Waterlogged soil, 96

Watlington Hill, Oxfordshire, 143
Wayland's Smithy, Berkshire, 34, 103, 148, 158, 262, **265**, 359, 362
Weathering, 207, **210**, 213, 219, 229, 257, 280, 324, 351
   ramp, 258
Weichselian, 137, 138, 141, 185, 351
   fauna, 352
Welland Valley, 176, 346
Wells, 31, 33, 36, 119, 346, 347
Wendover, Buckinghamshire, 232, 239
Wenlock Edge, 25, 27
West Country, 180
West Kennet Long Barrow, 123, 148, 154, 211, 213, **263**, 360, 362
West Kennett, Wiltshire, 263
West Overton, Wiltshire, 320
Wheatfen, 156
Whiteleaf Hill, 236, 246, 315, 316, 359, 361
White-lipped *Cepaea nemoralis*, 70, 174
Whitton, Glamorgan, 119, 176
   Roman villa, 303
   well, 347, 350
Wild places, 175, 176, 180, 191, 193, 201
Willerby Wold, Yorkshire, 226, 359, 363
Wilstone, Hertfordshire, 163, 184, 300, 305, 315
Wiltshire, 129, 137, 140, 143, 149, 164, 169, 171, 181, 183, 290, 304, 352, **359**
Winchester, Hampshire, 150, 154
Wind, 126
   erosion, 214, 284
   transport, 293
Windmill Hill, Wiltshire, 144, 147, 155, 211, 212, **242**, 261, 359, 360
Windmill Hill Culture/pottery, 332, 335
Wingham, Kent, 365
Winters, 361
Wisconsin Glaciation, 102
Wood, 350
Woodhenge, Wiltshire, 7, 140, 154
Woodland, 3, 87–89, 91, 94, 106, 111–115, 126–130, 133, 139–146, 151 *et seq.*, 196, 208, 219, 226, 235, 244, 252, 255–257, 274, 282, 295, 299, 314, 351, 356–359, 362 (*see also* individual tree species)
   clearance, 3, 98, 126, 127, 134, 142, 158, 172, 174, 177, 196, 226, 252, 261, 263, 274, 277, 284, 287, 334, 336, 341, **358** (*see also* Deforestation)
   faunas/species, 114, 116, 126, 128, 138 *et seq.*, 194, 197, 202, 213, 255, 264, 289, 305, 310, 326, 360
Woodstock, Oxfordshire, 305
Woodston, Hunts., 164, 166
Woodward, B. B., 7
Wychwood Forest, Oxfordshire, 138, 146, 154, 155, 163, 220
Wytham Wood, Berkshire, 153, 155

X

Xerophile faunas/species, 108, 115, 117, 119, 126, 130, 133, 140, 152, 156, 177, 179, 182

Y

Yarnbury Castle, Wiltshire, 7
Yorkshire, 137, 220, 277, 287, 305
   Wolds, 24, 278

Z

Zerna, 87, 113, 236
Zone
   I, 98, 125, 222, 224, 289, 352, 355
   II, 98, 353, 355 (*see also* Allerød Interstadial)
   III, 98, 125, 222–224, 229, 289, 352, 355
   IV, 357
   V, 165, 357
   VI, 176, 357
   VI/VIIa transition, 184, 185
   VIIa, 357
   VIIb, 300, 304, 357
   VIII, 191, 357
Zonitidae, 66, 74, 103, 143, **194**, 202, 249, 264, 310
Zonitoides
   excavatus, 74, 75*, 76*, **78**, **190**, 194
   nitidus, 74, 75*, 76*, **78**, 156, **190**, 199, 347